Industry 4.0 for SM

Dominik T. Matt · Vladimír Modrák ·
Helmut Zsifkovits
Editors

Industry 4.0 for SMEs

Challenges, Opportunities and Requirements

Editors
Dominik T. Matt
Faculty of Science and Technology
Free University of Bozen-Bolzano
Bolzano, Italy

Vladimír Modrák
Department of Manufacturing Management
Technical University of Košice
Prešov, Slovakia

Helmut Zsifkovits
Chair of Industrial Logistics
Montanuniversität Leoben
Leoben, Austria

ISBN 978-3-030-25427-8 ISBN 978-3-030-25425-4 (eBook)
https://doi.org/10.1007/978-3-030-25425-4

This Palgrave Macmillan imprint is published by the registered company Springer Nature Switzerland AG
The registered company address is: Gewerbestrasse 11, 6330 Cham, Switzerland

This project has received funding from the European Union's Horizon 2020 research and innovation program under the Marie Skłodowska-Curie grant agreement No. 734713 (Project title: SME 4.0—Industry 4.0 for SMEs).

Preface

The term Industry 4.0 describes the ongoing revolution of manufacturing industry around the world. Large companies in particular have rapidly embraced the challenges of Industry 4.0 and are currently working intensively on the introduction of the corresponding enabling technologies. Small- and medium-sized enterprises (SMEs) face the hurdle of possessing neither human nor financial resources to systematically investigate the potential and risks for introducing Industry 4.0. However, in most of the countries SMEs form the backbone of the economy, they account for the largest share of the gross domestic product and are also important employers. In this respect, the challenges, opportunities, and requirements of Industry 4.0 have to be examined specifically for SMEs, thus paving the way for the digital transformation of traditional SMEs into smart factories.

The central question in this book is therefore: Which opportunities arise from Industry 4.0, which challenges do SMEs face when introducing Industry 4.0, and which requirements are necessary for a successful and sustainable digital transformation of their company?

With this book the research consortium of the H2020 MSCA RISE project "SME 4.0—Industry 4.0 for SMEs" (grant agreement

No. 734713) encourages other researchers to conduct research in the field of Industry 4.0 specifically for SMEs and thus expanding the community in SME research. Practical methods, instruments, and best practice case studies are needed to support practitioners from SMEs in the introduction of Industry 4.0.

This book summarizes the research results of the first phase of the project “SME 4.0—Industry 4.0 for SMEs: Smart Manufacturing and Logistics for SMEs in an X-to-order and Mass Customization Environment,” which was conducted from 2017 to 2018. The project, started in January 2017 with a duration of four years and is funded by the European Union’s Horizon 2020 research and innovation program under the Marie Skłodowska-Curie grant agreement No. 734713.

In this initial book, that is being published within the framework of the above-mentioned research project, the editors and contributors focus their research results on possible challenges, opportunities, and requirements that arise from the introduction of Industry 4.0. A further book publication is planned for the final phase with the focus on research of methods for the introduction of Industry 4.0 to SMEs in addition to practical applications in SMEs.

A great opportunity for the future lies in the transfer of Industry 4.0 expertise and technologies in Small- and Medium-sized Enterprises (SMEs). This research project aims to close and overcome the gap in this transfer through the establishment of an international and interdisciplinary research network for this topic. This network has the objectives of identifying the requirements, the challenges, and the opportunities for a smart and intelligent SME factory, creating adapted concepts, instruments, and technical solutions for production and logistics systems in SMEs and developing suitable organisation and management models. The practical applicability of the results is guaranteed through a close collaboration of the network with small- and medium-sized enterprises from Europe, USA, Thailand, and India.

The book is structured into five parts with a total of 13 chapters:

Part I—Introduction to Industry 4.0 for SMEs

In the first part readers are introduced to the topic by reviewing the current state of the art of the transfer of Industry 4.0 in SMEs and the role of SMEs in the digital transformation.

Part II—Industry 4.0 Concepts for Smart Manufacturing in SMEs

In the second part the focus lies on manufacturing in SMEs. The first chapter in this part describes the main requirements, constraints, and guidelines for the design of smart and highly adaptable manufacturing systems. The second chapter reports how SMEs can implement an industrial internet of things and cyber-physical systems for achieving distributed and service-oriented control of their manufacturing system. The third chapter provides insights about potentials and challenges of automation through safe and ergonomic human–robot collaboration.

Part III—Industry 4.0 Concepts for Smart Logistics in SMEs

The third part concentrates on the introduction of Industry 4.0 in SME logistics. In the first chapter, requirements for the design of smart logistics in SMEs are summarized, while the second chapter shows how SMEs can implement identification and traceability of objects to enable automation. The third chapter gives an overview of the state of the art of the application and the potential of automation in logistics.

Part IV—Industry 4.0 Managerial, Organizational and Implementation Issues

The fourth part deals with organization and management models for smart SMEs. In the first chapter in this part, the contributors develop and test organizational models for smart SMEs in terms of mass customization. In the second chapter, a focus group study shows the main barriers that SMEs are facing when implementing Industry 4.0. As SMEs need to be guided and supported in the process of implementation of Industry 4.0, the third chapter provides an SME 4.0 implementation tool kit.

Part V—Case Studies and Methodical Tools for Implementing Industry 4.0 in SMEs

In this part, topics previously covered theoretically are described by means of practical case studies. The case studies describe both the underlying theoretical concepts as well as the practical implementation and validation in the laboratory environment. In the first chapter, the contributors report about a case study of automatic product identification and inspection by using tools of Industry 4.0. In the second chapter, readers can expect a laboratory case study for intuitive collaboration between man and machine in SME assembly.

In the third chapter we give an overview on Axiomatic Design as a design methodology pertinent to the introduction of Industry 4.0 to SMEs as this method can be found within some chapters of this book. This chapter explains the basic rules of Axiomatic Design: the different domains and levels used in Axiomatic Design, the independence axiom and the information axiom. Further, this chapter introduces how Axiomatic Design can be used for the design of complex systems including both products and manufacturing systems.

We would like to thank the authors for their refreshing ideas and interesting contributions to this topic.

Bolzano, Italy — Dominik T. Matt
Košice, Slovakia — Vladimír Modrák
Leoben, Austria — Helmut Zsifkovits
May 2019

Contents

Notes on Contributors

Christopher A. Brown earned his Ph.D. at the University of Vermont, and then spent four years in the Materials Department at the Swiss Federal Institute of Technology. Subsequently he was a senior research engineer working on product and process research and design at Atlas Copco's European research center. Since the fall of 1989, Chris Brown has been on the faculty at WPI. Chris Brown has published over a hundred and fifty papers on AD, manufacturing, surface metrology, and sports engineering. He has patents on characterizing surface roughness, friction testing, and sports equipment. He also developed software for surface texture analysis. He currently teaches graduate courses on Axiomatic Design of manufacturing processes, and on surface metrology, and undergraduate courses on manufacturing and on skiing technology.

Nilubon Chonsawat was born in Rayong, Thailand. She received a Bachelor of Engineering Program in Computer Engineering in 2012, Faculty of Engineering and a Master of Science Program in Technopreneurshipin 2015, Institute of Field Robotics (FIBO), King Mongkut's University of Technology Thonburi (KMUTT), Thailand. In 2016, she started her Doctor of Philosophy Program in Industrial

Engineering, Department of Industrial Engineering, Faculty of Engineering, Chiang Mai University (CMU), Thailand. Her research interests focus on advanced technology, industrial engineering, logistics and supply chain management, Smart SMEs, organizational performance measurement, and the Industry 4.0 principle.

Patrick Dallasega received degrees from the Free University of Bolzano, Bolzano, Italy, the Polytechnic University of Turin, Turin, Italy, and the Ph.D. degree from the University of Stuttgart, Stuttgart, Germany. He is an Assistant Professor of project management and industrial plants design with the Faculty of Science and Technology, the Free University of Bolzano. He was a Visiting Scholar with the Excellence Center in Logistics and Supply Chain Management Chiang Mai University, Chiang Mai, Thailand, and the Worcester Polytechnic Institute in Massachusetts, Worcester, MA, USA. His main research interests are supply chain management, Industry 4.0, lean construction, lean manufacturing and production planning, and control in MTO and ETO enterprises.

Luca Gualtieri is an Industrial Engineer specialized in manufacturing and logistics. He is also a certified Occupational Health and Safety Manager and Trainer, mainly focused on safety of machinery. He is working in the Smart Mini Factory lab as research fellow and doctoral candidate for the Industrial Engineering and Automation (IEA) group of the Faculty of Science and Technology, Free University of Bozen-Bolzano. He is responsible for research in the field of industrial collaborative robotics. In particular, he is focusing on human–machine interaction from the point of view of the operator's occupational health and safety conditions, ergonomics, and shared workplace organization. He is involved in the EU project "SME 4.0—Industry 4.0 for SMEs" as well as industry projects on collaborative robotics and workplace design.

Johannes Kapeller received the Dipl.-Ing. Degree in Industrial Logistics from the Montanuniversitaet Leoben, Austria. In 2018, he completed his Ph.D. in Industrial Engineering with special focus on the sequential combination of production control strategies within the line production area. Since 2018, he works as a management consultant at the Boston Consulting Group. Prior he has been a Postdoc at the Montanuniversitaet in Leoben and a visiting researcher within the research project "SME 4.0—Industry 4.0 for SMEs" at the University of Chiang Mai, Thailand. Johannes published his work in high-quality peer-reviewed journals.

Dominik T. Matt holds the Chair for Production Systems and Technologies and heads the research department "Industrial Engineering and Automation (IEA)" at the Faculty of Science and Technology at the Free University of Bozen-Bolzano. Moreover, Dominik is the Director of the Research Center Fraunhofer Italia in Bolzano. Dominik coordinates as Principal Investigator the Horizon 2020 research project SME 4.0 as Lead Partner. His research primarily focuses on the areas of Industry 4.0 and Smart Factory, Lean and Agile Production, on the planning and optimization of assembly processes and systems, as well as on organizational and technical aspects of in-house logistics. He has authored more than 200 scientific and technical papers in journals and conference proceedings and is member of numerous national and international scientific organizations and commitees (e.g., AITeM—Associazione Italiana di Tecnologia Meccanica | WGAB—Academic Society for Work and Industrial Organization | EVI—European Virtual Institute on Innovation in Industrial Supply Chains and Logistic Networks).

Vladimír Modrák is Full Professor of Manufacturing Technology at Faculty of Manufacturing Technologies of Technical University of Kosice. He obtained his Ph.D. degree at the same University in 1989. His research interests include cellular manufacturing systems design, mass customized manufacturing and planning/scheduling optimization. Since 2015 he is a Fellow of the European Academy for Industrial Management (AIM). He was the leading editor of three international books, *Operations Management Research and Cellular Manufacturing Systems*, *Handbook of Research on Design and Management of Lean Production Systems*, and *Mass Customized Manufacturing: Theoretical Concepts and Practical Approaches*. He is also active as editorial board member in several scientific journals and committee member of many international conferences. Moreover, Vladimiír is Vice-Rector for International Relations of Technical University of Kosice.

Guido Orzes received his M.Sc. in Management Engineering from the University of Udine (Italy) in 2011 with summa cum laude. In 2015, he obtained his Ph.D. degree in Industrial and Information Engineering (topic: operations management) from the University of Udine (Italy). Currently, he is an Assistant Professor in Management Engineering at the Free University of Bozen-Bolzano (Italy). He is also Honorary Research Fellow at the University of Exeter Business School (UK) and Visiting Scholar at the Worcester Polytechnic Institute (USA). His research focuses on international sourcing and manufacturing and their social and environmental implications. He has published over 70 scientific works on these topics in leading operations management and international business journals (e.g., *International Journal of Operations & Production Management*, *International Journal of Production Economics*, *International Business Review*, and *Journal of Purchasing and Supply Management*) as well as in conference proceedings and books. He is involved (as work package leader or coinvestigator) in various EU-funded research projects on global operations management and Industry 4.0. He is also Associate Editor of the *Electronic Journal of Business Research Methods* and member of the board of the European division of the Decision Science Institute.

Ilaria Palomba received the M.Sc. degree in Product Innovation Engineering from the University of Padua, in 2012. In 2016, she obtained the Ph.D. degree in Mechatronics and Product Innovation Engineering from the University of Padua with a dissertation on "State estimation in multibody systems with rigid or flexible links". Currently, she is an Assistant Professor in Applied Mechanics at the Free University of Bozen-Bolzano. Dr. Palomba's research activities chiefly concern theoretical and experimental investigations in the fields of mechanics of machines, mechanical vibrations, multibody dynamics, and robotics and automation. In particular, her research interests are focused on the following topics: nonlinear state estimation for multibody systems with rigid and flexible links; model reduction of vibrating systems; robotic grasping systems for soft and fragile bodies; structural modifications and model updating of vibrating systems; design of advanced mechatronic systems.

Ján Pitel is a Full Professor in automation at the Technical University of Košice (Slovakia) and he currently works at the Faculty of Manufacturing Technologies with a seat in Prešov as vice-dean and head of the Institute of Production Control. His research activities include modelling, simulation, automatic control and monitoring of machines and processes. He is author of more than 60 papers registered in databases WoS and Scopus with more than 200 SCI citations. He is inventor and coinventor of more than 50 patents and utility models. He has been leader of many national projects (e.g., EU Structural Funds projects) and currently he participates in 2 EC-funded projects (H2020, Erasmus+). As vice-dean for external relationships he is responsible for mobility programs under the framework of projects ERASMUS and CEEPUS.

Robert Poklemba is Ph.D. student in the department of automotive and manufacturing technologies at the Technical University of Kosice (Slovakia). His dissertation thesis is focused on composite material based on polymer concrete. He worked at the project as an early stage researcher and was part of a team that was focused on organizational and managerial models in a mass customized environment for small and medium enterprises.

Sakgasem Ramingwong is an Associate Professor in Industrial Engineering at the Faculty of Engineering at Chiang Mai University, Thailand. His research interests are in industrial logistics and supply chain management.

Erwin Rauch holds an M.Sc. in Mechanical Engineering from the Technical University Munich (TUM) and an M.Sc. in Business Administration from the TUM Business School and obtained his Ph.D. degree in Mechanical Engineering from the University of Stuttgart with summa cum laude. Currently he is an Assistant Professor for Manufacturing Technology and Systems at the Free University of Bolzano, where he is the Head of the Smart Mini Factory laboratory for Industry 4.0. His current research is on Industry 4.0, Social Sustainability in Production, Smart and Sustainable Production Systems, Smart Shopfloor Management and Engineer/Make to Order.

He has 10 years of experience as Consultant and later Associate Partner in an industrial consultancy firm operating in production and logistics. He is project manager of the EU-funded H2020 research project "SME 4.0—Industry 4.0 for SMEs" in an international partner consortium. Further, he is author and coauthor of more than 130 scientific and non-scientific books, chapters of books, articles, and other contributions and received several awards for scientific contributions.

Hermann Reiter received the Dipl.-Ing. Degree in Production and Management from the University of Applied Science Steyr, Austria. In 2009 he completed his Master's in Supply Chain Management and in 2017 his Executive MBA from the Danube University Krems, Austria. Since 2004, he was working in different positions, e.g., project manager, head of production planning and head of supply chain management in industrial manufacturing companies. He was working in the pharmaceutical area for 7 years. In 2012 he moved to the automotive supplier industry being responsible for the global supply chain and ERP-implementation in Austria and China. Since 2018 he is a member of the directors of global automotive supplier for headlamps.

Rafael A. Rojas received an M.Sc. degree in Mechanical Engineering at the Sapienza University of Rome. In 2016 he concluded his Ph.D. in theoretical and applied mechanics with ad dissertation on optimal control theory for semi-active actuators. Since 2016, he works as postdoc research fellow at the Free University of Bozen-Bolzano in the Smart Mini Factory laboratory, focusing in his research on cyber-physical production systems, smart manufacturing control systems, connectivity and interoperability, and collaborative robotics. As Visiting Scholar, he worked in the research project "SME 4.0—Industry 4.0 for SMEs" with the Faculty of Mechanical Engineering at Worcester Polytechnic Institute (WPI) in Massachusetts. Rafael published his work in high-quality peer-reviewed journals like *Robotics and Automation Letters* and *Mechanical Systems and Signal Processing*.

Manuel A. Ruiz Garcia received his Master's degree in Control Engineering in 2013 and his Ph.D. in Engineering in Computer Science in 2018, both from Sapienza University of Rome, Rome, Italy.

He is a postdoc research fellow in the Smart Mini Factory Laboratory of the Free University of Bozen-Bolzano, Bolzano, Italy. His research interests include reactive control of mobile manipulators, robotics perception, collaborative robotics, and human–robot cooperation. He had been involved in the EU Projects "NIFTi—Natural human-robot cooperation in dynamic environments" and "TRADR—Long-term human-robot teaming for disaster response" and is currently involved in the EU Project "SME 4.0—Industry 4.0 for SMEs" as well as in industry projects on collaborative robotics and robotics perception. Dr. Ruiz Garcia published his work in high-quality peer-reviewed journals and conference proceedings like *IEEE Robotics and Automation Letters*, Proceedings of the IEEE International Conference on Robotics and Automation and Proceedings of the International Conference on Computer Vision.

Zuzana Šoltysová completed her Bachelor's and Master's at the Technical University of Kosice, Faculty of Manufacturing Technologies with a seat in Presov, Department of Manufacturing Management in the field of Manufacturing Management. She completed her Ph.D. Study at Technical University of Kosice, Faculty of Manufacturing Technologies with a seat in Presov, Department of Manufacturing Management in the field of Industrial Technology. Her Ph.D. thesis was focused on the research of product and production complexity in terms of mass customization. Currently, she is an Assistant Professor at Faculty of Manufacturing Technologies with a seat in Presov at Department of Manufacturing Management. Moreover, her research activities include complexity, throughput, axiomatic design, and production line balancing rate.

Apichat Sopadang was born in Chiang Mai, Thailand. He graduated from Chiang Mai University, Thailand in 1987 with a degree in industrial engineer. For several years, he worked as a maintenance planning engineer in Electricity Generator Authority of Thailand (EGAT). He completed his Ph.D. from Clemson University, USA in 2001. Following the completion of his Ph.D., he is working for Chiang Mai University as an Associate Professor and head of Excellence Center in Logistics and Supply Chain Management (E-LSCM). He is a frequent speaker at industry and academic meetings. His current research areas are on Industry 4.0, Sustainability Supply Chain, Aviation Logistics, Lean Manufacturing System and Performance Measurement. Dr. Sopadang also served as a consultant in many private organizations in Thailand and international organizations such as the Asian Development Bank (ADB) and The Japan External Trade Organization (JETRO). He is author and coauthor of more than 100 academic papers that include book chapters and articles.

Korrakot Tippayawong graduated with B.Eng., M.Eng. and Ph.D. in Industrial Engineering from Chiang Mai University, Thailand, Swinburne University of Technology, Australia, and Tokyo Institute of Technology, Japan, respectively. She has over 20 years' experience in teaching, research, and industrial consultation. Korrakot has worked with more than 300 SMEs as well as a number of large public and private enterprises. She is currently an Assistant Professor at Department of Industrial Engineering, Chiang Mai University. Her research focuses on logistics & supply chain, industrial engineering & management. She has received many major grants, including those from Thai Ministry of Industry, Ministry of Science and Technology, and European Horizon 2020 MSCA-RISE programme.

Walter T. Towner teaches for the Foisie Business School at Worcester Polytechnic Institute, Worcester Massachusetts, USA. Courses include Achieving Effective Operations, Engineering Economics, Operations Management, Lean Process Design, Productivity Management, Design and Analysis of Manufacturing Processes, Six Sigma, and Axiomatic Design Theory. The courses were taught at Raytheon, UTC, Natick Labs, NE Utilities, NSTAR, ECNE, Public Service of NH, Electric Boat, UMass Hospital, and General Dynamics and GE Healthcare. These courses are interdisciplinary and combine elements of engineering, finance, and management. Prof. Towner is the former owner of a metal fabrication and laser cutting manufacturing company serving semiconductor equipment, medical equipment, and nuclear power. Degrees from WPI include BS Mechanical Engineering, MS Operations & Information Technology, MS Manufacturing Engineering, an M.B.A. from Babson College and alumnus of Owner/President

Management Program at Harvard Business School. The WPI Alumni Association awarded the John Boynton Award Young Alumni Award and the Herbert F. Taylor Award for service to the university. Awarded Provost's Undergraduate Capstone Project Award three times. Prof. Towner has completed 1/3 of the 2200-mile Appalachian Trail in the Eastern US.

Andrew R. Vickery is a Ph.D. student of manufacturing engineering in the mechanical engineering department of Worcester Polytechnic Institute (USA). He earned a Master's in materials science and engineering at WPI. His main research interests are the design of sustainable manufacturing systems for SMEs, systems design through axiomatic design, and the study and optimization of value chains for SMEs.

Renato Vidoni received his M.Sc. in Electronic Engineering—focus: industrial automation—from the University of Udine, Italy, in 2005. In 2009, he obtained his Ph.D. degree in Industrial and Information Engineering from the University of Udine, Italy. Currently, he is Associate Professor in Applied Mechanics at the Free University of Bozen-Bolzano (Italy) where he is responsible of the activities in robotics and mechatronics inner the Smart Mini Factory laboratory for industry 4.0 and he is the Head of the Field Robotics laboratory. He is course director of the M.Sc. in Industrial Mechanical Engineering and Rector's delegate at the CRUI (Conference of the Rectors of Italian Universities) Foundation's for the University-Business Observatory. His research activity is documented by more than 100 scientific contributions that deal with topics of the Applied Mechanics sector both in "classical" fields as well in new and emerging domains (e.g., industry and Agri 4.0). The recent research activity can be grouped in three different research areas that fall into the "Industrial Engineering and Automation" macro-area of the Faculty of Science and Technology: High-performance automatic machines and robots, Mechatronic applications for Energy Efficiency, Mechatronics and Robotics for field activities.

Erich J. Wehrle is an Assistant Professor for Applied Mechanics at the Free University of Bozen-Bolzano. He holds a Bachelor of Science in Mechanical Engineering from the State University of New York at

Buffalo (USA) and a Master of Science in Mechanical Engineering from the Technical University of Munich. He carried out research in structural design optimization with crash loading under uncertainty at Institute of Lightweight Structures leading to the Doktor-Ingenieur (doctor of engineering) degree. At his current appointment at the Free University of Bozen-Bolzano, he is deputy director of the Mechanical Lab. His current research includes applied mechanics, design optimization, multibody dynamics, topology optimization for compliant mechanisms, nonlinear mechanics, Industry 4.0, and engineering education. He is author and coauthor of more than 60 book chapters, conference papers, articles, and other contributions.

Christian Weichbold received the Dipl.-Ing. Degree in Industrial Logistics from the Montanuniversitaet Leoben, Austria. Since 2016, he works as a researcher at the Voestalpine Stahl Donawitz GmbH in the blast furnace area for reduction metallurgy with special focus on his Ph.D.-project material tracking of bulk material (iron ore sinter) and its properties.

Warisa Wisittipanich received a Master of Science in Systems Engineering from George Mason University, Virginia, USA in 2006 and obtained a Doctor of Engineering from Asian Institute of Technology, Thailand in 2012. Currently, she is working as an Assistant Professor at the Department of Industrial Engineering, Faculty of Engineering, Chiang Mai University, Chiang Mai, Thailand. Her areas of interests and research include operations research, production scheduling and sequencing, inbound and outbound truck scheduling, vehicle routing problem, supply chain and logistics management, and metaheuristic applications for real-world optimization problems.

Manuel Woschank is a Postdoc Senior Lecturer and Researcher at the Department of Economic and Business Management, Chair of Industrial Logistics at the University of Leoben. Manuel holds a Diploma in Industrial Management and a Master's Degree in International Supply Management from the University of Applied Sciences FH JOANNEUM, Austria and a Ph.D. in Management Sciences from the University of Latvia, Latvia. He has published a multitude of international, peer-reviewed papers and has conducted research projects with voestalpine AG, Stahl Judenburg—GMH Gruppe, Knapp AG, SSI Schaefer, MAGNA International Europe GmbH, Hoerbiger Kompressortechnik Holding GmbH, voestalpine Boehler Aerospace GmbH & Co KG, etc. His research interests include Logistics Systems Engineering, Logistics Process Optimization, Production Planning and Control Systems, and Behavioral Decision Making Theory. He is a member of BVL, WING, Logistikclub Leoben, ILA, and GfeW.

Kamil Židek is focused on the research in the area of image processing for manufacturing applications and knowledge extraction by algorithms of artificial intelligence. He completed his habitation thesis named "Identification and classification surface errors of mechanical engineering products by vision systems" at Faculty of Manufacturing Technologies of Technical University of Kosice. Currently, he is Associate Professor at this faculty. He published his research titled "Embedded vision equipment of industrial robot for in-line detection of product errors by clustering-classification algorithms" in the mentioned area in Current Contents (IF 2016 = 0.987). He is the lead author of one Current Contents article and coauthor of 3 other CC articles. He is inventor and coinventor of 6 published patents 4 and 8 utility models. He is also coauthor of 1 monograph, 2 university textbooks, and 3 scripts.

Helmut Zsifkovits holds the Chair of Industrial Logistics at the Department of Economics and Business Management at Montanuniversitaet Leoben, Austria. He graduated from the University of Graz, Austria and has professional experience in automotive industry, logistics consultancy, and IT. His research interests include logistics systems engineering, supply chain strategy, and operations management. He is a Board Member of the European Certification Board for Logistics (ECBL), Vice-President of Bundesvereinigung Logistik Austria (BVL), and President of Logistics Club Leoben. In 2018, he was appointed as an Adjunct Professor at the University of the Sunshine Coast, Australia. He has teaching assignments at various universities in Austria, Latvia, Colombia, and Germany, and is the author of numerous scientific publications and several books.

List of Figures

List of Tables

Part I

Introduction to Industry 4.0 for SMEs

1

SME 4.0: The Role of Small- and Medium-Sized Enterprises in the Digital Transformation

Dominik T. Matt and Erwin Rauch

1.1 Introduction

In recent years, the industrial environment has been changing radically due to the introduction of concepts and technologies based on the fourth industrial revolution (Sendler 2013). At the Hanover Fair 2011 for the first time, a synonym for such a new industrial revolution was mentioned, "Industry 4.0". The focus of Industry 4.0 is to combine production, information technology and the internet. Thus, newest information and communication technologies are combined in Industry 4.0 with traditional industrial processes (BMBF 2012).

In recent years, the global economy has become a strong competitor for industry in Europe. It is no longer enough to produce faster, cheaper, and with higher quality than the competitors, defending the achieved competitive advantage. The industry needs to introduce new types of innovative and "digital" production strategies to maintain

D. T. Matt · E. Rauch (✉)
Faculty of Science and Technology, Free University of Bozen-Bolzano, Bolzano, Italy
e-mail: erwin.rauch@unibz.it

D. T. Matt et al. (eds.), *Industry 4.0 for SMEs*,
https://doi.org/10.1007/978-3-030-25425-4_1

the current competitive advantage in the long term (Manhart 2013). The fourth industrial revolution should extend to the whole production and supply chain of components, and not only, as in past revolutions, to the mechanical manufacturing process of products and the associated process organization. The development of Industry 4.0 should provide a contribution to tackle global challenges, like sustainability, resource and energy efficiency and strengthen competitiveness (Kagermann et al. 2013). In the whole production life cycle, the data exchange should be improved leading to advantages for all involved parties. More functionalities and customization options are gained for the client and more flexibility, transparency, and globalization for the supply chain (Baum 2013). In addition, the return to uniqueness should be achieved by the fourth revolution (Hartbrich 2014). Therefore, to remain competitive, the ability to respond to customer requirements quickly and flexibly and to produce high version numbers at low batch sizes, must increase (Spath et al. 2013). Industry 4.0 aims to implement highly efficient and automated manufacturing processes, usually known from mass production, also in an industrial environment, where individual and customer-specific products are fabricated according to mass customization strategies (Modrak et al. 2014). Mass customization means the production of products customized by the customer, at production costs similar to those of mass-produced products. A production, based on the principle of Industry 4.0, creates the conditions to replace traditional structures, which are based on centralized decision-making mechanisms and rigid limits of individual value-added steps. These structures are replaced by flexible reconfigurable manufacturing and logistics systems, offering interactive and collaborative decision-making mechanisms (Spath et al. 2013).

In recent years, a growing number of authors have addressed the topic of Industry 4.0 for SMEs in their scientific works (Matt et al. 2016; Bär et al. 2018; Türkeş et al. 2019). In addition, the European Commission (EC) actively supports SMEs by providing direct financial support and indirect support to increase their innovation capacity through Horizon 2020. Thus, underpinning the Europe 2020 strategy for smart and sustainable growth, the EC supports research, development and innovation projects with the aim of creating a favorable ecosystem for SME innovation and development. Due to their flexibility, the entrepreneurial spirit, and the innovation capabilities, SMEs have proved to be more robust than large and multi-national enterprises, as the previous financial and economic crisis showed (Matt 2007; Matt et al. 2016). Typically, SMEs are not only adaptive and innovative in terms of their products, but also in terms of their manufacturing practices. Recognizing the continuing competitive pressures, small organizations are becoming increasingly proactive in improving their business operations (Boughton and Arokiam 2000), which is a good starting point for introducing new concepts like Industry 4.0. The successful implementation of Industry 4.0 has to take place not only in large enterprises but in particular, in SMEs (Sommer 2015). Various studies point out relevant changes and potential for SMEs in the context of Industry 4.0 (Rickmann 2014). Industry 4.0 technologies offer great opportunities for the SME sector to enhance its competitiveness. SMEs are most likely to be the big winners from the shift; they are often able to implement the digital transformation more rapidly than large enterprises, because they can develop and implement new IT structures from scratch more easily (Deloitte 2015). Many small- and medium-sized companies are already focusing on digitized products in order to stand out in the market (PWC 2015). The integration of information and communication technology (ICT) and modern Industry 4.0 technologies would transform today's SME factories into smart factories with significant economic potential (Lee and Lapira 2013; Gualtieri et al. 2018).

Industry 4.0 represents a special challenge for businesses in general and for SMEs in particular. The readiness of SME adapted Industry 4.0 concepts and the organizational capability of SMEs to meet this challenge exist only in part. The smaller SMEs are, the higher the risk that they will not be able to benefit from this revolution. European SMEs are conscious about the knowledge in adaption deficits. This opens the need for further research and action plans for preparing SMEs in a technical and organizational direction (Sommer 2015). The introduction of Industry 4.0 often shows difficulties and leads to headlines such as "most SME production companies are currently not yet ready for Industry 4.0", "SMEs are missing the trends of the future" or "Industry

4.0 has not arrived at SMEs" (Olle and Clauß 2015). Today, most SMEs are not prepared to implement Industry 4.0 concepts (Brettel et al. 2014; Orzes et al. 2018).

Therefore, special research and investigations are needed for the implementation of Industry 4.0 technologies and concepts in SMEs. SMEs will only achieve Industry 4.0 by following SME-customized implementation strategies and approaches and realizing SME-adapted concepts and technological solutions. Otherwise, actual effort for sensitization and awareness building among SMEs for Industry 4.0 will not show the expected success and results. According to this identified gap, this book considers and investigates the specific requirements of SMEs introducing Industry 4.0 and reflects on opportunities and difficulties in the digital transformation of manufacturing, logistics and organizational processes in SMEs.

1.2 Industry 4.0 as the Fourth Industrial Revolution

1.2.1 Origin and Characterization of Industry 4.0

In 2011, the German group of scientists Acatech (Deutsche Akademie der Technikwissenschaften) presented the term "Industrie 4.0" for the first time during the Hannover Fair, symbolizing the beginning of the fourth industrial revolution. The fourth industrial revolution can be described as the introduction of modern ICT in production. At the end of the eighteenth century, the first industrial revolution was initiated by inventing the machine and thus replacing muscle force. With the development of the industrial nations around the year 1870, the second industrial revolution began. The second industrial revolution was determined by the introduction of the division of labor and mass production with the help of electrical energy (Kagermann et al. 2013). The third industrial revolution referred to the multiplication of human brainpower to the same extent as human muscle power had been multiplied in the first and second industrial revolutions (Balkhausen 1978). The fourth industrial revolution describes a further step ahead where people, machines, and products are directly connected with each other and their environment (Plattform Industrie 4.0 2014). Figure 1.1 illustrates the fourth industrial revolutions and how each revolution has affected manufacturing.

In 2013, the working group "Arbeitskreis Industrie 4.0", consisting of representatives of industry, research, the "Forschungsunion" and associations, presented their report of recommendations for the introduction of Industry 4.0 to the government at the Hannover Fair. In addition, in 2013, the German associations BITKOM, VDMA, and ZVEI created the "Plattform Industrie 4.0" as a reference platform for a further promotion of Industry 4.0 in German politics and industry (Acatech 2013). In the first few years after the presentation and introduction of this new term, the characterization and description of Industry 4.0 varied greatly and a concrete, generally accepted definition of Industry 4.0 did not exist at that time (Bauer et al. 2014).

The main objectives of Industry 4.0 include individualization of customer requirements, flexibility, and adaptability of manufacturing and logistics systems, improved decision-making, the integration of ICT and Cyber-Physical Systems (CPS), the introduction of advanced production technologies (additive manufacturing, precise manufacturing,...), intelligent automation concepts, adapted business

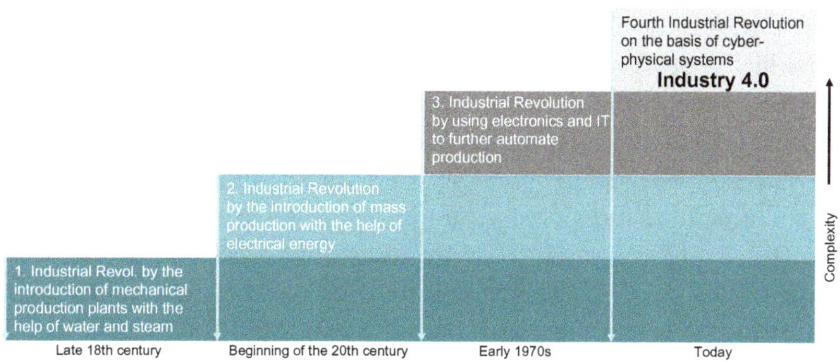

Fig. 1.1 The four industrial revolutions (Adapted from Kagermann et al. 2013)

and organizational models as well as concepts for more sustainable production and logistics processes (Spath et al. 2013).

One of the biggest opportunities of Industry 4.0 is expected in capabilities of CPS for self-organization and self-control in so-called Cyber-Physical Production Systems (CPPS) (Monostori 2014). CPS are systems of collaborating computational entities which are in intensive connection with the surrounding physical world and its on-going processes, providing and using, at the same time, data-accessing and data processing services available on the internet. In other words, CPS can be generally characterized as "physical and engineered systems whose operations are monitored, controlled, coordinated, and integrated by a computing and communicating core" (Rajkumar et al. 2010; Laperrire and Reinhart 2014). Here the physical and digital world are combined and interact through the so-called "Internet of things" (Lee and Seshia 2016). Production data is provided of a completely new quality and with real-time information on production processes. This is possible through the comprehensive placement of production with sensors and the continuous integration of intelligent objects (Spath et al. 2013; Zhou et al. 2018). Future data models will work in real time and the production will be more transparent (Rauch et al. 2018a). An image of the production occurs as essential to new, decentralized and real-time production control. Such a production control system can cope in the future with uneven workloads in the short term and make complex decisions decentralized in a customized production environment (Spath et al. 2013; Meissner et al. 2017). With CPS, the pursuit of economies of scale as a means to reduce costs loses significance because many individual process steps can be combined more flexibly by the use of computer-based modularization of production runs. This means that with networked production technologies, an individualized production at low costs will become possible. In a traditional production system, the fullfilment of individual customer requirements is possible only by frequent change of variants or by individual execution of individual production steps, which often means higher costs because of higher production costs (BMBF 2012; Lee et al. 2015).

According to a definition of the European Commission (EC 2015), Industry 4.0 consists of a number of new and innovative technologies:

- **Information and communication technology (ICT)** to digitize information and integrate systems at all stages of product creation and use (including logistics and supply), both inside companies and across company boundaries.
- **Cyber-physical systems** that use ICTs to monitor and control physical processes and systems. These may involve embedded sensors, intelligent robots that can configure themselves to suit the immediate product to be created, or additive manufacturing (3D printing) devices.
- **Network communications** including wireless and internet technologies that serve to link machines, work products, systems, and people, both within the manufacturing plant, and with suppliers and distributors.
- **Simulation**, modeling, and virtualization in the design of products and the establishment of manufacturing processes.
- **Big data analysis** and exploitation, either immediately on the factory floor, or through cloud computing.
- **Digital assistance systems for human workers**, including robots, augmented reality, and intelligent aid systems.

1.2.2 Industry 4.0—A Challenge for Europe and Beyond

The EU supports industrial change through its industrial policy and through research and infrastructure funding. Member States are also sponsoring national or trans-national initiatives. The need for investment, changing business models, data issues, legal questions of liability and intellectual property, standards, and skill mismatches are among the challenges that must be met if benefits are to be gained from new manufacturing and industrial technologies. If these obstacles can be overcome, Industry 4.0 may help to reverse the past decline in industrialization and increase total value added from manufacturing to a targeted 20%. From 2014 to 2020, the Horizon 2020 research programme's industrial leadership pillar provides almost €80 billion for research and innovation, including support for developing key enabling technologies. In addition, "Factories of the Future" is a public-private partnership (PPP), launched initially under the earlier Seventh Framework Programme but continuing under Horizon 2020, that centers on advanced, smart, digital, collaborative, human-centred, and customer-focused manufacturing (EC 2015). The EU is undertaking a wide range of additional measures to support and connect national initiatives that focus on Industry 4.0 and the digitalization of industry. An overview of the European Commission shows that there are more than 30 national and regional initiatives at European level: e.g., Plattform Industrie 4.0 in Germany Catapult in UK, Fabbrica Digitale in Italy, Made Different in Belgium, Industry du Futur in France, Produktion 2030 in Sweden, Made in Denmark, Smart Industry in Netherlands, Produtech in Portugal, Industria Conectada 4.0 in Spain, Production of the Future in Austria, Průmysl 4.0 in Czech Republic, Smart Industry SK in Slovakia and many others (Plattform Industrie 4.0 2019). Also, in the new EU flagship funding program for Research and Innovation, "Horizon Europe", research for Industry 4.0 technologies, sustainable production and artificial intelligence (AI) will be addressed in the Cluster "Digital and Industry" (€15 billion) of the "Global Challenges and Industrial Competiveness pillar" (€52.7 billion) (EC 2018).

Also, in the United States, the National Science Foundation (NSF) offers funding programs for topics like Industry 4.0 technologies, in the United States better known as "Smart Manufacturing", "Smart Factory" or "Internet of Things". The Division of Civil, Mechanical and Manufacturing Innovation (CMMI) addresses research topics like additive manufacturing, CPS, computational engineering (NSF 2019). Further, the "Manufacturing USA" program, a PPP of 14 manufacturing institutes, connect member organizations, work on major research, and development collaboration projects and train people on advanced manufacturing skills (Manufacturing USA 2019).

The trend of Industry 4.0 has also reached and influenced Asian nations in their political programs and future strategies. China wants to catch up with the strongest economic powers in the world. The government has therefore drawn up the program "Made in China 2025", an ambitious plan to bring the country to the technological forefront. Where manual labor currently accounts for the majority of value added, automated production should dominate in the future—and turn the country into an "industrial superpower" In the Made in China 2025 strategy, the government has identified ten industries in which it wants to form leading Chinese global companies among which also are high-end automation, robotics, artificial intelligence, and robotics (FAZ 2019).

Whilst China has its Made in China 2025 strategy for upgrading its manufacturing sector, India has its "Make in India" initiative to boost investments and improvements across its various industrial sectors. The Indian automotive industry occupies a prominent place within the Indian economy and therefore India is trying to remain attractive and ready for Industry 4.0 as well as the Internet of Things (Moreira 2017).

In addition, other Asian countries like Thailand are dealing with Industry 4.0. "Thailand 4.0" is an economic model that aims to unlock the country from several economic challenges resulting from past economic development models which place emphasis on agriculture (Thailand 1.0), light industry (Thailand 2.0), and advanced industry (Thailand 3.0). The objective of Thailand 4.0 is to become attractive for innovative and value-based industry. Besides a strong emphasis on agriculture, food health and medical technologies, research topics like robotics, mechatronics, artificial intelligence, Internet of Things, and smart devices are also a focus of the Thailand 4.0 initiative (ThaiEmbDC 2019).

1.3 The Contribution of Small and Medium Enterprises to Economic Development

1.3.1 The Role of SMEs in the European Economy

Small- and medium-sized enterprises (SMEs) are a focal point in shaping enterprise policy in the European Union. The European Commission considers SMEs and entrepreneurship as key to ensuring economic growth, innovation, job creation, and social integration in the EU (Eurostat 2018).

SMEs—defined by the European Commission as having fewer than 250 employees, an annual turnover of less than €50 million, or a balance sheet total of no more than €43 million—are the backbone of the European economy (Kraemer-Eis and Passaris). According to the SEM statistics published on the Eurostat database in 2015, there were 23.4 million SMEs in the European Union's non-financial business economy. Together they employed 91 million people and generated €3934 billion of value added. The economic contribution from SMEs was particularly apparent in Malta, Cyprus, and Estonia, with SMEs providing more than three quarters of the total value-added generated in each of their non-financial business economies. Large enterprises (0.2% of total enterprises in EU-28) generate 43.5% of value added and count for around one third of the number of employees (33.7%). Two third of employees in the non-financial sector are employed by SMEs, with 29.1% in micro enterprises with less than 10 persons employed and 20.2% in small enterprises with less than 49 persons employed and 17.1% in medium enterprises with less than 250 employees. All three sizes of SMEs are contributing nearly equally to value added in the EU-28 with 20.3% for micro enterprises, 17.6% for small enterprises and 18.5% for medium-sized enterprises (Eurostat 2018). Until the next SME statistics report update, planned for May 2019, recent statistics confirm that, in 2017, there were 24.6 million SMEs in the EU-28 non-financial business sector, of which 22.9 (~93%) million were micro SMEs, 1.4 million were small SMEs (~6%) and 0.2 million were medium-sized SMEs (~1%). In contrast, there were only 47,000 large enterprises (EU 2018).

The economic contribution from SMEs was particularly apparent in Malta, Cyprus, Estonia, Greece, Latvia, Lithuania, and Italy with SMEs providing more than two thirds of the total value added generated in each of their non-financial business economies. In the overall EU-28, the value-added generated by SMEs is 56.5% (Eurostat 2018).

Furthermore, the number of SMEs in the EU-28 increased by 13.8% between 2008 and 2017. SMEs represented 88.3% of all EU-28 enterprises exporting goods. The rest of the world accounted for only 30% of all SME exports. In 2016, 80% of all exporting SMEs were engaged in intra-EU trade, while less than half of exporting SMEs sold to markets outside the EU-28, and slightly more than a quarter of exporting SMEs sold to both markets (Brusselsnetwork 2018).

1.3.2 The Role of SMEs in the United States

In the United States, the definition of SMEs is dependent not only on the number of employees, annual sales, assets, or any combination of these, it also varies from industry to industry, based on the North American Industry Classification System (NAICS). In manufacturing, for example, an SME is defined as having 500 employees or less, whereas in wholesale trades, it is typically 100 employees or less (Madani 2018). It has to be stated here that all the following numbers refer to SMEs including also financial businesses (different to the EU definition) as the US definition makes no distinction. Small to medium enterprises also make up the vast majority of businesses in the United States. According to the U.S. Census Bureau Data from 2016, of the 5.6 million employer firms in the United States (Ward 2018):

- 99.7% had fewer than 500 employees
- 98.2% had fewer than 100 employees
- 89.0% had fewer than 20 employees.

SMEs contributed 46% of the private nonfarm GDP in 2008 (the most recent year for which the source data are available), making them hugely important for economic growth, innovation, and diversity (Ward 2018). Regarding the employed workforce, small businesses in the US employed 58.9 million people, or 47.5% of the private workforce, in 2015. SMEs contribute to economic growth in the United States as small businesses created 1.9 million net jobs in 2015. Firms employing fewer than 20 employees experienced the largest gains, adding 1.1 million net jobs. The smallest gains were in firms employing 100–499 employees, which added 387,874 net jobs (SBA 2018).

1.3.3 The Role of SMEs in Asia

SMEs represent a significant proportion of enterprises, especially in developing countries, as many Asian countries are (ADB 2018). SMEs are also the backbone of the Asian economy. They make up more than 96% of all Asian businesses, providing two out of three private-sector jobs on the continent. Therefore, it is vital for Asian economies' economic success that they have fully functioning support measures for SMEs (Yoshino and Taghizadeh-Hesary 2018).

For the 20 countries in Asia and the Pacific with available data during 2011–2014, on average, SMEs represented 62% of national employment, ranging from 4 to 97%, and accounting for 42% of gross domestic product (GDP), ranging from 12 to 60% (ADB 2018). Also, in Asia the definition of SMEs is heterogeneous varying from country to country. For example, SMEs are defined as having up to 1000 employees in some sectors in the People's Republic of China (PRC), but the cutoff is up to 200 workers for some sectors in Thailand. Malaysia considers manufacturing firms to be SMEs if they have fewer than 200 workers or revenue of less than RM50 million (about $12 million) (ADB 2018). In Japan, the definition of SMEs depends not only on the sector of activity but also on the value of capital and the number of employees (Madani 2018). Furthermore, very often, government agencies within the same country may use different definitions. For example, a ministry may use one definition while the national statistics office uses another, and a lending policy may adopt yet another (ADB 2018).

In the Association of Southeast Asian Nations (ASEAN) countries—Singapore, Brunei, Malaysia, Thailand, Philippines, Indonesia, Vietnam, Laos, Cambodia, Myanmar—micro, SMEs represent around 97–99% of the enterprise population. The SME sector tends to be dominated by micro enterprises, which typically account for 85–99% of enterprises (where data are available). There is a relatively low share of medium-sized enterprises across the region as a whole, which may be indicative of a "missing middle" in the region's productive structure. In most ASEAN countries, SMEs are predominantly found in labor-intensive and low value-added sectors of the economy, particularly retail, trade, and agricultural activities. As such, they continue to account for a high share of employment but a low share of gross value added in most countries. In the ASEAN region, SMEs account for around 66.3% of employment (based on the median) and 42.2% of gross value added (OECD/ERIA 2018).

1.3.4 The Role of SMEs in the World

For statistical purposes, the following numbers refer to the European definition of SMEs as the firms employing up to 249 persons, with the following breakdown: micro (1–9), small (10–49), and medium (50–249).

According to WTO calculations based on World Bank Enterprise Surveys covering over 25,000 SMEs in developing countries, direct exports represent just 7.6% of total sales of SMEs in the manufacturing sector, compared to 14.1% for large manufacturing enterprises. Among developing regions, Africa has the lowest export share at 3%, compared to 8.7% for developing Asia. Participation by SMEs indirect exports of services in developing countries is negligible, representing only 0.9% of total services sales compared to 31.9% for large enterprises (WTO 2016).

In many countries, and in particular Organization for Economic Cooperation and Development (OECD) countries, SMEs are key players in the economy and the wider eco-systems of firms. In the OECD area, SMEs are the predominant form of enterprise, accounting for approximately 99% of all firms. They provide the main source of employment, accounting for about 70% of jobs on average, and are major contributors to value creation, generating between 50 and 60% of value added on average. In emerging economies, SMEs contribute up to 45% of total employment and 33% of GDP. When taking the contribution of informal businesses into account, SMEs contribute to more than half of employment and GDP in most countries irrespective of income levels. In addition, SME development can contribute to economic diversification and resilience and therefore to a more sustainable economy. This is especially relevant for resource-rich countries that are particularly vulnerable to commodity price fluctuations (OECD 2017).

1.4 Current State of the Transition of Industry 4.0 to SMEs

1.4.1 State of the Art of Industry 4.0 for SMEs in Scientific Literature

For an analysis of the state of the art in the field of Industry 4.0 for SMEs, the scientific literature with the keywords "Industry 4.0" as well as "Smart Manufacturing" in the title, keywords and abstract is linked with the occurrence of the keywords "small- and medium-sized enterprises" as well as "SME" in the title of the papers. For the analysis, the database Scopus is used, which is known as a high-quality and comprehensive scientific database in the field of engineering. The results of the search are a list of 55 papers, of which 25 are conference contributions, 23 are journal articles, 5 book chapters, and 2 are reviews. The search results show an important increase in scientific papers starting from 2017 (see Table 1.1). Also, the found literature at the date of the search leads to the hypothesis that the topic Industry 4.0 in SMEs is of increasing importance for scholars in engineering and production research.

In the following, we will summarize the main findings in the current state analysis based on the identified papers. At the beginning of the scientific discussions of "Industry 4.0", most of the papers describe the big challenge that SMEs will face with the new hype of Industry 4.0 (Färber 2013; Matt et al. 2016). Later Reuter (2015) addresses the future importance of Industry 4.0 and IT in business continuity management of SMEs, but remains very vague about the use of the term

Table 1.1 Search results in SCOPUS

Year	Papers
2019[a]	9
2018	27
2017	12
2016	5
2015	1
2013	1
Sum	55

[a]28 March 2019

"Industry 4.0". Decker (2017) analyzes the readiness of Danish SMEs from the metal processing sector for Industry 4.0 using case study research. Up to this point, there was no maturity or readiness model available and thus the analysis was conducted basically on a qualitative level. The basic outcome is that SMEs at this time were not sure if, when and how they should start to introduce Industry 4.0 in their firms. Bollhöfer et al. (2016) describe the potential of service-based business models for SMEs although many uncertainties and barriers (see also Müller and Voigt 2016) discourage or limit SMEs from starting an implementation process for Industry 4.0. Later Seidenstricker et al. (2017), Safar et al. (2018), Müller (2019), and Bolesnikov et al. (2019) describe specific frameworks for introducing new and innovative as well as digital business models in SMEs. In 2016, Ganzarain and Errasti (2016) are the first to discuss the adoption of maturity models in SMEs to support the implementation of Industry 4.0 in SMEs. Later also, other authors like Wiesner et al. (2018) and Jones et al. (2018) came up with SME-specific maturity and readiness models. Mittal et al. (2018a) reviewed different smart manufacturing and Industry 4.0 maturity models and their implications for SMEs. Matt et al. (2016) describe a very early first attempt at a methodical approach to how SMEs can introduce Industry 4.0. Later in 2018, they refine the approach to a five-step methodology for SMEs (Matt et al. 2018a). Jørsfeldt and Decker (2017) as well as Jun et al. (2017) promote the concept of digitally enabled platforms generating entrepreneurial opportunities for smart SMEs but up to now such platforms do not really exist. Bakkari and Khatory (2017) as well as Schlegel et al. (2017) encourage SMEs in their works to put a certain emphasis on Industry 4.0 strategies in the integration of concepts for a more sustainable and ecological manufacturing environment. At this time, the concepts proposed are still on a very rough and abstract level without any clear and tangible recommendations about how to achieve the proposed goals. Several works try also to combine the advantages of Lean and Industry 4.0 or discuss why both principles are complementary and do not exclude each other (Matt et al. 2016; Müller et al. 2017; Rauch et al. 2017). Other papers are presenting competence centers, learning factories, and laboratories for Industry 4.0 specific research, training offers or knowledge transfer to smaller firms (Müller and Hopf 2017; Scheidel et al. 2018; Gualtieri et al. 2018). Rauch et al. (2018b) used Axiomatic Design in their research to develop a methodology for SMEs in order to introduce flexible and agile manufacturing systems. Goerzig and Bauernhansl (2018) propose a framework architecture for SMEs adopting Industry 4.0 in their firm. Similarly, other researchers are working on the development of Industry 4.0 tool kits and roadmaps in order to simplify the introduction of Industry 4.0 in SMEs (Mittal et al. 2018b; Modrak et al. 2019). Further investigation has also been started on social sustainability in SME manufacturing in the context of Industry 4.0 (Matt et al. 2018b). Moica et al. (2018) address the need of Industry 4.0 also for shop floor management and Menezes et al. (2017) the need for adapted Manufacturing Execution Systems (MES) for SMEs. Other works discuss the need for retrofitting old machinery to be prepared for Industry 4.0 (Pérez et al. 2018). Although the discussion about artificial intelligence had already started, Sezer et al. (2018) address the technologies to be used for low-cost predictive maintenance in SMEs. As Industry 4.0 is not limited to SMEs working in the manufacturing sector, Nowotarski and Paslawski (2017) started also to investigate the use of Industry 4.0 methods and technologies on-site in the construction industry, Weiß et al. (2018) in the textile industry and Zambon et al. (2019) in agriculture.

Analyzing the content of the identified scientific papers about Industry 4.0 in SMEs, the following hypotheses can be derived:

- Several works are dealing with innovative and digital business models, also putting SMEs in a position to take advantage of Industry 4.0 as a business model.
- Researchers propose that digital platforms are interesting opportunities for SMEs to increase their business, but there are no relevant implementations of such SME platforms up to now.
- Several researchers are working on readiness, assessment, or maturity models to help SMEs understand their actual status.
- There is a rising number of works on frameworks, tool sets as well as roadmaps to guide SMEs in the implementation of Industry 4.0.

- It seems that Industry 4.0 is more and more of interest for other industries rather than manufacturing (construction, textile, agriculture).
- Although there is still little research about artificial intelligence in manufacturing in general, there are first attempts to introduce low cost and easy approaches also in SMEs.
- Sustainability (in the sense of ecological as well as social sustainability) is gaining attention.

1.4.2 Current EU Research Initiatives on Industry 4.0 for SMEs

As identified in the current state of research and scientific literature, there is still need to further investigate the transition to Industry 4.0 in SME firms. The European Commission is already financing a few projects related to this important niche topic. A first small H2020 research project with around €70,000 of project volume with the title "Industrial FW 4.0 – Internet 4.0 based MES for the SME sector" (Grant agreement ID: 710130) was awarded in 2015 and ended in 2016. The first important research initiative in this direction started in January 2017 with the project "SME 4.0 – Industry 4.0 for SMEs" (Grant agreement ID: 734713) funded by the EC H2020 program. The details regarding this project will be explained in the following Sect. 1.5. In April 2017, the H2020 project "IoT4Industry – Towards smarter means of production in European manufacturing SMEs through the use of the Internet of Things technologies" (Grant Agreement 777455) started. Later, in October 2017, the H2020 project "L4MS – Logistics for Manufacturing SMEs" (Grant Agreement 767642) was awarded with funding of nearly €8 mio to investigate IoT platforms and smartization services for logistics in SMEs from manufacturing. Another related project is "ENIT – Agent 2.0 – The world's first edge computing solution for SMEs enabling energy efficiency, Industry 4.0 and new business models for the energy sector" (Grant agreement ID: 811640) with a project volume of €2 mio and EU funding of €1.4 mio which started in 2018 and focuses on the adoption of Industry 4.0 technologies in SMEs for energy management. A recently started small scale (project volume of around €70,000) H2020 project is "Katana – Bringing Industry 4.0 to the hands of small manufacturers: Feasibility study for scaling up Katana smart workshop software" (Grant agreement ID: 855987) which aims to realize smart workshop software for SMEs (Cordis 2019).

1.4.3 Summary of the State-of-the-Art Analysis

Both in the analysis of scientific works and of EU projects to research the introduction of Industry 4.0 in SMEs, it can be seen that the topic is still relatively "young" and unexplored as it has basically only been dealt with seriously since 2017. Therefore, it will require further efforts to provide appropriate instruments for SMEs introducing Industry 4.0 in practice. What is striking is that there are no best practice examples of SMEs where a big part of the Industry 4.0 technologies have been successfully adopted. For this reason, more attention should be paid to this in future work and pilot introductions in SMEs should be accompanied and documented by scientists. In addition, the overview of current research initiatives in Europe showed that the "SME 4.0" project is currently the only initiative that considers the problem holistically and from an international point of view and is not limited to partial aspects of Industry 4.0 in SMEs.

1.5 SME 4.0—Industry 4.0 for SMEs

1.5.1 "SME 4.0" Project Key Data and Objectives

The research project "SME 4.0" with the full title "Industry 4.0 for SMEs—Smart Manufacturing and Logistics for SMEs in an X-to-order and Mass Customization Environment" and a project volume of €954,000 received €783,000 funding from the European Commission H2020 MSCA Research and Innovation Staff Exchange (RISE) program. The project has a duration of four years starting in 2017 and ending at the end of 2020. The project consortium consists of the following

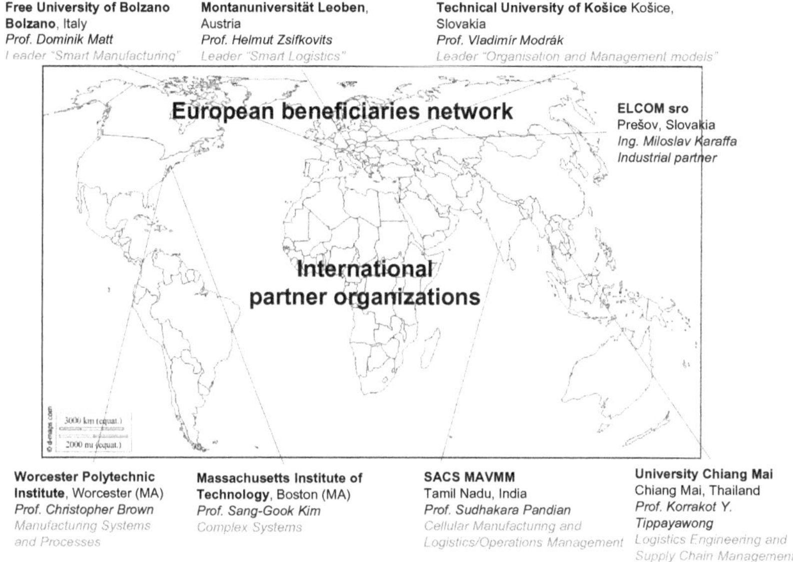

Fig. 1.2 European beneficiaries and international partner organizations in the project (*Source of the map* www.d-maps.com)

four European beneficiaries, both from academia and industry, as well as four academic partner organizations from the USA, Thailand, and India (see also Fig. 1.2):

- Free University of Bolzano-Bozen, Italy (project coordinator and EU beneficiary)
- Technical University of Kosice, Slovakia (EU beneficiary)
- Montanuniversität Leoben, Austria (EU beneficiary)
- Elcom sro Presov, Slovakia (EU beneficiary)
- MIT Massachusetts Institute of Technology, USA (partner organization)
- WPI Worcester Polytechnic Institute, USA (partner organization)
- Chiang Mai University, Thailand (partner organization)
- SACS MAVMM Engineering College, India (partner organization).

As the first large research initiative in Europe to investigate the introduction of Industry 4.0 in SMEs, the project covers the following specific research questions:

A. Identification of requirements for Industry 4.0 applications and implementation in SME manufacturing and logistics:

- What are the actual known concepts and technologies of Industry 4.0?
- What are the main opportunities/risks for the use of these concepts in SMEs?
- How suitable are the different concepts for application in SMEs?
- What are SME-specific requirements for the adaptation of the most promising concepts and technologies?

B. Development of SME-specific concepts and strategies for smart and intelligent SME manufacturing and logistics:

- What are possible forms or migration levels for realizing smart and intelligent manufacturing systems for x-to-order and mass customization production?
- How can automation, advanced manufacturing technologies, ICT, and CPS improve productivity in SME manufacturing and logistics?
- What are suitable models for smart and lean supply chains in SME logistics?

C. Development of specific organization and management models for smart SMEs:

- What are innovative and promising new business models for smart SMEs?
- What are optimal implementation strategies for the introduction of Industry 4.0 in SMEs?
- What are ideal organizational models for smart SMEs or SME networks?

In addition to the above described research questions, the project aims to achieve the following general objectives in relation to the European Community:

- Ensure the transfer of Industry 4.0 to SMEs through adapted template models;
- Maintain and develop the competitive level of European SMEs;
- Accelerate the transition of Industry 4.0 from research to practice;
- Maintain the prosperity of the European population by securing jobs;
- Develop and progress the careers of European experts and qualified young scientists in SME research for Industry 4.0.

1.5.2 Project Structure

The project is organized into three research fields (RF) (Fig. 1.3): (i) smart manufacturing in SMEs, specific solutions for (ii) smart logistics in SMEs and (iii) adapted organization and management models for the introduction of Industry 4.0 and the management of smart SMEs. These research fields are further decomposed into nine research topics (RT) to investigate specific concepts.

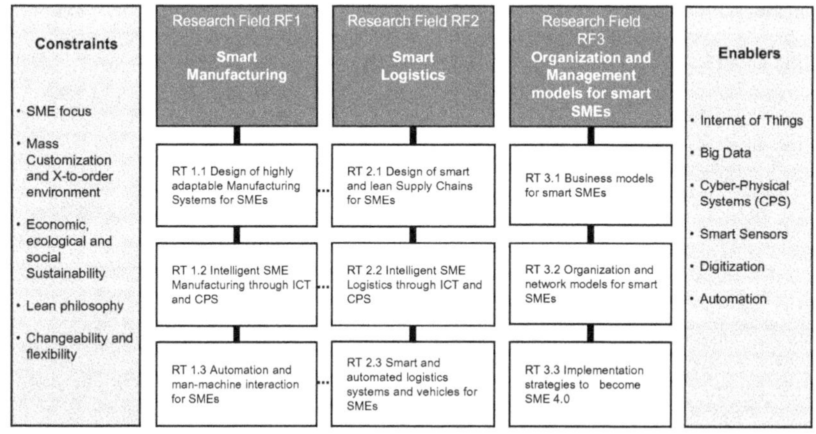

Fig. 1.3 Research fields and topics in the SME 4.0 project

The research project, with a duration of four years, is divided into two equal phases each of two years (see Fig. 1.4).

The first phase from 2017 to 2018 was dedicated to analyzing existing Industry 4.0 concepts and to investigating their suitability for SMEs and/or their need to be adapted for the specific requirements of SMEs. Therefore, the first phase was used also to collect and analyze the requirements of SMEs to introduce Industry 4.0 in small- and medium-sized firms. In the first operative work package (WP3), the research team analyzed the Industry 4.0 requirements in manufacturing with a special focus on the need for adaptable manufacturing systems, the potential of ICT and CPS in manufacturing and the potential of automation in small and medium firms. In WP4, a similar approach is used to investigate the requirements and opportunities of Industry 4.0 in SME logistics. In WP5, the research team analyzed the suitability of SMEs strategies for the introduction of Industry 4.0; they collected organizational requirements for smart SMEs and analyzed the current state of methods and tools for the implementation of Industry 4.0 in SMEs. The content of this book is directly related to this first project phase and summarizes the scientific findings and results.

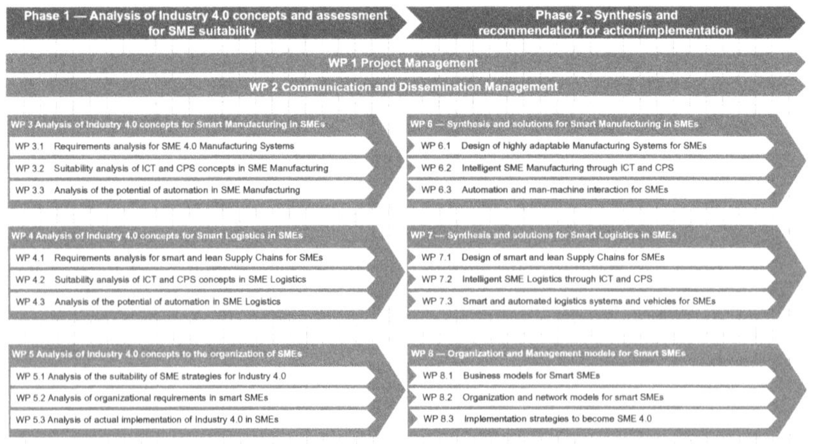

Fig. 1.4 Work packages, tasks, and project phases

The second phase from 2019 to 2020 is dedicated to the synthesis of the results of the first phase and the elaboration of methods, tools, and recommendations for actions to support SMEs introducing Industry 4.0. Similarly to the first phase, we have three work packages (WP) for the derivation of measures for manufacturing, logistics, and organizational issues. The main expected results of the WP6 are to design guidelines and templates for highly adaptable SME manufacturing systems, ICT and CPS-based solutions for more intelligent manufacturing as well as enabling easy implementation of solutions for automation in collaboration with the human on the shop floor. In WP7, related to logistics, the main expected results are templates and guidelines for smart and lean supply chains, the use of ICT and CPS for a smarter supply chain as well as easy and low-cost solutions for automation in intralogistics and transport. In WP8, the expected results are the elaboration of a generally applicable business model (or business models) for smart SMEs, smarter processes and organizational models as well as Industry 4.0 implementation strategies, roadmaps and supporting assessment models for SMEs. The findings of this second project phase will be summarized in a second book project planned to be published by the end of the project.

1.5.3 Research Methodology in the First Project Phase

In the first project phase, the research team adopted a dual strategy using primary research methods (direct collection of data and information from SMEs) as well as secondary research methods (analysis of data collected/published from previously undertaken research or other sources). This kind of research strategy was chosen, as in the scientific literature (especially before 2017) only a few works dealt with Industry 4.0 and the specific needs of SMEs.

In order to analyze the current state of the art as well as the existing Industry 4.0 technologies and concepts, secondary research was used by using literature research methods like the Systematic Literature Review (SLR) (Booth et al. 2016). For secondary research, the research team used explorative field studies with focus groups (Becker et al. 2009;

Wölfel et al. 2012) organizing four SME workshops in 2017. In each of the four workshops held in Italy, Austria, USA, and Thailand, a number of around 10 SMEs participated (total number of 37 SMEs in the workshop series). The workshops had a standardized structure in order to guarantee comparable results. Based on the results of the literature search and the SME workshops, the research consortium worked on the results of this first project phase.

In addition, the research team also visited different SMEs to get a practical understanding of the problems and requirements of SMEs. Through the industry partner Elcom sro, the voice of SMEs is also represented in the research consortium to monitor that research is characterized by a practical approach.

1.6 Conclusion and Structure of the Book

This chapter has shown how important SMEs are in Europe as well as in all other economies of the world. Industry 4.0 is also a major challenge for large enterprises, which in turn, have the necessary financial and human resources to introduce industry into their enterprises. For SMEs, it is not a question of whether they should introduce Industry 4.0 or not, but rather how they can do so as quickly as possible in order to maintain or achieve a large competitive advantage. As the analysis of the research has shown, the research on Industry 4.0 for SMEs is still in its infancy.

This book represents a first major step in this direction by analyzing and describing the challenges, opportunities, and requirements of SMEs in terms of manufacturing, logistics, and organization. Therefore, this book is organized into several sections, which summarize the results of the research according to the topic. The second section focuses on Industry 4.0 concepts for smart manufacturing in SMEs. In addition to the detailed analysis of the requirements for a highly adaptable manufacturing system, it also describes how an Industrial IoT and CPS can be implemented and what kind of potential and challenges SMEs expect to face when they start to implement manufacturing automation and human-robot collaboration. In the third section, the focus is on Industry 4.0 concepts for smart logistics in SMEs. Here the authors provide detailed information about the requirements for designing smart supply chains, they analyze the use of ICT and embedded systems for tracking and tracing in SME logistics and they review the state of the art of automation in logistics. In the fourth section, the work is focused on the organization of smart SMEs. Innovative organizational models for mass customization are developed and tested and the major limitations and barriers to SMEs introducing Industry 4.0 are studied. Finally, a toolkit for the implementation of Industry 4.0 is proposed in order to support SMEs in this challenging initial phase. The fifth section is dedicated to practical case studies on how Industry 4.0 has been applied in a laboratory environment or in real industrial situations. This section is a valuable addition to the previously discussed chapters on smart manufacturing, smart logistics, and organization in smart SMEs. In the case studies, the authors describe both the underlying theoretical concepts and their practical implementation and validation. Readers can expect practical insights into automatic product identification and inspection by means of Industry 4.0 as well as into intuitive collaboration between man and machine in SME assembly. The last chapter of this book explains the basics of Axiomatic Design. As several of the chapters use Axiomatic Design as a research method, the chapter is intended to present the necessary basics and to make it possible to refer to these basics in the individual chapters.

References

Acatech. 2013. *Bundeskanzlerin Angela Merkel nimmt Bericht des Arbeitskreises Industrie 4.0 entgegen.* https://www.acatech.de/allgemein/bundeskanzler-in-angela-merkel-nimmt-bericht-desarbeitskreises-industrie-4-0-entgegen/. Accessed on 23 Mar 2019.

ADB. 2018. Support for Small and Medium-Sized Enterprises, 2005–2017: Business Environment, Access to Finance, Value Chains, and Women in Business, Linked Document, A Small and Medium-Sized Enterprises in Asia and the Pacific: Context and Issues. https://www.adb.org/sites/default/files/linked-documents/A-SME-Context-and-Issues.pdf. Accessed on 21 Feb 2019.

Bakkari, M., and A. Khatory. 2017. Industry 4.0: Strategy for More Sustainable Industrial Development in SMEs. In *Proceedings of the IEOM 7th International Conference on Industrial Engineering and Operations Management*, 11–13. Rabat, Morocco. https://pdfs.semanticscholar.org/ca4b/140f92a0bbbeee7dd0604f5ea33c77636ad4.pdf. Accessed on Aug 2019.

Balkhausen, D. 1978. *Die dritte industrielle Revolution, wie die Mikroelektronik unser Leben verändert*. Düsseldorf und Wien: Econ Verlag.

Bär, K., Z.N.L. Herbert-Hansen, and W. Khalid. 2018. Considering Industry 4.0 Aspects in the Supply Chain for an SME. *Production Engineering* 12 (6): 747–758. https://doi.org/10.1007/s11740-018-0851-y.

Bauer, W., S. Schlund, D. Marrenbach, and O. Ganschar. 2014. Industrie 4.0 – Volkswirtschaftliches Potenzial für Deutschland. BITKOM, 18. Berlin: Das Fraunhofer-Institut für Arbeitswirtschaft und Organisation. https://doi.org/10.1007/978-3-642-36917-9.

Baum, G. 2013. Innovationen als Basis der nächsten Industrierevolution. In *Industrie 4.0*, ed. U. Sendler, 37–53. Berlin and Heidelberg: Springer Vieweg. https://doi.org/10.1007/978-3-642-36917-9_3.

Becker, J., D. Beverungen, M. Matzner, and O. Müller. 2009. Design Requirements to Support Information Flows for Providing Customer Solutions: A Case Study in the Mechanical Engineering Sector. In *Proceedings of the First International Symposium on Services Science*, Leipzig, Germany.

BMBF. 2012. Zukunftsbild Industrie 4.0. https://www.plattform-i40.de/PI40/Redaktion/DE/Downloads/Publikation/zukunftsbild-industrie-4-0.pdf?__blob=publicationFile&v=4. Accessed on 23 Mar 2019.

Bolesnikov, M., M. Popović Stijačić, M. Radišić, A. Takači, J. Borocki, D. Bolesnikov, and J. Dzieńdziora. 2019. Development of a Business Model by Introducing Sustainable and Tailor-Made Value Proposition for SME Clients. *Sustainability* 11 (4): 1157. https://doi.org/10.3390/su11041157.

Bollhöfer, E., D. Buschak, and C. Moll. 2016. Dienstleistungsbasierte Geschäftsmodelle für Industrie 4.0–aktueller Stand und Potenziale für KMU. In *Multikonferenz Wirtschaftsinformatik*, 1287–1298. Ilmenau: Universitätsverlag Ilmenau.

Booth, A., A. Sutton, and D. Papaioannou. 2016. *Systematic Approaches to a Successful Literature Review*, 2nd ed. London: Sage.

Boughton, N.J., and I.C. Arokiam. 2000. The Application of Cellular Manufacturing: A Regional Small to Medium Enterprise Perspective. *Proceedings of the Institution of Mechanical Engineers, Part B: Journal of Engineering Manufacture* 214 (8): 751–754. https://doi.org/10.1243/0954405001518125.

Brettel, M., N. Friederichsen, M. Keller, and M. Rosenberg. 2014. How Virtualization, Decentralization and Network Building Change the Manufacturing Landscape: An Industry 4.0 Perspective. *International Journal of Mechanical, Industrial Science and Engineering* 8 (1): 37–44. https://doi.org/10.5281/zenodo.1336426.

Brusselsnetwork. 2018. Annual Report 2018 on European SMEs. https://www.brusselsnetwork.be/annual-report-2018-on-european-smes/. Accessed on 27 Dec 2018.

Cordis. 2019. Cordis EU Research Results. https://cordis.europa.eu. Accessed on 27 Mar 2019.

Decker, A. 2017. Industry 4.0 and SMEs in the Northern Jutland Region. In *Value Creation in International Business,* 309–335. Cham: Palgrave Macmillan. http://dx.doi.org/10.1007/978-3-319-39369-8_13.

Deloitte. 2015. Industry 4.0—Challenges and Solutions for the Digital Transformation and Use of Exponential Technologies. Study of Deloitte Consulting. http://www2.deloitte.com/content/dam/Deloitte/ch/Documents/manufacturing/ch-en-manufacturing-industry-4-0-24102014.pdf. Accessed on 7 Jan 2016.

EC. 2015. Industry 4.0—Digitalisation for Productivity and Growth. http://www.europarl.europa.eu/RegData/etudes/BRIE/2015/568337/EPRS_BRI(2015)568337_EN.pdf. Accessed on 27 Mar 2019.

EC. 2018. Commission Proposal for the Next Research and Innovation Programme (2021–2027). Horizon Europe. https://ec.europa.eu/info/sites/info/files/horizon-europe-presentation_2018_en.pdf. Accessed on 27 Mar 2019.

EU. 2018. Small and Medium-Sized Enterprises: An Overview. Annual Report on European SMEs 2017/2018. Special Background Document on the Internationalisation of SMEs. SME Performance Review 2017/2018. eurostat. https://ec.europa.eu/eurostat/web/products-eurostat-news/-/EDN-20181119-1. Accessed on 3 Dec 2019.

Eurostat. 2018. Statistics on Small and Medium-Sized Enterprises. https://ec.europa.eu/eurostat/statistics-explained/index.php?title=Statistics_on_small_and_medium-sized_enterprises&oldid=451334. Accessed on 29 Sept 2019.

Färber, R. 2013. Hype trifft Mittelstand: Relevanz von Industrie 4.0 für kleine und mittelständische Unternehmen. *Mechatronik* 121 (12): 45–47.

FAZ. 2019. https://www.faz.net/aktuell/wirtschaft/infografik-made-in-china-2025-15936600.html. Accessed on 27 Mar 2019.

Ganzarain, J., and N. Errasti. 2016. Three Stage Maturity Model in SME's Toward Industry 4.0. *Journal of Industrial Engineering and Management* 9 (5): 1119–1128. https://doi.org/10.3926/jiem.2073.

Goerzig, D., and T. Bauernhansl. 2018. Enterprise Architectures for the Digital Transformation in Small and Medium-Sized Enterprises. *Procedia CIRP* 67 (1): 540–545. https://doi.org/10.1016/j.procir.2017.12.257.

Gualtieri, L., R. Rojas, G. Carabin, I. Palomba, E. Rauch, R. Vidoni, and D.T. Matt. 2018. Advanced Automation for SMEs in the I4. 0 Revolution: Engineering Education and Employees Training in the Smart Mini Factory Laboratory. In *2018 IEEE International Conference on Industrial Engineering and Engineering Management (IEEM),* 1111–1115. https://doi.org/10.1109/IEEM.2018.8607719.

Hartbrich, I. 2014. Industrie 4.0-in der Zukunftsfabrik. *Die Zeit* 5 (1): 1–3.

Jones, M., L. Zarzycki, and G. Murray. 2018. Does Industry 4.0 Pose a Challenge for the SME Machine Builder? A Case Study and Reflection of Readiness for a UK SME. In *International Precision Assembly Seminar*, 183–197. Cham: Springer. https://doi.org/10.1007/978-3-030-05931-6_17.

Jørsfeldt, L.M., and A. Decker. 2017. Digitally Enabled Platforms: Generating Innovation and Entrepreneurial Opportunities for SMEs. In *Motivating SMEs to Cooperate and Internationalize*, 93–111. Routledge. https://doi.org/10.4324/9781315412610.

Jun, C., J.Y. Lee, J.S. Yoon, and B.H. Kim. 2017. Applications' Integration and Operation Platform to Support Smart Manufacturing by Small and Medium-Sized Enterprises. *Procedia Manufacturing* 11: 1950–1957. https://doi.org/10.1016/j.promfg.2017.07.341.

Kagermann, H., J. Helbig, A. Hellinger, and W. Wahlster. 2013. Recommendations for Implementing the Strategic Initiative Industrie 4.0: Securing the Future of German Manufacturing Industry. Final report of the Industrie 4.0 Working Group. Forschungsunion.

Laperrire, L., and G. Reinhart. 2014. *CIRP Encyclopedia of Production Engineering*. Springer. http://dx.doi.org/10.1007/978-3-642-20617-7.

Lee, J., B. Bagheri, and H.A. Kao. 2015. A Cyber-Physical Systems Architecture for Industry 4.0-Based Manufacturing Systems. *Manufacturing Letters* 3: 18–23. https://doi.org/10.1016/j.mfglet.2014.12.001.

Lee, J., and E. Lapira. 2013. Predictive Factories: The Next Transformation. *Manufacturing Leadership Journal* 20 (1): 13–24. https://doi.org/10.3182/20130522-3-BR-4036.00107.

Lee, E.A., and S.A. Seshia. 2016. *Introduction to Embedded Systems: A Cyber-Physical Systems Approach*. Cambridge: MIT Press.

Madani, A.E. 2018. SME Securitization in SME Policy: Comparative Analysis of SME Definitions. *International Journal of Academic Research in Business and Social Sciences* 8 (8): 103–114. https://doi.org/10.6007/IJARBSS/v8-i8/4443.

Manhart, K. 2013. Industrie 4.0 könnte schon bald Realität sein. https://computerwelt.at/knowhow/industrie-4-0-konnte-schon-bald-realitat-sein/. Accessed on 15 Mar 2015.

Manufacturing USA. 2019. https://www.manufacturingusa.com/. Accessed on 25 Mar 2019.

Matt, D.T. 2007. Reducing the Structural Complexity of Growing Organizational Systems by Means of Axiomatic Designed Networks of Core Competence Cells. *Journal of Manufacturing Systems* 26 (3–4): 178–187. https://doi.org/10.1016/j.jmsy.2008.02.001.

Matt, D.T., E. Rauch, and D. Fraccaroli. 2016. Smart Factory for SMEs: Designing a Holistic Production System by Industry 4.0 Vision in Small and Medium Enterprises (SMEs). *ZWF Zeitschrift für wirtschaftlichen Fabrikbetrieb* 111: 2–5. https://doi.org/10.3139/104.111471.

Matt, D.T., E. Rauch, and M. Riedl. 2018a. Analyzing the Impacts of Industry 4.0 in Modern Business Environments. *Knowledge Transfer and Introduction of Industry 4.0 in SMEs: A Five-Step Methodology to Introduce Industry 4.0*, 256–282. IGI Global. https://doi.org/10.4018/978-1-5225-3468-6.

Matt, D.T., G. Orzes, E. Rauch, and P. Dallasega. 2018b. Urban Production–A Socially Sustainable Factory Concept to Overcome Shortcomings of Qualified Workers in Smart SMEs. *Computers & Industrial Engineering*. https://doi.org/10.1016/j.cie.2018.08.035.

Meissner, H., R. Ilsen, and J.C. Aurich. 2017. Analysis of Control Architectures in the Context of Industry 4.0. *Procedia CIRP* 62: 165–169. https://doi.org/10.1016/j.procir.2016.06.113.

Menezes, S., S. Creado, and R.Y. Zhong. 2017. Smart Manufacturing Execution Systems for Small and Medium-Sized Enterprises. *Decision Making* 13: 14. https://doi.org/10.1016/j.procir.2018.03.272.

Mittal, S., M.A. Khan, D. Romero, and T. Wuest. 2018a. A Critical Review of Smart Manufacturing & Industry 4.0 Maturity Models: Implications for Small and Medium-Sized Enterprises (SMEs). *Journal of Manufacturing Systems* 49: 194–214. https://doi.org/10.1016/j.jmsy.2018.10.005.

Mittal, S., D. Romero, and T. Wuest. 2018b. Towards a Smart Manufacturing Toolkit for SMEs. In *IFIP International Conference on Product Lifecycle Management*, 476–487. Cham: Springer. https://doi.org/10.1007/978-3-030-01614-2_44.

Modrak, V., D. Marton, and S. Bednar. 2014. Modeling and Determining Product Variety for Mass-Customized Manufacturing. *Procedia CIRP* 23: 258–263. https://doi.org/10.1016/j.procir.2014.10.090.

Modrak, V., Z. Soltysova, and R. Poklemba. 2019. Mapping Requirements and Roadmap Definition for Introducing I 4.0 in SME Environment. In *Advances in Manufacturing Engineering and Materials*, 183–194. Cham: Springer. http://dx.doi.org/10.1007/978-3-319-99353-9_20.

Moica, S., J. Ganzarain, D. Ibarra, and P. Ferencz. 2018. Change Made in Shop Floor Management to Transform a Conventional Production System into an "Industry 4.0": Case Studies in SME Automotive Production Manufacturing. In *2018 7th International Conference on Industrial Technology and Management (ICITM)*, 51–56. https://doi.org/10.1109/icitm.2018.8333919.

Monostori, L. 2014. Cyber-Physical Production Systems: Roots, Expectations and R&D Challenges. *Procedia CIRP* 17: 9–13. https://doi.org/10.1016/j.procir.2014.03.115.

Moreira, C.F. 2017. The Make in India Initiative and the Role of Industry 4.0. https://www.enterprisetv.com.my/the-make-in-india-initiative-and-the-role-of-industry-4-0/. Accessed on 22 Mar 2019.

Müller, J.M. 2019. Business Model Innovation in Small- and Medium-Sized Enterprises: Strategies for Industry 4.0 Providers and Users. *Journal of Manufacturing Technology Management*, in press. https://doi.org/10.1108/JMTM-01-2018-0008.

Müller, E., and H. Hopf. 2017. Competence Center for the Digital Transformation in Small and Medium-Sized Enterprises. *Procedia Manufacturing* 11: 1495–1500. https://doi.org/10.1016/j.promfg.2017.07.281.

Müller, J., and K.I. Voigt. 2016. Industrie 4.0 für kleine und mittlere Unternehmen. Welche spezifischen Probleme werden bei der Einführung von Industrie 4.0 von kleinen und mittleren Unternehmen gesehen? *Productivity Management* 3: 28–30.

Müller, R., M. Vette, L. Hörauf, C. Speicher, and D. Burkhard. 2017. Lean Information and Communication Tool to Connect Shop and Top Floor in Small and Medium-Sized Enterprises. *Procedia Manufacturing* 11: 1043–1052. https://doi.org/10.1016/j.promfg.2017.07.215.

NSF. 2019. https://www.nsf.gov/div/index.jsp?div=CMMI. Accessed on 24 Mar 2019.

Nowotarski, P., and J. Paslawski. 2017. Industry 4.0 Concept Introduction into Construction SMEs. *IOP Conference Series: Materials Science and Engineering* 245 (5), 52043. http://dx.doi.org/10.1088/1757-899X/245/5/052043.

OECD. 2017. Enhancing the Contributions of SMEs in a Global and Digitalized Economy. https://www.oecd.org/mcm/documents/C-MIN-2017-8-EN.pdf. Accessed on 20 Mar 2019.

OECD/ERIA. 2018. *SME Policy Index: ASEAN 2018: Boosting Competitiveness and Inclusive Growth*. Paris/Economic Research Institute for ASEAN and East Asia, Jakarta: OECD Publishing. https://doi.org/10.1787/9789264305328-en.

Olle, W., & D. Clauß. (2015). Industrie 4.0 braucht den Mittelstand–Kurzstudie. Chemnitz Automotive Institute. http://cati.institute/wp-content/uploads/2015/03/Kurzstudie_Endfassung.pdf. Accessed on 15 Aug 2019.

Orzes, G., E. Rauch, S. Bednar, and R. Poklemba. 2018. Industry 4.0 Implementation Barriers in Small and Medium Sized Enterprises: A Focus Group Study. In *2018 IEEE International Conference on Industrial Engineering and Engineering Management (IEEM)*, 1348–1352. https://doi.org/10.1109/IEEM.2018.8607477.

Pérez, J.D.C., R.E.C. Buitrón, and J.I.G. Melo. 2018. Methodology for the Retrofitting of Manufacturing Resources for Migration of SME Towards Industry 4.0. In *International Conference on Applied Informatics*, 337–351. Cham: Springer. https://doi.org/10.1007/978-3-030-01535-0_25.

Plattform Industrie 4.0. 2014. Plattform Industrie 4.0. http://www.plattform-i40.de/. Accessed on 2 Nov 2014.

Plattform Industrie 4.0. 2019. Industrie 4.0—A Competitive Edge for Europe. https://www.plattform-i40.de/I40/Navigation/EN/InPractice/International/EuropaeischeEbene/europaeische-ebene.html. Accessed on 26 Mar 2019.

PWC. 2015. Industry 4.0—Opportunities and Challenges of the Industrial Internet. Study of Pricewaterhouse Coopers PWC. http://www.strategyand.pwc.com/media/file/Industry-4-0.pdf. Accessed on 22 Dec 2015.

Rajkumar, R., I. Lee, L. Sha, and J. Stankovic. 2010. Cyber-Physical Systems: the Next Computing Revolution. In *Design Automation IEEE International Conference on Industrial Engineering and Engineering Management*, 731–736. https://doi.org/10.1145/1837274.1837461.

Rauch, E., P. Dallasega, and D.T. Matt. 2017. Critical Factors for Introducing Lean Product Development to Small and Medium Sized Enterprises in Italy. *Procedia CIRP* 60: 362–367. https://doi.org/10.1016/j.procir.2017.01.031.

Rauch, E., P. Dallasega, and D.T. Matt. 2018a. Complexity Reduction in Engineer-to-Order Industry Through Real-Time Capable Production Planning and Control. *Production Engineering* 12 (3–4): 341–352. https://doi.org/10.1007/s11740-018-0809-0.

Rauch, E., P.R. Spena, and D.T. Matt. 2018b. Axiomatic Design Guidelines for the Design of Flexible and Agile Manufacturing and Assembly Systems for SMEs. *International Journal on Interactive Design and Manufacturing (IJIDeM)*, 1–22. http://dx.doi.org/10.1007/s12008-018-0460-1.

Reuter, C. 2015. Betriebliches Kontinuitätsmanagement in kleinen und mittleren Unternehmen–Smart Services für die Industrie 4.0. In *Mensch und Computer 2015–Workshopband*. https://doi.org/10.1515/9783110443905-006.

Rickmann, H. 2014. Verschläft der deutsche Mittelstand einen Megatrend? http://www.focus.de/finanzen/experten/rickmann/geringer-digitalisierungsgrad-verschlaeft-der-deutschemittelstand-einen-megatrend_id_3973075.html. Accessed on 10 Jan 2016.

Safar, L., J. Sopko, S. Bednar, and R. Poklemba. 2018. Concept of SME Business Model for Industry 4.0 Environment. *TEM Journal* 7 (3), 626. https://doi.org/10.18421/TEM73-20.

Schlegel, A., T. Langer, and M. Putz. 2017. Developing and Harnessing the Potential of SMEs for Eco-Efficient Flexible Production. *Procedia Manufacturing* 9: 41–48. https://doi.org/10.1016/j.promfg.2017.04.028.

SBA. 2018. Small Business Profile. https://www.sba.gov/sites/default/files/advocacy/2018-Small-Business-Profiles-US.pdf. Accessed on 12 Jan 2019.

Scheidel, W., I. Mozgova, and R. Lachmayer. 2018. Teaching Industry 4.0–Product Data Management For Small and Medium-Sized Ebterprises. *Proceedings of the 20th International Conference on Engineering and Product Design Education (E&PDE 2018)* DS 93, 151–156. Dyson School of Engineering, Imperial College, London, 6th–7th September 2018.

Seidenstricker, S., E. Rauch, and P. Dallasega. 2017. Industrie-4.0-Geschäftsmodell-innovation für KMU. *ZWF Zeitschrift für wirtschaftlichen Fabrikbetrieb* 112 (9), 616–620. https://doi.org/10.3139/104.111776.

Sendler, U. (ed.). 2013. *Industrie 4.0: Beherrschung der industriellen Komplexität mit SysLM*. Berlin and Heidelberg: Springer Vieweg. https://doi.org/10.1007/978-3-642-36917-9_1.

Sezer, E., D. Romero, F. Guedea, M. Macchi, and C. Emmanouilidis. 2018. An Industry 4.0-Enabled Low Cost Predictive Maintenance Approach for SMEs. In *2018 IEEE International Conference on Engineering, Technology and Innovation (ICE/ITMC)*, 1–8. https://doi.org/10.1109/ICE.2018.8436307.

Sommer, L. 2015. Industrial Revolution-Industry 4.0: Are German Manufacturing SMEs the First Victims of This Revolution? *Journal of Industrial Engineering and Management* 8 (5): 1512–1532. https://doi.org/10.3926/jiem.1470.

Spath, D., O. Ganschar, S. Gerlach, T.K. Hämmerle, and S. Schlund. 2013. *Produktionsarbeit der Zukunft – Industrie 4.0*. Stuttgart: Fraunhofer Verlag.

ThaiEmbDC. 2019. https://thaiembdc.org/thailand-4-0-2/. Accessed on 10 Mar 2019.

Türkeş, M.C., I. Oncioiu, H.D. Aslam, A. Marin-Pantelescu, D.I. Topor, and S. Căpuşneanu. 2019. Drivers and Barriers in Using Industry 4.0: A Perspective of SMEs in Romania. *Processes* 7 (3): 153. https://doi.org/10.3390/pr7030153.

Ward. 2018. SME Definition (Small to Medium Enterprise). https://www.thebalancesmb.com/sme-small-to-medium-enterprise-definition-2947962. Accessed on 13 Mar 2019.

Weiß, M., M. Tilebein, R. Gebhardt, and M. Barteld. 2018. Smart Factory Modelling for SME: Modelling the Textile Factory of the Future. Lecture Notes. *Business Information Processing* 319: 328–337. https://doi.org/10.1007/978-3-319-94214-8_24.

Wiesner, S., P. Gaiardelli, N. Gritti, and G. Oberti. 2018. Maturity Models for Digitalization in Manufacturing-Applicability for SMEs. In *IFIP International Conference on Advances in Production Management Systems*, 81–88. Cham: Springer. https://doi.org/10.1007/978-3-319-99707-0_1.

Wölfel, C., U. Debitz, J. Krzywinski, and R. Stelzer. 2012. Methods Use in Early Stages of Engineering and Industrial Design—A Comparative Field Exploration. In *Proceedings of DESIGN 2012, the 12th International Design Conference*, Dubrovnik, Croatia DS 70.

WTO. 106. World Trade Report 2016. Levelling the Trading Field for SMEs. https://www.wto.org/english/res_e/booksp_e/world_trade_report16_e.pdf. Accessed on 15 Feb 2019.

Yoshino, N., and F. Taghizadeh-Hesary. 2018. The Role of SMEs in Asia and Their Difficulties in Accessing Finance. ADBI Working Paper 911. Tokyo: Asian Development Bank Institute. https://www.adb.org/publications/role-smes-asia-and-their-difficulties-accessing-finance. Accessed on 11 Feb 2019.

Zambon, I., M. Cecchini, G. Egidi, M.G. Saporito, and A. Colantoni. 2019. Revolution 4.0: Industry vs. Agriculture in a Future Development for SMEs. *Processes* 7 (1): 36. https://doi.org/10.3390/pr7010036.

Zhou, J., P. Li, Y. Zhou, B. Wang, J. Zang, and L. Meng. 2018. Toward New-Generation Intelligent Manufacturing. *Engineering* 4 (1): 11–20. https://doi.org/10.1016/j.eng.2018.01.002.

Part II

Industry 4.0 Concepts for Smart Manufacturing in SMEs

2

SME Requirements and Guidelines for the Design of Smart and Highly Adaptable Manufacturing Systems

Erwin Rauch, Andrew R. Vickery, Christopher A. Brown and Dominik T. Matt

2.1 Introduction

The industrial environment has undergone a radical change with the introduction of new technologies and concepts based on the fourth industrial revolution, also known as Industry 4.0 (I4.0) (Sendler 2013), the Fourth Industrial Revolution (Kagermann et al. 2013) or Smart

E. Rauch (✉) · D. T. Matt
Faculty of Science and Technology,
Free University of Bozen-Bolzano, Bolzano, Italy
e-mail: erwin.rauch@unibz.it

D. T. Matt
e-mail: dominik.matt@unibz.it

A. R. Vickery · C. A. Brown
Department of Mechanical Engineering,
Worcester Polytechnic Institute, Worcester, USA
e-mail: andrewv@wpi.edu

C. A. Brown
e-mail: brown@wpi.edu

D. T. Matt et al. (eds.), *Industry 4.0 for SMEs*,
https://doi.org/10.1007/978-3-030-25425-4_2

Manufacturing (Kang et al. 2016). The concept of I4.0 is building on the integration of information and communication technologies and advanced industrial technologies in so-called Cyber-Physical Systems (CPS) to realize a digital, intelligent, and sustainable factory (Zhou et al. 2015). The basic meaning of I4.0 lies in connecting products, machines, and people with the environment and combining production, information technology, and the internet (Kagermann et al. 2013). Industry, especially in high-wage countries, must introduce these types of smart production strategies to maintain the current competitive advantage in the long-term competing on a global market (Manhart 2017). To remain competitive, lead times, flexibility, and the ability to produce many individual kinds of products in low batch sizes or batch sizes of one, must improve (Spath et al. 2013; Matt and Rauch 2013a). In a mass customization and "design for x" environment, more functionality and customization options have to be provided to the client and more flexibility, transparency, and globalization for the supply chain (Baum 2013). On the other hand, this also leads to a more difficult and complex situation for manufacturing companies. Quickly responding to the expectations and requirements of customers is not easy and requires agile and highly adaptable manufacturing systems (Zawadzki and Żywicki 2016). The introduction of I4.0 in manufacturing companies contributes exactly to tackling these global challenges for strengthening competitiveness of high-wage countries (Kagermann et al. 2013).

Manufacturing companies, and especially SMEs, struggle with the introduction of I4.0 and to gain from its potential to increase productivity on the shop floor (Matt et al. 2014). Very often, they do not know how to face the challenge of I4.0 or how to start introducing and implementing I4.0 concepts (Ganzarain and Errasti 2016). A recent 2017 study (Wuest et al. 2018) conducted with manufacturing SMEs in West Virginia, USA, confirmed the struggle for SMEs to adopt Smart Manufacturing (Mittal et al. 2018). According to their literature review, only a few studies specifically focus on supporting SMEs' evolutionary path and paradigm shift toward "Smart Manufacturing (SM)" or "Industry 4.0". SMEs often face complications in such innovative processes due to the continuous development of innovations and technologies. Therefore, further research is needed to provide specific instruments and models for SMEs introducing I4.0 in their companies and production shop floors. In addition, policy makers should propose strategies with the aim of supporting SMEs to invest in these technologies and make them more competitive in the marketplace (Zambon et al. 2019).

The objective of this chapter is to analyze and evaluate the specific needs and requirements of SMEs with the aim of defining guidelines for the design of highly adaptable and smart manufacturing systems for SMEs in a dynamic environment using I4.0. After a brief introduction on I4.0 and its impact for SMEs, Sect. 2.2 summarizes the state of the art in I4.0 and its transfer to SMEs based on a literature review. The following Sect. 2.3 gives an overview of the problem formulation and the system limits of this research. Section 2.4 illustrates the research methodology, which is grounded in Axiomatic Design (AD) theory to transform user needs into functional requirements and finally into design guidelines for highly adaptable manufacturing systems. Section 2.5 describes in detail the analysis of the user needs of SMEs to introduce I4.0 in their factories. The collection of these user needs is based on an explorative study, while the derivation of functional requirements and design parameters is based on AD theory. Results of this main section are a final list of SME requirements as well as constraints to introducing I4.0 in manufacturing and a set of coarse design parameters for their implementation. In Sect. 2.6, there follows a critical discussion of the obtained results and in Sect. 2.7, the conclusion and outlook for further necessary research are presented.

2.2 Background and Literature Review

2.2.1 Industry 4.0—The Fourth Industrial Revolution

The term I4.0 was introduced in 2011 by a German group of scientists during the Hannover Fair, which symbolized the beginning of the fourth industrial revolution (Lee 2013). After mechanization, electrification, and computerization, the fourth stage of industrialization aims to introduce concepts like CPS, Internet of Things (IoT), Automation, and Human–Machine Interaction (HMI) as well as Advanced Manufacturing Technologies in a factory environment (Zhou et al. 2015). Since then, the term I4.0 has become one of the most popular manufacturing topics among industry and academia in the world and has been considered the fourth industrial revolution with its impact on future manufacturing (Kagermann et al. 2013; Qin et al. 2016). Based on the principle of I4.0, traditional structures can be replaced, which are based on centralized decision-making mechanisms and rigid limits on individual value added steps. These structures are replaced by highly adaptable and agile manufacturing systems, offering interactive, collaborative decision-making mechanisms (Spath et al. 2013).

A key element in I4.0 for manufacturing companies is CPS with capabilities for self-organization and self-control. CPS are computers with networks of small sensors and actuators installed as embedded systems in materials, equipment and machine parts and connected via the Internet (Kagermann et al. 2013; Broy and Geisberger 2012; VDI/VDE 2013). CPSs positively affect manufacturing in the form of Cyber-Physical Production Systems (CPPS) in process automation and control (Monostori 2014). There is still a need for further research on CPS (Wang et al. 2015). In the future, CPS and networks of CPS, better known as CPPS, as well as all the technologies behind them, may act as enablers for new business models, which have the potential to be disruptive (Rauch et al. 2016).

Furthermore, the "Internet of Things" (IoT) is also one key element of I4.0, when the physical and the digital world are combined (Federal Ministry of Education and Research 2013). In its origins, the IoT means an intelligent connectivity of anything, anytime, anywhere (Atzori et al. 2010). IoT has developed into the combination and integration of information and physical world addresses to create the "4Cs" (Connection, Communication, Computing, and Control) (Tao et al. 2014). Production data are provided in a new way with real-time information on production processes, through sensors, and continuous integration of intelligent objects (Spath et al. 2013; Gneuss 2014). With connected production technologies, individualized production at low costs will become possible (Kraemer-Eis and Passaris 2015).

The potential benefits from the successful implementation of IoT in the context of I4.0 are immense and research is still important.

Other key elements of I4.0 are Automation, HMI, and Advanced Manufacturing. Automation needs to become more flexible allowing manufacturing processes to be automated with changing products or volumes (Rüßmann et al. 2015). To achieve a symbiosis between automation and operators, HMI plays a major role in providing adequate technological assistance as well as intelligent user interfaces (Gorecky 2014). Advanced manufacturing technologies like high-precision machining, reconfigurable manufacturing units, additive manufacturing, and others are changing production strategies, but also processes and manufacturing systems (Chen et al. 2018; Frank et al. 2019). A prominent example of such advanced technologies in Additive Manufacturing (AM), also known as 3D printing (Rauch et al. 2018). It is defined by the American Society for Testing and Materials (ASTM) as "the process of joining materials to make objects from 3D-model data, usually layer upon layer, as opposed to subtractive manufacturing methodologies, such as traditional machining" (ASTM 2013).

2.2.2 State of the Art in the Introduction of Smart and Highly Adaptable Manufacturing Systems in SMEs

However, challenges arise for companies due to the immense financial resources required to acquire new I4.0 technologies, which makes it difficult for SMEs to introduce I4.0 (Erol et al. 2016). Despite these difficulties, SMEs will not be able to ignore the trend toward I4.0 and therefore, it will be a major challenge for them in the near future (Matt et al. 2018). I4.0 is particularly interesting for these companies, as this term promises the enabling of intelligent automation toward batch size one (Matt et al. 2016). SMEs are the backbone of the EU and many other economies (Federal Ministry of Education and Research 2013). European SMEs provide around 45% of the value added by manufacturing while they provide around 59% of manufacturing employment (Vidosav 2014). In the United States, SMEs account for nearly two-thirds of net new private sector jobs (USTR 2017). Programs like the European Horizon 2020 research and innovation program actively support SMEs by providing direct financial support and indirect support to increase their innovation capacity, although, the number of publications and research activities related to I4.0 for SMEs is still limited (Mittal et al. 2018). New technologies and ideas related to I4.0 need to be further researched and adapted to make it possible to use them in SMEs (Nowotarski and Paslawski 2017).

According to a survey, many SMEs struggle with increasing product variety and individualization in a mass customization environment. Price competition, high quality requirements, and short delivery times are becoming increasingly important (Spena et al. 2016). Due to their flexibility, entrepreneurial spirit, and innovation capabilities, SMEs have proved to be more robust than large and multinational enterprises, as the previous financial and economic crisis showed (Matt 2007). SMEs are not only adaptive and innovative in terms of their products, but also in their manufacturing practices. Recognizing rising competitive pressure, small organizations are becoming proactive in improving their business operations (Boughton and Arokiam 2000), which is a good starting point for introducing new concepts of I4.0 like smart and highly adaptable manufacturing systems.

Successful implementation of such intelligent manufacturing systems must take place not only in large enterprises but also in SMEs (Sommer 2015). Various studies point out relevant changes and potential for SMEs in the context of I4.0 (Rickmann 2017). I4.0 technologies offer opportunities for SMEs to enhance their competitiveness. The integration of ICT and CPS with production, logistics, and services in current industrial practices would transform today's SME factories into smarter and more adaptable factories with significant economic potential (Lee and Lapira 2013). Previous works have shown a limited but positive impact of Industry 4.0 in SME operational performance, with little investment and little expertise when it relates to cloud computing (Radziwon et al. 2014).

The introduction of new technologies and practices is always risky in SMEs (Moeuf et al. 2017) and represents a big challenge for them. SMEs are only partly ready to adapt to I4.0 concepts due to their current organizational capabilities. The smaller the SME, the greater the risk that they will not be able to benefit from this revolution. Many SMEs are not prepared to implement I4.0 concepts. This opens the need for further research and action plans to support SMEs in introducing I4.0 concepts (Sommer 2015) like smart and highly adaptable manufacturing systems.

Only a few works address the specific requirements of SMEs for introducing such intelligent manufacturing systems and most of them do not provide a complete list of them. In the work of Rauch et al. (2019) the authors present a study regarding requirements for the design of flexible and agile manufacturing systems for SMEs. This work does not consider the introduction of I4.0 concepts, but highlights the need for research into SME specific I4.0 solutions. The work of Mittal et al. (2018) is one of the only works that provides a list of SME requirements regarding the design of smart manufacturing systems by introducing I4.0. The work is based on literature research as well as a survey-based study with US SME companies. According to this work, the main SME requirements for I4.0 in manufacturing are (a) the need for financial resources, (b) the need for advanced manufacturing technologies, (c) the need for industrial standards, (d) the need to include I4.0 in the organizational culture, (e) the need to develop and include employees in I4.0 related changes, (f) the need for alliances with universities and research institutions, and (g) the need for collaboration with customers and suppliers. Although the results show a good starting point for further considerations, they are formulated very generally, they do not address the specific requirements for designing an SME manufacturing system and most of them are typical requirements of any kind of companies introducing I4.0. Therefore, we conclude that there is still a need to investigate the specific requirements of SMEs for smart manufacturing system design.

2.3 Problem Formulation

As previously identified in the literature review, there is a need for research and investigations for the implementation of I4.0 technologies and concepts in SME manufacturing. The authors compare these challenges with the introduction of Lean Management in small- and medium-sized enterprises over the past 20 years. While most large companies have introduced or integrated Lean, at least in part, into their corporate strategy, most SMEs have gradually addressed this topic. Carrying out an analysis in Scopus with the keywords "lean" and "SME," for example, shows research on this topic was carried out from 2001 onwards. There are several papers recommending specific strategies for the introduction of lean (Medbo et al. 2013) and specific lean methods for SMEs (Dombrowski et al. 2010; Matt and Rauch 2013b). As a result, Lean has now been implemented in many SMEs in practice. A similar approach is therefore also expected for SME manufacturing companies introducing I4.0, even as large companies have already addressed this goal.

As with the introduction of Lean, the success rate for introducing I4.0 in SME manufacturing can be increased by developing SME-customized implementation strategies, SME-adapted concepts and technologically feasible solutions. Otherwise, the current efforts for awareness-building of SMEs for I4.0 are at risk of failing to achieve the expected results and benefits. As mentioned in the conclusion of the literature review in Sect. 2.2, we can state that there is still a lack of scientific literature regarding detailed analysis of the needs and requirements of smart SME factories for a better understanding of the necessities and problems involved in the introduction of I4.0 in SME manufacturing. In addition, there are already no clear design guidelines available about how SMEs can implement I4.0 in their manufacturing facilities and processes. Another important question is what kind of limitations or barriers could hinder the successful implementation of I4.0 in manufacturing. Knowing these barriers, SMEs can better define the constraints for I4.0 implementation strategies and actions.

For this reason, we define the aim of our research with the following research questions:

- What are the current needs of SMEs when I4.0 is being introduced into manufacturing?
- What are the functional requirements of SMEs based on their specific user needs for smart manufacturing?

- What are coarse design guidelines to facilitate the introduction of I4.0 in SME manufacturing systems?
- What are the possible limitations and barriers for SMEs introducing I4.0 in manufacturing?

2.4 Research Methodology

In order to obtain direct input from the beneficiaries of smart manufacturing systems, we selected a primary research approach to collect specific user needs by interviewing SMEs. Another approach to get this information could also be to conduct a survey like in the work of Mittal et al. (2018). Due to the novelty of I4.0, many SMEs have not yet dealt with the topic at all or only to a limited extent, thus a survey might not produce any usable results. Therefore, the approach of an explorative field study (see also Becker et al. 2009; Wölfel et al. 2012) based on SME workshops was chosen, which allowed direct contact to be made with SMEs in order to better understand their real requirements. In the exploratory study, the researchers preferred discussion in smaller workshop groups. Such workshops allow a common exchange of experiences and stimulate discussion among the participants, thus creating a more creative atmosphere.

The workshops themselves were structured as follows. A total of four SME workshops were held in Europe (Italy and Austria), USA (Massachusetts), and Asia (Thailand) to investigate specific requirements for SME (see Fig. 2.1). The implementation of SME workshops in different countries/continents should also help to identify cultural or country-specific differences, thus avoiding local needs having a strong influence on the final design guidelines for the introduction of I4.0 in SMEs. A limit of 10–12 participating companies (owner, general manager, operations manager) facilitated a productive interaction in the workshops. The workshops had a standardized structure starting with an initial introduction and overview of I4.0, then presenting some practical applications and best practice examples in SMEs. This should help raise awareness that I4.0 will be an important topic for SMEs in the future and prove that even smaller companies can implement I4.0. Afterward,

Fig. 2.1 Explorative field study through SME workshops (Map reproduced from D-maps.com: https://d-maps.com/carte.php?num_car=3267&lang=en)

the participants were asked to express their needs and requirements for introducing I4.0 concepts in their companies and share their experiences with the other participants. They were then asked about the main barriers and limitations for the implementation of I4.0. The inputs were collected in the form of sticky-notes on pinboards and categorized by topic. Before starting the evaluation of the collected inputs, several company visits were carried out at participating SMEs to gain a better practical understanding of the requirements and barriers on site.

For the evaluation of the collected inputs from the SME workshops, the research team applied Axiomatic Design (AD) theory (see also Fig. 2.2). AD is a method used for the systematic design of complex systems (Suh 2001). In AD so-called Customer Needs (CNs) are translated into Functional Requirements (FRs) because not all customer "wishes" can be considered as functional. In addition, some of the CNs are translated into Constraints (Cs) as some of them limit design space. Once the needs and requirements have been determined, the next step starts with a decomposition and mapping process selecting appropriate solutions or Design Parameters (DPs) for individually fulfilling each FR. So-called Process Variables (PVs) are then the real process parameters

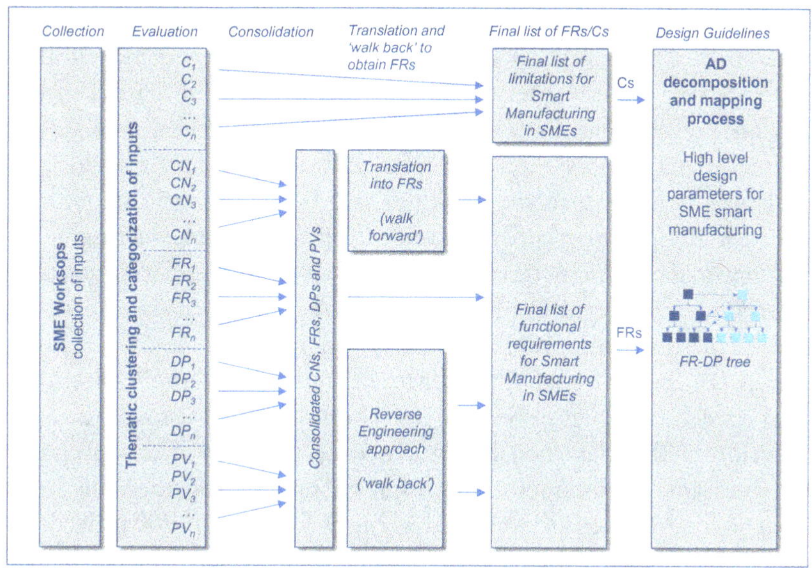

Fig. 2.2 Axiomatic Design based methodology for the analysis of SME requirements and design guidelines for smart manufacturing

in the phase of realization of the DPs. Chapter 6 gives a detailed overview about the AD methodology used in this chapter and explains the application of AD for the design of complex products, processes, and systems.

Although people in the workshop were asked about their needs and requirements for introducing I4.0, the experience of the authors showed that, often, people do not express their thoughts in the form of solution-neutral CNs or FRs, but rather in the form of physical solutions in the sense of DPs or PVs. Thus, the research team categorized the inputs from the SME workshops into Cs, CNs, FRs, DPs, and PVs. Cs are collected and build a final list of constraints that must be considered when realizing a smart manufacturing system in SMEs. The other inputs had to be further processed and interpreted to create a final list of solution-neutral FRs as a basis for the later definition of DPs. CNs were translated into FRs by analyzing the expressed needs and deriving the functional requirements by which the needs can be fulfilled.

FRs were added directly to the final list of FRs. DPs and PVs needed to be further processed to create "true FRs." Users had most difficulties expressing solution-neutral CNs or FRs, proposing partial physical solutions, rather than basic needs. According to Girgenti et al. (2016), such a mixing of CNs and FRs with DPs or PVs can introduce personal bias, forestall creative thinking, and further complicate and constrain the design process. Therefore, we applied a Reverse Engineering (RE) approach, which starts from DPs/PVs from the SME workshops to derive solution-neutral FRs and CNs. This idea of using reverse engineering to solve this problem is based on previous research (Girgenti et al. 2016; Sadeghi et al. 2013). More details on the application of the RE approach is shown in Sect. 2.5. To build the final list of FRs, a consolidation of the identified FRs was needed as many of the inputs deal with the same requirement and could be merged together. In the last step, the final list of FRs was used as input for the top-down decomposition and mapping process in AD to derive coarse design guidelines for the design of smart manufacturing systems for SMEs.

2.5 Analysis of Requirements for SME 4.0 Manufacturing Systems and Coarse Design Guidelines

2.5.1 Collection of User Needs Through an Explorative Study

As explained in the previous section, the research team conducted four SME workshops in Italy, Austria, USA, and Thailand in order to collect inputs for the analysis of needs and requirements of SMEs regarding the introduction of I4.0. To ensure a uniform collection of requirements, a standardized procedure, and presentation for the conduct of the workshops were defined in advance. Table 2.1 illustrates the standardized structure of the workshops, where inputs for smart manufacturing were collected in three categories defined previously by the research team: (i) adaptable manufacturing system design, (ii) smart manufacturing through ICT and CPS, and (iii) automation and man-machine interaction. In each brainstorming round, participants were also asked to express the main barriers and difficulties of introducing I4.0 concepts in manufacturing, which they had experienced, or foresaw experiencing as they planned on implementing I4.0 within their firms.

Table 2.1 Structure of SME workshops

No	Agenda point	Duration	Objective	Method
1	Introduce project presentation	15 min	Explanation of the project and research objectives	Opening presentation
2	Concept and origin of I4.0	30 min	Introduction to I4.0 for a common understanding	Opening presentation
3	Best practice examples	20 min	Awareness raising for implementation	Case studies, pictures, videos
4	Overview AD	15 min	Understanding of the research method and of the difference of CNs, FRs, DPs	Introductory presentation, examples
5	Introduction brainstorming session	10 min	Understanding of the brainstorming method	Introductory presentation
6	Brainstorming "adaptable manufacturing systems design"	30 min	Creative brainstorming with sticky-notes and subsequent discussion	Sticky-notes method
7	Brainstorming "smart manufacturing through ICT and CPS"	30 min	Creative brainstorming with sticky-notes and subsequent discussion	Sticky-notes method
8	Brainstorming "automation and man-machine interaction"	30 min	Creative brainstorming with sticky-notes and subsequent discussion	Sticky-notes method
9	Discussion and closure	30 min	Summary and closure	Open discussion

Table 2.2 Categories used in the workshop brainstorming sessions

No	Category	Brainstorming session	Sticky-notes
1	Adaptable manufacturing systems design	Session 1—smart manufacturing	58
2	Smart manufacturing through ICT and CPS	Session 1—smart manufacturing	64
3	Automation and man-machine interaction	Session 1—smart manufacturing	41
4	Main barriers and difficulties for SMEs—manufacturing	Session 1—smart manufacturing	60
Sum			223

Participants of the SME workshops who could speak well to the needs of SMEs in the manufacturing sector were invited through contact databases of the research team and professional associations. To allow an open discussion, the number of participants was limited to around a dozen companies in each workshop. Only owners, general managers, and production or logistics managers were invited. A total of 67 people from 37 SME companies attended and contributed to collect 163 user needs and 60 inputs regarding barriers/difficulties in the form of sticky-notes (see Table 2.2). Participants came from a variety of fabrication backgrounds, such as metal fabricators, wood processors, and many other industries.

2.5.2 Thematic Clustering and Categorization of Inputs

The workshop results built the basis for the definition of FRs and a subsequent AD decomposition and mapping process to derive DPs for the design of smart manufacturing systems for SMEs. The evaluation of the workshop results showed that the participants did not always write down Cs, CNs, or FRs as desired, but replied partly in the form of DPs or PVs. As this is a common behavior of people when they are asked to express their basic needs and requirements, the research team categorized all sticky-note responses.

The results were interpreted using the following procedure to define the AD domain:

- Each category was discussed during the brainstorming session and notes were taken to ensure the intent of the inputs when final collation of data was to be done after the workshop. The open discussion of participants' feedback on post-its ensures a correct interpretation of the statements. The moderator needed to check if the respondents understood the concepts of I4.0 correctly and used them in a correct way according to what they intended to express. In addition, this confirmed the alignment between their understanding and the interpretation of the research team.
- After the workshop, inputs, and notes were collected in an Excel spreadsheet and inputs were categorized into thematic "clusters" (see Table 2.3), which were used to identify subjects of interest for several categories.
- Each piece of input was then categorized as a C, CN, FR, DP, or PV based on AD grammar, notes, and interpreted design space.

Table 2.3 Thematic clustering of workshop inputs

No	Cluster	Sticky-notes	No	Cluster	Sticky-notes
1	Agility	23	15	Production planning and control	10
2	Automation	16	16	Preventive and predictive maintenance	5
3	Connectivity	12	17	Real-time status	10
4	Culture	14	18	Remote control	3
5	Design for manufacturing	4	19	Resource management	14
6	Digitization	22	20	Safety	2
7	Ease of use	8	21	Security	4
8	Implementation	12	22	Strategy	2
9	Inspection	5	23	Sustainability	4
10	Lean	8	24	Tracking and tracing	5
11	Machine learning	3	25	Transport	1
12	Mass customization	9	26	Upgrade	3
13	Network	4	27	Warehouse management	1
14	People	16	28	Virtual reality	3

Table 2.4 Breakdown of categorization of workshop outputs

Abbreviation	AD domain	Sticky-notes	%	Check
C	Constraints	47	21.08	✓
CN	Customer Needs	65	29.15	✓
FR	Functional Requirements	34	15.25	✓
DP	Design Parameters	76	34.08	✗
PV	Process Variables	1	0.45	✗

Table 2.4 summarizes the result of the categorization. 21.08% of the inputs were constraints. In particular, the inputs regarding limitations and barriers for the introduction of I4.0 were good sources for the collection of constraints. Overall, 29.15% of the inputs were categorized as CNs and another 15.25% as FRs. CNs could be translated by the research team and companies into real FRs. However, nearly 35% of the inputs were categorized as DP and PVs and need a reverse engineering interpretation to be used for further AD design studies.

2.5.3 Reverse Engineering of Inputs Categorized as DPs and PVs

DPs and PVs were derived to functional requirements (see Table 2.5) applying the reverse engineering approach (hereinafter called FR_{RE}s). Through logical regression, the research team then "walked back" each input to make it a functional requirement. For this purpose, these were analyzed in detail and discussed together with companies from the workshops in order to identify the real needs.

The grammatical rules of AD were applied for this "walk back." A look at the first example will show that "automate a current manual loading…" is a physical solution, and that the true FR would be to "mitigate highly repetitive tasks." This gives a larger solution space as the design team is no longer constrained to using automation, but whatever solution is deemed best by the design team and customer.

Table 2.5 shows an excerpt of the complete list of 43 derived FR_{RE}s. Due to repetition of similar DPs in the various workshops, many DPs have been consolidated into single inputs to make reading the FR list easier to digest. This means that the original 77 non-satisfactory inputs from sticky-notes have been reduced to 43 FR_{RE}s.

Table 2.5 Excerpt from the list of the reverse engineering approach

No	Inputs (DPs and PVs)	Reverse engineered FR (FR_{RE}s)
1	Automate a current manual loading process using a robot to load and process	Mitigate highly repetitive manual tasks
2	Augmented reality in service, maintenance and after sales, augmented reality for information provision at assembly	Allow user-friendly "smart" representation of information for production, maintenance, design, and service
3	Machine driven SPC and adaptive tool path generation	Identify and adjust parameter deviations in the manufacturing process influenced by environmental variance
4	Automation for billing, order management for correct priorities, and workflow optimization	Automate and digitize internal workflows and report generation
5	Simulation of components before production	Avoid cost and time for physical prototyping
6	Data acquisition of machines, workstations, warehouses, and buildings	Collect real-time data of machines, warehouses, and facilities to keep production under control
7	Optimal utilization of space thanks to flexible working systems, with shortened distances through flexible workstations	Create compact production lines and work stations
8	Automated time recording of staff presence	Create data-driven resource and process capability monitoring system for all relevant resources
9	Computational design and engineering as well as simulation for products can save cost and test process, etc.	Create data-driven system for product development, improvement, and management
10	Use of sensors on the machine for data acquisition, real-time data collection, machine reports capacity usage, digital feedback of work steps	Create a digital feedback system, and infrastructure, which monitors real-time status of production

A limitation of the reverse engineering approach is a possible misunderstanding of the user input by the research team. However, the risk of making a misjudgment through the reverse engineering approach is lower than the limitation one would accept if one continued to work with inputs that are not solution-neutral. Furthermore, as the case study in this research confirms, many user inputs can be categorized often as

DPs or PVs (in the described case study nearly 35%). Therefore, simply ignoring these inputs is not a recommended way forward. Thus, the presented reverse engineering approach represented a good possibility to transfer "false CNs" into useful requirements for further design studies.

2.5.4 Final List of Functional Requirements and Constraints Regarding the Introduction of Industry 4.0 in SMEs

FRs (directly collected in the workshops or translated from CNs) and FR_{RE}s (obtained from DPs and PVs using the previously explained RE approach) were consolidated, and redundancies removed by combining similar FRs and FR_{RE}s and merging them into one. Due to the high number of inputs from SME workshops and many similar inputs from different workshops, this was necessary and reasonable to make the document and the final FR list more workable and useful. The same was also done for the identified constraints to achieve a list of the main limitations and barriers that SMEs are facing to introducing I4.0 in their companies. This final FR list, together with the final list of Cs is used in a next step to derive coarse design guidelines for the design of smart manufacturing systems for SMEs (see also Sect. 2.5.5).

Table 2.6 shows the consolidated list of functional requirements for SMEs based on the procedure discussed throughout Sect. 2.4 of this chapter.

In addition, Table 2.7 shows the consolidated list of the main limitations and barriers (deduced from the identified Cs) for SMEs introducing I4.0. This list serves as a starting point for measures to minimize the listed barriers or also to set SME specific limits in the design of smart manufacturing systems.

2.5.5 Derivation of Coarse Design Guidelines for Smart Manufacturing in SMEs

The consolidated final list of FRs builds the basis for the next step to derive coarse design guidelines for the design of smart manufacturing systems. According to AD, this can be achieved through a top-down

Table 2.6 Final list of SME functional requirements for smart manufacturing

Cluster	No	(Functional) Requirements for the design of smart manufacturing systems in SMEs
Agility	1	Build or improve production lines and work stations to be more compact
	2	Ensure flexible, scalable, customizable production systems
	3	Minimize set up time for new configurations
	4	Enable the ability to produce a wide variety of products and a wide range of volumes without significant reconfiguration of costs and time
	5	Create self-adjusting processes
	6	Enable easy to use and change systems of new manufacturing technologies
	7	Take advantage of rapid prototyping technologies to make product development easier, and reduce requirements for stock
Automation	8	Mitigate repetitive tasks with quick payback time
	9	Enable on demand customizable packaging
	10	Reduce labor and cost of all production and logistics processes
	11	Implement self-maintaining processes
Connectivity	12	Ensure the ability to easily and efficiently communicate on a sufficiently real-time basis with internal and external customers
	13	Standardize and simplify security and interoperability of information and communication technologies
	14	Create standardized easy to use systems for connectivity, communication, and transparency
	15	Enable internal and external information connectivity to enable better forecasting, inventory management, current demand measuring, internal material requirements, etc.
Culture	16	Understand the culture of customers to interpret preferences for cost and quality
Design for manufacturing	17	Enable the use of advanced manufacturing technologies in the design phase
Digitization	18	Implement automation and digitization of internal workflows and report generation
	19	Avoid cost of physical prototyping
	20	Implement clear data gathering, management, analysis, and visualization to both internal and external customers

(continued)

Table 2.6 (continued)

Cluster	No	(Functional) Requirements for the design of smart manufacturing systems in SMEs
	21	Collect real-time data of machines, warehouses, and facilities to keep production under control
	22	Enable data flow to be consistent through the whole product life cycle and in the whole supply chain
	23	Enable fast measurement on-site and immediate delivery of data to production facility
	24	Provide and visualize information everywhere and every time to reduce waiting times and unnecessary delays
Ease of use	25	Simplify maintenance of newly adopted manufacturing technologies
	26	Minimize informational barrier, complexity of entry to new manufacturing technologies
	27	Enable user-friendly robot programming for "normal" workers
Implementation	28	Manage legal and bureaucratic hurdles for introducing I4.0 technologies
	29	Measure the impact of I4.0 on the company's sustainable success
	30	Provide an overview of existing I4.0 instruments and their suitability for SMEs or industry sectors
	31	Gain access to knowledge needed to implement I4.0
Inspection	32	Identify a defect as early as possible with little to no worker intervention needed
	33	Mitigate the human element in otherwise tedious or low information content tasks, such as delicate maintenance, equipment calibration, etc.
	34	Identify defects through in line inspection of process and material to avoid non-quality at the customer side
Lean	35	Eliminate non-value adding activities in production and logistics
	36	Produce on demand and deliver just in time
	37	Move product individualization as late as possible in the value chain
Machine learning	38	Automatically identify and adjust parameter deviations in the manufacturing process influenced by environmental variance
	39	Implement fast and automated design-based generation of tool path, part processing plan, and quotation
Mass customization	40	Gain the ability to produce small lot sizes (lot size 1) without losing efficiency

(continued)

Table 2.6 (continued)

Cluster	No	(Functional) Requirements for the design of smart manufacturing systems in SMEs
Network	41	Ensure that SME has a culture which includes the needs of the customer and workers through discourse and communication to enable full and productive integration of SME 4.0
	42	Gain the ability to communicate and/or share capacity, materials, infrastructure, and information with internal and external customers, and suppliers
People	43	Enable ergonomic support for physically difficult tasks
	44	Manage internal knowledge and staff development for Industry 4.0
Production planning and control	45	Enable a decentralized and highly reactive production planning and control
	46	Create system which can forecast demand changes quickly and interact with systems for planning, control, and logistics
Preventive and predictive maintenance	47	Ensure maintenance costs are minimized while maximizing value added time of machines
	48	Proactively maintain to ensure availability and decrease downtime of machines
	49	Predict data-based probability of machine stops or machine downtime
Real-time status	50	Create digital feedback system, and infrastructure, which monitors status of production, storage, shipping, risk, and crisis management
	51	Gather real-time status and visualize these data for operators and management
Remote control	52	Enable location independent control of maintenance, facilities, and products
Resource management	53	Create data-driven material, and process capability monitoring system for all relevant resources
	54	Ensure machines are capable for prospective jobs, and are able to be repurposed for a variety of other jobs
	55	Minimize time investment for I4.0 implementation and throughout life cycle
Safety	56	Provide workers with ergonomic workplace
	57	Provide safe working environment
Sustainability	58	Minimize energy consumption and environmental cost
	59	Measure and optimize energy, material, and time usage on processes

(continued)

Table 2.6 (continued)

Cluster	No	(Functional) Requirements for the design of smart manufacturing systems in SMEs
Tracking and tracing	60	Implement easy tracking of products from origin through the value chain
	61	Ensure supply chain has capability to digitally trace, and allow localization of systems
Transport	62	Create easy to use, worker independent material transport system for inside plant
Upgrade	63	Reuse and upgrade of existing manufacturing equipment
Virtual reality	64	Allow user-friendly "smart" representation of systems for production, maintenance, design, and service
	65	Create data-driven system for product development, improvement, management, and security to ensure product is more profitable for SME and customer through product life

decomposition and mapping approach of FR-DP pairs applied to decompose first level FR-DP pairs from an initially abstract level toward more tangible design guidelines (see also Fig. 2.3). To conduct such a decomposition, the two basic Axioms of AD will be considered (see also Chapter 13 in the Appendix). The application of the first Axiom, the Independence Axiom, favors DPs which are independent of FRs other than the one they were selected to fulfill. The second Axiom, the Information Axiom, ensures that in case of alternative solutions (alternative DPs), the best DP has the lowest information content, or greatest probability of success (Suh 2001):

- Axiom 1—Independence Axiom: the design of a system is considered ideal if all functional requirements are independent of the others to avoid any kind of interaction among them. Each defined design parameter is only related to one functional requirement and has no influence on other functional requirements.
- Axiom 2—Information Axiom: The Information Axiom helps the designer to choose among multiple possible solutions. The design parameter should be part of the physical domain with the smallest information content, to ensure a higher probability to satisfy a requirement. The aim is to minimize the information content or complexity of the design.

Table 2.7 Constraints (limitations and barriers) of SMEs introducing smart manufacturing

No	Cluster	Limitations and barriers for the design of smart manufacturing systems in SMEs
1	Culture	Lack of cooperation, openness, and trust between firms
2		Lack of employee acceptance of new operational processes and technologies
3		Company needs a well-entrenched top-down culture which allows continual improvement and mitigation of silo syndrome
4		Regulations and culture of the sphere within which the SME and parent organization functions must be such that proliferation of I4.0 is enabled, rather than disabled
5		Lack of visibility of I4.0 among professionals who would otherwise champion the implementation of I4.0
6	Implementation	Lack of experience in project management and budgeting for implementation of I4.0
7	People	Lack of training and qualification of personnel for systems to encourage communication, flexibility, education of I4.0, and soft skills
8		SMEs lack access to the financial, informational, digital, physical, and educational resources to ensure I4.0 is fully realized
9	Resource management	Lack of easy access to thought leaders and talent (relative to multinational companies)
10		Buildings are not designed for automating internal transports or processes or for new manufacturing technologies
11		High financial barrier to new manufacturing technologies
12	Security	Lack of and need for better, data security for operations such that potentially unforeseen dangers can be mitigated or blocked entirely
13	Strategy	Current lack of knowledge transfer from experts to SMEs for the implementation of I4.0
14		Lack of risk management tools for investments in new processes

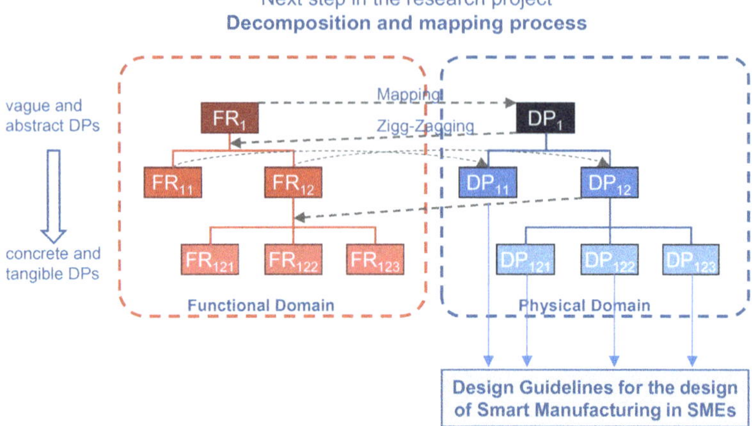

Fig. 2.3 AD approach to deduce design parameters for smart manufacturing in SMEs

By using the previously explained Axiomatic Design approach and examining the final list of FRs in Table 2.6, we identified the following top-level (Level 0 and Level 1) and upper-level FRs as well as their related design solutions (DPs).

FR_0 Create a smart and highly adaptable manufacturing system for SMEs
DP_0 Design guidelines for a smart and highly adaptable manufacturing system for SMEs

The abovementioned highest level FR-DP pair (Level 0) can be further decomposed into the following top-level FR-DP pairs.

FR_1 Adapt the manufacturing system very quickly in a flexible way
DP_1 Changeable and responsive manufacturing system
FR_2 Make the manufacturing system smarter
DP_2 Industry 4.0 technologies and concepts

The top-level FR-DP pairs, describing the main goals in sense of a highly adaptive and a more intelligent manufacturing system, can again be further decomposed into a set of upper-level FR-DP pairs.

For FR_1/DP_1 (Adaptability of the manufacturing system), the decomposition is as follows.

$FR_{1.1}$ Change and reconfigure the system with low effort
$DP_{1.1}$ Changeable SME manufacturing system
$FR_{2.1}$ React immediately to changes
$DP_{2.1}$ Responsive SME manufacturing system

For FR_2/DP_2 (Smartness of the manufacturing system), the decomposition is as follows.

$FR_{2.1}$ Enable the manufacturing system for Industry 4.0
$DP_{2.1}$ Digitalization, Smart Sensors, and Cyber-Physical Systems
$FR_{2.2}$ Connect all elements in the system to get real-time data
$DP_{2.2}$ Connectivity and Interoperability in SME Cyber-Physical Production Systems
$FR_{2.3}$ Take advantage of available data in the system
$DP_{2.3}$ SME-adapted Big Data Analytics and Artificial Intelligence
$FR_{2.4}$ Make automation in SME manufacturing easier
$DP_{2.4}$ SME Automation and Human–Machine Interaction
$FR_{2.5}$ Prepare typically low qualified people in SMEs for Industry 4.0
$DP_{2.5}$ SME specific Industry 4.0 qualification programs
$FR_{2.6}$ Provide appropriate protection against cyber attacks
$DP_{2.6}$ Cyber Security systems for SMEs
$FR_{2.7}$ Reduce ecological impact of manufacturing
$DP_{2.7}$ Sustainable and Green Manufacturing for SMEs

Once the decomposition and mapping process is finalized, the lowest level DPs of every branch in the FR-DP tree build a list of coarse guidelines for the design of smart manufacturing systems for SMEs. This final list of design guidelines will support researchers to develop specific I4.0 implementation strategies and solutions for SMEs and should guide practitioners from SMEs in their work to design smart manufacturing systems.

2.6 Discussion

The derivation procedure described previously in this chapter and the results summarized in Tables 2.6 and 2.7 give a good overall list of needs and constraints for SMEs to introduce I4.0. In the following, we try to use all these inputs to describe a picture of a smart SME manufacturing firm using the concepts of I4.0.

The needs discussed by the SME workshop participants desire a rapidly evolving manufacturing facility, where machines are easy to set up, and quick to adhere to the steps of ever-changing product configurations. These processes track themselves such that the personnel running the facility can concentrate on progressive improvement and upgrades to the system rather than acting as firefighters keeping the production working from day to day. Furthermore, these processes nondestructively inspect themselves. This would give operators the ability to be the first line of defense in quality control by giving them the tools to understand what the implications of process variations are, to lower their workload and increase the efficiency of the firm. Such an SME facility is also highly digitized with the ability for workplace user interfaces to be connected vertically and laterally within the organization. This allows for the destruction of silo syndrome through meaningful connectivity both within and without the organization and interoperability between single machines or processes. This allows the SME to better communicate within itself to ensure the manufacturing floor is always pushing the edge of productivity and adaptability. In addition, there is also the possibility for SMEs to achieve higher efficiency in higher-level supply chain management by connecting the company with suppliers and customers. The management in such a smart SME manufacturing firm has real-time numbers on the outputs of different machines, problems on the shop floor, potential upcoming costs, through predictive maintenance, or tracking the manufacturing environment and resources needed to ensure that all the needs of the floor workers are met, enabling increases in profitability. Furthermore, the leaders of these firms have access to experts, thought leaders as well as cognitive assistance systems that can give guidance on decisions which would otherwise have lasting costs. These leaders also engender an empowered workforce which is highly encouraged to bring possible improvements of the process to the fore, even when everything is working as expected.

These needs were not found to change much from culture to culture, or sector to sector, which lead us to believe that SMEs worldwide and from different sectors face similar challenges and problems. The lists in Tables 2.6 and 2.7 are general needs and constraints for most small- and medium-sized companies. The authors believe that these final list of FRs and Cs do give a good initial list of subjects to be pursued for implementation in SMEs throughout the world, due to the repetition of similar needs across these multinational workshops.

Possible limitations of this research include that the derived requirements and constraints using the reverse engineering approach are subject to the interpretation of the authors, as well as the initial company leaders, who communicated these needs. The authors attempted to hedge against this by taking notes on the intent behind the inputs, as well as diversifying the backgrounds, and geographical locations, of the participants of the workshops and by intensive discussions with SMEs during the phase of evaluation of the workshop results. It is believed by the authors that this did mitigate possible misinterpretations of needs, as well as incomplete needs for SMEs for implementing I4.0.

A current limitation of the presented decomposition of FRs into DPs is the fact that the design guidelines derived describe coarse design parameters. Manufacturing engineers receive a good basis for the appropriate design of their manufacturing system, but they do not yet find a very detailed, so-called "leaf-level" of design guidelines in order to be supported in the very detailed levels for machine design or the design of single processes. This would need a much more detailed investigation regarding the low-level decomposition of FRs and DPs defined in this work.

2.7 Conclusions

In this chapter, a comprehensive list of SME specific requirements and limitations regarding the introduction and implementation of I4.0 was proposed using an explorative field study as well as Axiomatic

Design theory. These lists are based on multinational workshops, which brought together leaders from manufacturing organizations from a variety of manufacturing spaces. To better organize the inputs of these workshops, they were broken down according to the subject matter of the session being discussed, then broken down further by "Clusters." These clusters allowed for an efficient manner for categorizing and further refining the requirements and constraints.

Upon initial processing of the content from the international workshops, the authors found that almost 35% of the input given was not solution-neutral. With the use of AD, this is a requirement to ensure the best solution is reached. The authors thus concluded that the inputs would need refinement to derive the "true FRs" behind the input from the workshops. The FR derivation technique, which was discussed, is a good methodology to derive solution-neutral requirements from these organizational leaders. These requirements and constraints show the basis for further research on the subject matter, giving a starting point for researchers to begin investigating, developing, and delivering tools for SMEs to fully realize the advantages which I4.0 is believed to offer them.

The decomposition and mapping process was used to derive coarse design guidelines for manufacturing system designers implementing I4.0 in SMEs. Together with the list of requirements and constraints, these guidelines form the main result of this research and a useful tool set for practitioners to design manufacturing systems for SMEs that are not only flexible and reconfigurable, but also smart and innovative as described in the picture in the previous section.

Further research will be needed now to investigate lower-level design guidelines and to develop techniques, methods, tools, and techniques as well as organizational solutions for SMEs to satisfy the functional requirements and to apply the defined coarse design guidelines. It is believed that this will deliver a suite of instruments for SMEs to take full advantage of I4.0 such that they do not lose their competitive advantage to large enterprises.

References

ASTM. 2013. *Standard Terminology for Additive Manufacturing Technologies*. F2792.

Atzori, L., A. Iera, and G. Morabito. 2010. The Internet of Things: A Survey. *Computer Networks* 54 (15): 2787–2805. https://doi.org/10.1016/j.comnet.2010.05.010.

Baum, G. 2013. *Innovationen als Basis der nächsten Industrierevolution. Industry 4.0 – Beherrschung der industriellen Komplexität mit SysLM*. Munich: Springer.

Becker, J., D. Beverungen, M. Matzner, and O. Müller. 2009. Design Requirements to Support Information Flows for Providing Customer Solutions: A Case Study in the Mechanical Engineering Sector. In *Proceedings of the First International Symposium on Services Science*, Leipzig, Germany.

Boughton, N.J., and I.C. Arokiam. 2000. The Application of Cellular Manufacturing: A Regional Small to Medium Enterprise Perspective. *Proceedings of the Institution of Mechanical Engineers, Part B: Journal of Engineering Manufacture* 214 (8): 751–754. https://doi.org/10.1243/0954405001518125.

Broy, M., and E. Geisberger. 2012. *Agenda CPS—Integrierte Forschungsagenda Cyber-Physical Systems*. Berlin and Heidelberg: Springer.

Chen, B., J. Wan, L. Shu, P. Li, M. Mukherjee, and B. Yin. 2018. Smart Factory of Industry 4.0: Key Technologies, Application Case, and Challenges. *IEEE Access* 6: 6505–6519.

Dombrowski, U., I. Crespo, and T. Zahn. 2010. Adaptive Configuration of a Lean Production System in Small and Medium-Sized Enterprises. *Production Engineering* 4 (4), 341–348. https://doi.org/10.1007/s11740-010-0250-5.

Erol, S., A. Schumacher, and W. Sihn. 2016. Strategic Guidance Towards Industry 4.0—A Three-Stage Process Model. In *International Conference on Competitive Manufacturing*, 495–501.

Federal Ministry of Education and Research. 2013. Zukunftsbild Industry 4.0. https://www.bmbf.de/pub/Zukunftsbild_Industrie_4.0.pdf. Accessed on 11 Mar 2018.

Frank, A.G., L.S. Dalenogare, and N.F. Ayala. 2019. Industry 4.0 Technologies: Implementation Patterns in Manufacturing Companies. *International Journal of Production Economics* 210: 15–26.

Ganzarain, J., and N. Errasti. 2016. Three Stage Maturity Model in SME's Toward Industry 4.0. *Journal of Industrial Engineering and Management* 9 (5): 1119–1128. https://doi.org/10.3926/jiem.2073.

Girgenti, A., B. Pacifici, A. Ciappi, and A. Giorgetti. 2016. An Axiomatic Design Approach for Customer Satisfaction Through a Lean Start-Up Framework. *Procedia CIRP* 53: 151–157. https://doi.org/10.1016/j.procir.2016.06.101.

Gneuss, M. 2014. *Als die Werkstücke laufen lernten, Industrie 4.0*. Berlin: Reflex.

Gorecky, D., M. Schmitt, M. Loskyll, and D. Zühlke. 2014. Human-Machine-Interaction in the Industry 4.0 Era. In *12th IEEE International Conference on Industrial Informatics (INDIN)*, 289–294. http://dx.doi.org/10.1109/INDIN.2014.6945523.

Kagermann, H., W. Wahlster, and J. Helbig. 2013. Recommendations for Implementing the Strategic Initiative Industrie 4.0: Securing the Future of German Manufacturing Industry. Final report of the Industrie 4.0 Working Group. Frankfurt: Acatech.

Kang, H.S., J.Y. Lee, S. Choi, H. Kim, J.H. Park, J.Y. Son, B.H. Kim, and S. Do Noh. 2016. Smart Manufacturing: Past Research, Present Findings, and Future Directions. *International Journal of Precision Engineering and Manufacturing-Green Technology* 3 (1): 111–128. https://doi.org/10.1007/s40684-016-0015-5.

Kraemer-Eis, H., and G. Passaris. 2015. SME Securitization in Europe. *The Journal of Structured Finance* 20 (4): 97–106. https://doi.org/10.3905/jsf.2015.20.4.097.

Lee, J. 2013. Industry 4.0 in Big Data Environment. *German Harting Magazine* 26, 8–10.

Lee, J., and E. Lapira. 2013. Predictive Factories: The Next Transformation. *Manufacturing Leadership Journal* 20 (1): 13–24.

Manhart, K. 2017. Industrie 4.0 könnte schon bald Realität sei. http://www.computerwelt.at/news/wirtschaft-politik/infrastruktur/detail/artikel/99076-industrie-40-koennte-schon-bald-realitaet-sein/. Accessed on 10 Aug 2017.

Matt, D.T. 2007. Reducing the Structural Complexity of Growing Organizational Systems by Means of Axiomatic Designed Networks of Core

Competence Cells. *Journal of Manufacturing Systems* 26: 178–187. https://doi.org/10.1016/j.jmsy.2008.02.001.
Matt, D.T., and E. Rauch. 2013a. Design of a Network of Scalable Modular Manufacturing Systems to Support Geographically Distributed Production of Mass Customized Goods. *Procedia CIRP* 12: 438–443. https://doi.org/10.1016/j.procir.2013.09.075.
Matt, D.T., and E. Rauch. 2013b. Implementation of Lean Production in Small Sized Enterprises. *Procedia CIRP* 12: 420–425. https://doi.org/10.1016/j.procir.2013.09.072.
Matt, D.T., E. Rauch, and P. Dallasega. 2014. Mini-Factory—A Learning Factory Concept for Students and Small and Medium Sized Enterprises. *Procedia CIRP* 17: 178–183. https://doi.org/10.1016/j.procir.2014.01.057.
Matt, D.T., E. Rauch, and D. Fraccaroli. 2016. Smart Factory für den Mittelstand. *ZWF Zeitschrift Für Wirtschaftlichen Fabrikbetrieb* 111 (1–2): 52–55. https://doi.org/10.3139/104.111471.
Matt, D.T., E. Rauch, and M. Riedl. 2018. Knowledge Transfer and Introduction of Industry 4.0 in SMEs: A Five-Step Methodology to Introduce Industry 4.0. In *Analyzing the Impacts of Industry 4.0 in Modern Business Environments*, ed. R. Brunet-Thornton and F. Martinez, 256–282. Hershey, PA: IGI Global.
Medbo, L., D. Carlsson, B. Stenvall, and C. Mellby. 2013. Implementation of Lean in SME, Experiences from a Swedish National Program. *International Journal of Industrial Engineering and Management* 4 (4): 221–227.
Mittal, S., M.A. Khan, D. Romero, and T. Wuest. 2018. A Critical Review of Smart Manufacturing & Industry 4.0 Maturity Models: Implications for Small and Medium-Sized Enterprises (SMEs). *Journal of Manufacturing Systems* 49: 194–214.
Moeuf, A., R. Pellerin, S. Lamouri, S. Tamayo-Giraldo, and R. Barbaray. 2017. The Industrial Management of SMEs in the Era of Industry 4.0. *International Journal of Production Research* 56 (3): 1118–1136. https://doi.org/10.1080/00207543.2017.1372647.
Monostori, L. 2014. Cyber-Physical Production Systems: Roots, Expectations and R&D Challenges. *Procedia CIRP* 17: 9–13. https://doi.org/10.1016/j.procir.2014.03.115.
Nowotarski, P., and J. Paslawski. 2017. Industry 4.0 Concept Introduction into Construction SMEs. *IOP Conference Series: Materials Science and Engineering* 245 (5): 052043. Bristol, UK: IOP Publishing.

Qin, J., Y. Liu, and R. Grosvenor. 2016. A Categorical Framework of Manufacturing for Industry 4.0 and Beyond. *Procedia CIRP* 52: 173–178. https://doi.org/10.1016/j.procir.2016.08.005.

Radziwon, A., A. Bilberg, M. Bogers, and E.S. Madsen. 2014. The Smart Factory: Exploring Adaptive and Flexible Manufacturing Solutions. *Procedia Engineering* 69:1184–1190. https://doi.org/10.1016/j.proeng.2014.03.108.

Rauch, E., S. Seidenstricker, P. Dallasega, and R. Hämmerl. 2016. Collaborative Cloud Manufacturing: Design of Business Model Innovations Enabled by Cyberphysical Systems in Distributed Manufacturing Systems. *Journal of Engineering*, 1308639. http://dx.doi.org/10.1155/2016/1308639.

Rauch, E., P.R. Spena, and D.T. Matt. 2019. Axiomatic Design Guidelines for the Design of Flexible and Agile Manufacturing and Assembly Systems for SMEs. *International Journal on Interactive Design and Manufacturing (IJIDeM)* 13 (1): 1–22. https://doi.org/10.1007/s12008-018-0460-1.

Rauch, E., M. Unterhofer, and P. Dallasega. 2018. Industry Sector Analysis for the Application of Additive Manufacturing in Smart and Distributed Manufacturing Systems. *Manufacturing Letters* 15: 126–131. https://doi.org/10.1016/j.mfglet.2017.12.011.

Rickmann, H. 2017. Verschläft der deutsche Mittelstand einen Megatrend? http://www.focus.de/finanzen/experten/rickmann/geringer-digitalisierungsgrad-verschlaeft-der-deutschemittelstand-einen-megatrend_id_3973075.html. Accessed on 11 Aug 2017.

Rüßmann, M., M. Lorenz, P. Gerbert, M. Waldner, J. Justus, P. Engel, and M. Harnisch. 2015. Industry 4.0: The Future of Productivity and Growth in Manufacturing Industries. Boston Consulting Group. http://www.inovasyon.org/pdf/bcg.perspectives_Industry.4.0_2015.pdf. Accessed on 17 Aug 2018.

Sadeghi, L., L. Mathieu, N. Tricot, L. Al Bassit, and R. Ghemraoui. 2013. Toward Design for Safety Part 1: Functional Reverse Engineering Driven by Axiomatic Design. In *7th ICAD International Conference on Axiomatic Design*, 27–28.

Sendler, U. (ed.). 2013. *Industrie 4.0: Beherrschung der industriellen Komplexität mit SysLM*. Berlin and Heidelberg: Springer Vieweg. https://doi.org/10.1007/978-3-642-36917-9_1.

Sommer, L. 2015. Industrial Revolution Industry 4.0: Are German Manufacturing SMEs the First Victims of This Revolution? *Journal of Industrial Engineering and Management* 8 (5): 1512–1532. https://doi.org/10.3926/jiem.1470.

Spath, D., O. Ganschar, S. Gerlach, T.K. Hämmerle, and S. Schlund. 2013. *Produktionsarbeit der Zukunft – Industrie 4.0*. Stuttgart: Fraunhofer Verlag.

Spena, P.R., P. Holzner, E. Rauch, R. Vidoni, and D.T. Matt. 2016. Requirements for the Design of Flexible and Changeable Manufacturing and Assembly Systems: A SME-Survey. *Procedia CIRP* 41: 207–212. https://doi.org/10.1016/j.procir.2016.01.018.

Suh, N.P. 2001. *Axiomatic Design: Advances and Applications*. New York: Oxford University Press.

Tao, F., Y. Cheng, L. Da Xu, L. Zhang, and B.H. Li. 2014. CCIoT-CMfg: Cloud Computing and Internet of Things-Based Cloud Manufacturing Service System. *IEEE Transactions on Industrial Informatics* 10 (2): 1435–1442. https://doi.org/10.1109/TII.2014.2306383.

VDI/VDE. 2013. *Cyber-Physical Systems: Chancen und Nutzen aus Sicht der Automation*. Düsseldorf: VDE Gesellschaft Mess- und Automatisierungstechnik.

Vidosav, D.M. 2014. Manufacturing Innovation and Horizon 2020—Developing and Implement New Manufacturing. *Proceedings in Manufacturing Systems* 9 (1): 3–8.

Wang, L., M. Törngren, and M. Onori. 2015. Current Status and Advancement of Cyber-Physical Systems in Manufacturing. *Journal of Manufacturing Systems* 37 (2), 517–527. https://doi.org/10.1016/j.jmsy.2015.04.008.

Wölfel, C., U. Debitz, J. Krzywinski, and R. Stelzer. 2012. Methods Use in Early Stages of Engineering and Industrial Design—A Comparative Field Exploration. *Proceedings of DESIGN 2012* DS 70. The 12th International Design Conference, Dubrovnik, Croatia.

Wuest, T., P. Schmid, B. Lego, and E. Bowen. 2018. Overview of Smart Manufacturing in West Virginia. WVU Bureau of Business & Economic Research. Morgantown, WV, USA.

USTR. 2017. Office of the United States Trade Representative, Small- and Medium-Sized Enterprises (SMEs). https://ustr.gov/trade-agreements/free-trade-agreements/transatlantic-trade-and-investment-partnership-t-tip/t-tip-12. Accessed on 12 Sep 2017.

Zambon, I., M. Cecchini, G. Egidi, M.G. Saporito, and A. Colantoni. 2019. Revolution 4.0: Industry vs. Agriculture in a Future Development for SMEs. *Processes* 7 (1): 36. https://doi.org/10.3390/pr7010036.

Zawadzki, P., and K. Żywicki. 2016. Smart Product Design and Production Control for Effective Mass Customization in the Industry 4.0 Concept.

Management and Production Engineering Review 7 (3): 105–112. https://doi.org/10.1515/mper-2016-0030.

Zhou, K., T. Liu, and L. Zhou. 2015. Industry 4.0: Towards Future Industrial Opportunities and Challenges. In *12th International Conference on Fuzzy Systems and Knowledge Discovery (FSKD)*, 2147–2152. https://doi.org/10.1109/fskd.2015.7382284.

3

Implementation of Industrial Internet of Things and Cyber-Physical Systems in SMEs for Distributed and Service-Oriented Control

Rafael A. Rojas and Manuel A. Ruiz Garcia

3.1 Introduction

The trend toward Industry 4.0 intends to populate traditional shop floors with digitalized systems that are able to share their process parameters, their operative status and express their availability for collaboration with other machines or workers. In other words, this new industrial philosophy foresees each instance of the value chain of a product as a smart one, that is, endowed with decision-making capabilities and the means of communication for valuable information sharing between instances (Kagermann et al. 2013). In this sense, the situational awareness of the manufacturing environment greatly relies on connectivity solutions, like IoT or Internet of Services (IoS) (Gilchrist 2016).

R. A. Rojas (✉) · M. A. Ruiz Garcia
Faculty of Science and Technology, Free University of Bozen-Bolzano, Bolzano, Italy
e-mail: rafael.rojas@unibz.it

M. A. Ruiz Garcia
e-mail: ManuelAlejandro.RuizGarcia@unibz.it

D. T. Matt et al. (eds.), *Industry 4.0 for SMEs*,
https://doi.org/10.1007/978-3-030-25425-4_3

Although comparable connectivity solutions are not new, thanks to the qualitative change in the processing capacity of modern embedded systems (ES) and the introduction of CPS (Lee 2008) a new kind of networked control system for factory automation is now possible. When integrated in a connected manufacturing environment, CPS replace the traditional automation pyramid and, by combining IoT and automation systems, merge two domains that had been traditionally been separate in industrial systems (Monostori et al. 2016): the IT domain and the operation technology (OT) domain. The former, related to the processing of data to obtain valuable information. The latter, related to the support of physical value creation in manufacturing processes. This integration could be achieved through the digital integration of traditional software systems as enterprise resource systems (ERP) and manufacturing execution systems (MES) with CPS. Following this idea, CPS will become the building blocks of the smart factory, the central CPPS of Industry 4.0.

The major challenges of achieving digital integration are related to the natural software heterogeneity in industrial systems. The effectiveness of modern enterprises depends on hundreds if not thousands of custom-built digital applications that can be acquired from a third party belong to a legacy system or a combination thereof. In fact, programming business applications is a challenging task and creating a single application capable of running a complete business is next to impossible (Hohpe and Woolf 2004). Although ERP systems are the most popular integrations points, they only provide a fraction of the functionalities required in an enterprise. Regardless, heterogeneity of components in modern industrial systems is a necessary or even favorable condition (Lin and Miller 2016). For example, acquiring components produced by different vendors may exploit the benefits of each distributor. Also, the continuous evolution of technology introduces new components that need to be integrated with legacy systems. Moreover, different norms and standards may require specific solutions that are not scalable or convenient for adoption in every application.

This chapter is devoted to the integration between OT and IT, in terms of the commutation network infrastructure necessary to define an IoT industrial solution. In particular, we present the design tools of a manufacturing service bus (MSB) to overcome many of the aforementioned integration issues, that is, the software infrastructure defining a homogeneous information channel among disparate, possibly event-driven, platforms.

3.2 Fundamentals of Connectivity

3.2.1 The OSI Model

The open systems interconnection reference model (OSI model) defines a seven-layer framework to describe the information flow between digital systems (Zimmermann 1980). It was developed to introduce a mechanisms abstraction to transfer information between a pair of digital systems. This model allows the commutation logic to be decoupled from the actual implementation, which is subject to the particularities of each application.

Figure 3.1 shows how the information flow between a CPS and other system travels through the OSI layers to reach both ends. Each block represents a digital system. At the bottom of each block, it is possible

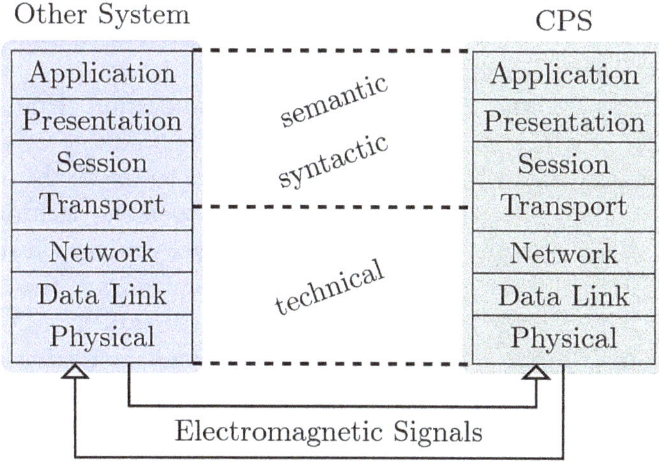

Fig. 3.1 The OSI model

to observe the physical layer, or layer 1, which describes the underlying mechanical, physical, optical, or electrical platform of the communication channel. On top of it, the data link layer or layer 2, provides the means to link two directly connected nodes, to access the physical communication medium and check possible errors induced by it. Next, the network layer, or layer 3, resolves the complex paths that may exist between the origin and destination nodes, thanks to specified target addresses. Layer 4 or the transport layer, sets up an end-to-end connection providing a means of transferring data sequences from a source to a destination host regardless of the routing paths. Layers 5 and 6 are often subject to criticism and are not of interest in this chapter. However, they are often integrated inside the application layer or layer 7, which is of special interest as represents the interface between the software applications running on the digital system and the communication system.

For this reason, the application layer encloses the major challenges of seamless integration between heterogeneous digital systems. On the one hand, this is because implementations of layers 1–4 are rapidly converging to ethernet-based TCP/IP technologies. On the other, it is because each digital system provides its own abstract representations of common sources of information and data therein. For example, consider a hybrid manufacturing station where a collaborative manipulator is integrated with a 3D camera to keep track of the human operator's activities. The manipulator receives data directly from the 3D camera to monitor the pose of the operator, so to collaborate with him in avoiding collisions. Both systems are attached with a direct communication channel based on USB or ethernet. As these digital systems are produced by different manufacturers, they offer different software abstractions of the information of common interest, i.e., the pose of the operator. Such data may be defined, for example, in different units or data structures inside each system. Although some standards may be implemented in the communication channel made available by the manufacturer of the 3D camera, there is a limit in the level of detail that a standard can offer. Moreover, the software inside the robotic application may be designed to represent the position of the operator in a way convenient to the programmer or follow the company's internal standards. Therefore, the data of the

3D camera shared through the communication channel cannot be used without the necessary representation transformation.

3.2.2 CPS Architecture

Heterogeneity is a fundamental characteristic of CPS. In fact, they are composed of three fundamental layers (Sztipanovits et al. 2012) that provide a division of their heterogeneous components. First, the physical layer refers to the material components and their physical interactions. The platform layer refers to the hardware electronics supporting the digital systems comprising the communication infrastructure. Finally, the software layer comprises the operating system and the different digital processes which actually control the CPS and provide means to implement intelligent or complex tasks. The relation between these layers and the OSI model is represented in Fig. 3.2.

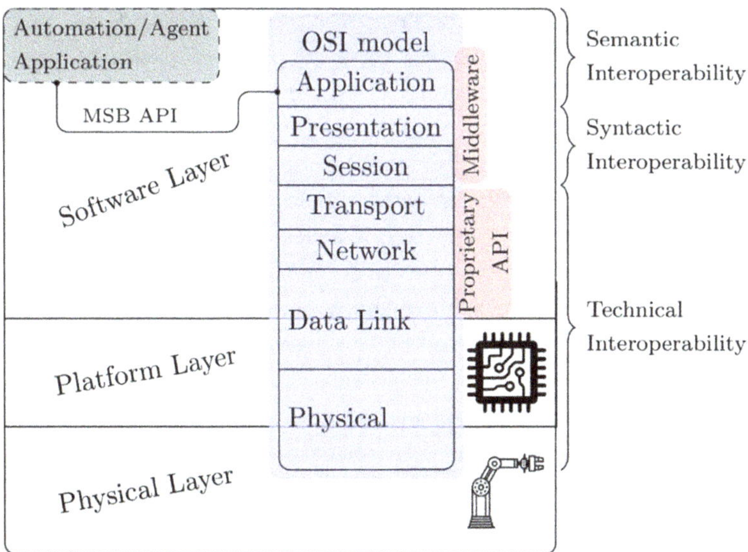

Fig. 3.2 Relation between the layered architecture of CPS and the OSI model

CPS are part of the common trend of pervasive computing, where distributed computing systems represent a dominant paradigm (Wooldridge and Jennings 1995). This concept has been tackled by the idea that intelligent behaviors "emerge" from the interaction of many simple entities. Together, the concept of CPS and "emergent" smart behaviors resonate with the idea of CPPS: a body of autonomous entities that smartly interact to achieve global objectives. Two main paradigms have been proposed to enable smart behaviors of autonomous entities in production systems: multi-agent systems (MAS) and holonic manufacturing systems (HMS). Agents and holons can be defined as self- and ambient-aware entities that can adapt to ambient variations, exhibit goal-oriented behaviors and interact with their peers.

MAS were proposed in the field of artificial intelligence (AI) to characterize such distributed computing systems. An agent may be defined as a system that is situated in some environment, capable of exerting autonomous actions on such an environment to meet its design objectives (Wooldridge and Jennings 1995). For leveraging of such characteristics, authors in Vogel-Heuser et al. (2015), Ji et al. (2016) and Monostori et al. (2016) propose the MAS technology as the main enabler of smart collective behaviors in CPPS. Beyond the fact that agents do not necessarily have a physical part, we desire to underline some important features which CPS share with agents in MAS (Weiss 1999): (i) are self-aware and ambient-aware, (ii) react in a timely way to ambient variations, (iii) exhibit goal-oriented behaviors, and (iv) interact with peers. On the other hand, HMS are constituted by autonomous entities that interact through a variety of hierarchic or egalitarian relations to achieve similar objectives of CPPS. In contrast to MAS, rooted in AI methods to achieve smart emergent behaviors, HMS is a conceptual paradigm motivated by the need to optimize manufacturing systems.

It is worth noticing that both paradigms share the common vision of manufacturing systems defined in terms of autonomous and cooperative units. As a consequence, the integration of a heterogeneous digital system is, in turn, an enabler of smart behaviors in CPPS. Therefore, as depicted in Fig. 3.2, the software layer integration represents a key milestone to enable the "agent behavior" of the CPS regardless of the underlying data heterogeneity contained inside the CPPS.

3.2.3 The Concept of Interoperability

In our context, integration implies a set of steps allowing a body of disparate systems to be treated as a whole (Bellman and Landauer 2000). As such, this conglomerate unicum of entities can be understood, monitored, reasoned about, configured or controlled without requiring explicit knowledge of its enclosed systems. As remarked upon in Gössler and Sifakis (2005), achieving integration requires that the system meets the following conditions: (i) compositionality, that the behavior of the system is predictable from the behavior of its components, (ii) composability, that the attributes of each component do not depend on other components nor on their interactions. Among all concerns that must be addressed to integrate heterogeneous systems, in this chapter, we are only interested in those related to the information sharing between software layers of different CPS. Under this delimitation, by integration of CPPS, we refer to the necessary steps to create a coherent and seamless information exchange between CPS software layers (Vernadat 2007). In other words, we limit our analysis to the interoperability aspects of the integration.

Interoperable systems provide understood interfaces for message exchange and functionality sharing. Therefore, three levels of interoperability can be identified: technical, syntactic, and semantic. Technical interoperability represents the capacity to exchange a raw sequence of bites. Syntactic interoperability is associated with data formats and structures, i.e., the symbols represented by such sequences of bits. Finally, semantic interoperability is the capacity to exchange meaning between systems. As a consequence, semantic interoperability necessarily depends on syntactic interoperability, which in turn, depends on technical interoperability. Unlike the wide concept of integration, interoperability is related to the coherence and uniformization of data and is a prerequisite to integrability itself. In fact, integration must assure composability, compositionality, and flexibility (Lin and Miller 2016).

3.2.4 Loosely Coupled Systems and SOA

Achieving loose coupling is to reduce the assumptions two software applications make about each other when they exchange information. The more assumptions they make about each other, the less tolerant is the connectivity solution to changes in the system. Common assumptions that lead to a tightly coupled system are (i) about the platform technology (internal bit representations), (ii) location (hard-coded addresses), (iii) time (availability), and (iv) data formats.

Integration of CPPS becomes familiar with a long history of effort to integrate disparate digital platforms that begins when computers and software applications become pervasive in office and business (Chappell 2004). In these early days, it was noted that a large number of small distributed software procedures allow for flexibility and reuse. Those approaches achieved simple remote communications by packaging a remote data exchange into the same semantics as a local method call (a traditional function of programming languages). This strategy resulted in the notion of a remote procedure call (RPC). Common implementations of RPC are CORBA, Microsoft DCOM, .NET Remoting, Java RMI, XML-RPC and SOAP. This marked the evolution from point-to-point integration solutions to the so-called service-oriented architectures (SOA) (Chen et al. 2008). Service is a common name for a functionality that is executed in distributed systems and SOA is an approach to encapsulate functional components in services. However, one of the main challenges to implementing a SOA is that remote communication invalidates many of the assumptions that a local method call is based on. To achieve a well-designed SOA requires management, a centralized service directory and effective documentation.

In parallel, communication solutions in industrial automation systems also began by using point-to-point solutions that were expensive and bulky (Felser 2002). In the two "worlds," business/office and industrial automation, the following metaphor appears as the ideal solution to achieve seamless connectivity: a single cable or bus of communication where a message can transit and every communication transaction may appear as a point-to-point exchange of data.

In the business/office realm, the term ESB was introduced to describe the software infrastructure capable of emulating such a bus and, in the field of industrial automation, the term "fieldbus" was coined. Unlike ESB, fieldbuses were originally developed for ES with reduced resources designed to exchange small amounts of data with stringent timing requirements. With the introduction of CPS factory shop floor, the quality of the exchanged information becomes more similar to office systems. Instead of exchange positions of servos, a CPS may receive the complete model of a product to be assembled.

3.2.5 The Publish and Subscribe Pattern

"Publish" and "Subscribe" is a king of messaging pattern designed to achieve a network of loosely coupled nodes. To reduce the number of assumptions made about peers, messages are sent through specialized channels often called topics. The middleware infrastructure hides the details of complex message routing between nodes and permits access to a channel using a semantic identifier instead of complex addresses such as IP number and port. To describe the type of messages that a channel conveys and which channels are available in the systems, it is possible to implement centralized systems to retrieve the desired information. Thanks to this middleware, it is possible to send messages only assuming that the receiver is listening at a specific channel. Nodes which send messages through a channel are called publishers of that channel. On the other hand, nodes which listen to a channel are called subscribers of that channel. Generally, channels have only one direction, but several implementations are possible. Publishers do not program messages to be sent directly to specific subscribers, but instead characterize published messages into classes using the available channels without making any assumptions about the subscribers' routing requirements. Similarly, subscribers express interest in one or more classes or channels and only receive messages that are of interest, without knowledge of which publishers, if any, there are.

Often publish-subscribe systems are constructed around a central entity called a broker that is responsive to managing channels and its publishers and subscriber. Brokers contain the list of the available channels and the IP addresses of every node in the publish-subscribe network. Each node performs client-server style operations with the broker to subscribe to channels or create new channels to publish data. Publish-subscribe is a sibling of the message queue paradigm, and is typically one part of a larger message-oriented middleware system. As publisher can "publish" information without regard for subscribers and a subscriber can "subscribe" to information without regard for publishers; the result is a loosely coupled network where each node can be replaced independently of one another. This king of messaging patterns is well suited to exchanging data updates between components and allows the communication paths to be optimized based on their requirements. It provides scalability for an evolving number of data sources and contributes to IIoT reliability, maintenance, and resilience by the decoupling of publishing and subscribing components in both location (location transparency) and time (asynchronous delivery). This decreases the likelihood of fault-propagation and simplifies incremental updating and evolution. The concept of channels allows flexibility in the interaction between nodes allowing periodic (time-driven) or responsive (event-driven) behaviors depending on the needs of the user. Asynchronous data transfers can be implemented in different publish-subscribe solutions making the IIoT more robust to component failures and unexpected delays.

3.2.6 Service Discovery, Zero Configuration, and Plug-and-Play/Work Networks

IIoT requires a flexible method for service composition, such that different functionalities can be dynamically integrated at run-time. Therefore, allowing dynamic networks to address services, without affecting the end users, represents a desirable quality. In this regard, the numerical-based addressing mechanisms used by IIoT platforms have two principal drawbacks. First, they are intrinsically not human-friendly. Second, IIoT networks require dynamic address assignment on hardware variations or software migrations. To overcome such limitations, a stable name can be associated with each service through a uniform resource identifier (URI), a string of characters with a predefined set of syntax rules that unambiguously identifies a particular resource. A URI does not refer to a particular piece of hardware or software, but a logical service with which any end user can communicate using a specified protocol. In the past, human-meaningful names (URI) were bound to computer-meaningful network addresses by manual configuration of the network. Today, they are several technologies to automate this procedure allowing each network entity to be interrogated about its services and protocols. As a result, any device or application can share its virtual representation or manifest (Monostori et al. 2016). In the context of desktop computers, the ability to automatically add devices to a network without having to manually register its services and configuration has been called zero configuration networking (Steinberg and Cheshire 2005). Zero configuration networking is based on the following three elements:

- Automatic IP address selection for networked devices (without DHCP server).
- Translation between names and IP addresses without a DNS server (Multicast DNS).
- Automatic location of network services through DNS service discovery (which eliminates the need for a directory server).

On the other hand, in the context of manufacturing, the term *plug-and-work* appears with the capability of a production system to automatically identify a new or modified component and to correctly integrate it into the running production process without manual efforts and changes within the design or implementation of the remaining production system (Schleipen et al. 2015). Schleipen et al. (2015) identify the following requirements for the application of the plug-and-work technology:

- *Component description*: the ability to get a complete description of an entity in the system.
- *Component selection*: the ability to compare all entities capabilities and choose the one which is able to perform some task.

- *Component access*: the ability to communicate with the entity.
- *Component control*: the ability to provide a control structure to the entity.

Under the perspective of IIoT, we recognize both concepts of zero configuration networks and plug-and-work as different names for a method of achieving dynamic CPPS composition. This composition is characterized by a loose coupling between the SOA and messaging model (OSI layers 4–7) and low-level network implementation (OSI layers 1–3). In general, the zero configuration infrastructure should be agnostic to application protocol design and advertise any kind of application protocol. At the lowest level, zero configuration networks or plug-and-work IIoT may be achieved through the implementation of the following technologies:

- *mDNS*: Multicasting DNS protocol resolves host names to addresses within small networks that do not include a local name server. It uses IP multicast user datagram protocol (UDP) packets.
- *UPnP*: Universal plug-and-play is a set of networking protocols on the top the internet protocol (IP), leveraging on HTTP to provide device/service description, actions, data transfer, and eventing. Device search requests and advertisements are supported by running HTTP on top of UDP (port 1900) using multicast (known as HTTPMU).

However, it is worth noticing that application-layer protocols are subject to constant evolution and changes.

3.2.7 Ethernet-Based Connectivity Technologies for SME

IT system integration costs may be a factor able to outweigh all other technical considerations (Wagner et al. 2015). Such costs include materials, software licenses, hardware installation, and other technical labor. Although there are some efforts of providers to adapt their offerings to the needs of small- and medium-sized enterprises (SMEs), usually their primary customers are large organizations able to afford such costs (Cruz-Cunha 2009). However, in the case of SMEs, the deployment of IIoT requires considering several trade-offs. Among others, we highlight the trade-offs between outsourcing and in-house software development and between open-source software (OSS) and proprietary software tools. OSS is a viable alternative in terms of costs and independence to vendor-locked applications (Olson et al. 2018). In fact, OSS does not require vendor-specialized consultants and tools, and allows the use of in-house available hardware and software rather than proprietary ones, subject to maintenance contracts. According to Weber (2004), OSS enhance software reliability and quality through independent peer review and rapid evolution of source code. Following these ideas, together with Lin and Miller (2016) and Gilchrist (2016), we suggest that OSS and open standards are also key enablers of IIoT in SME. To support this statement, we observe that SMEs have a relatively small amount of digital systems and processes to be integrated, thus they may be able to afford to develop in-house software applications. Moreover, thanks to the source code availability, OSS gives SMEs the possibility of developing highly customized and lean solutions based on their know-how. As a last remark, we emphasize that a growing amount of companies are providing support to open source IT solutions (Olson et al. 2018).

The following publish-subscribe and SOA messaging protocols, implemented on top of the TCP/IP protocol, are useful OSS for IoT applications:

- Java message service *(JMS), part of the Java EE, defines a generic and standard application programming interface (API) for the implementation of message-oriented middlewares. It does not provide any concrete implementation of a messaging engine.*
- The advanced message queuing protocol *(AMQP) is an open standard application-layer protocol for messaging and publish-subscribe messaging. AMQP is often used with RabbitMQ, a free and complete AMQP broker implementation and API for Java, C# and Erlang.*
- RabbitMQ *is written in Erlang and it is available in Ubuntu through the rabbitmq-server package. It offers a flexible broker implementation based on open standards.*

- Message queue telemetry transport *(MQTT) is an ISO standard (ISO/ IEC PRF 20922). It is broker-based and designed for networks with limited resources.*
- Extensible messaging and presence protocol *(XMPP) is a communication protocol for message-oriented middleware based on XML (extensible markup language).*

3.3 The Integration Drivers

The challenges of CPPS integration span far across operational and technical issues. In fact, beyond classical interoperability between digital systems, IIoT also implies the integration between organizational units and IT systems. Such an integration may require a significant shift in corporate politics and defining clear separations between inherent modules is not an easy task. In Lin et al. (2015) a set of different viewpoints are given to achieve such a modularity. Following this idea, we distinguish between organizational and technical drivers for the IIoT integration effort. On the one hand, organization drivers refer to those required for OT, i.e., to create physical value for a global market. On the other, technical drivers are those localized at the digital platforms, specifically related to the available or required digital systems.

3.3.1 Organizational Drivers

To define the necessary information flows inside a CPPS, it is also necessary to map business/organization processes into data requirements (see Fig. 3.3). The business vision, values, practices, key objectives, and capabilities are fundamental inputs to clearly define such system requirements. As a consequence, it is of fundamental regard to identify the OT activities, their inherent tasks and the roles of each comprising party on the accomplishment of such activities. Either tasks dependencies, constraints, and workflows need to be mapped into data requirements and data flows.

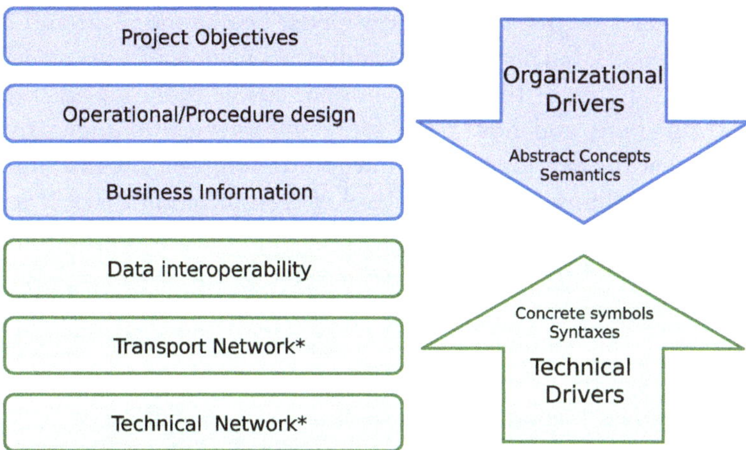

Fig. 3.3 The drivers of the integration

In Lin et al. (2015), the concept of domains is introduced, to simplify the functional separation between building blocks of business and organization processes. Each functional domain lies on a specific granularity and time scale, and can be hierarchically organized from high-level intelligence, to low-level control. The control domain directly deals with low-level OT. The operations domain contains the functional elements enabling operability of the hardware on the control domain (monitoring, register, track, deploy, and retrieve assets). The information domain provides the collection of data from all domains to high-level analysis. The application domain applies and defines coarse-grained logic, rules, and models for workflows. In other words, it provides a high-level abstraction of functionalities that could lie in a specific domain or be distributed among many of them.

3.3.2 Technical Drivers

The implementation viewpoint of Lin et al. (2015) is concerned with the technical representation of the IIoT. This representation takes into account the technologies and system components required to implement all activities and functions prescribed by the usage and functional viewpoints. Such viewpoints address all technical issues associated with the future use of the system, how to enable those usages through specific functions and how these functions interact. It might also be observed that the development of integration solutions is constrained to the limited or almost nonexistent level of customization that participating applications may have. In most cases, applications belong to external providers or legacy systems and cannot be changed or upgraded. This often leaves the integration developers in a situation where they need to make up for deficiencies or idiosyncrasies inside the participating applications. Often, it would be easier to implement part of the solution inside the application endpoints, but for political or technical reasons that option may not be available.

3.4 Connectivity Architecture

3.4.1 Ethernet-Based Automation System

In the last decade, the ethernet has become increasingly popular and pervasive inside the mid-level of the automation pyramid (Sauter 2014). Ethernet implements the OSI layers 1 and 2 and it is generally deployed with the complete IP protocol stack (OSI layers 3 and 4). Such a combination of ethernet and IP provides important building blocks to achieve technical interoperability between disparate digital systems. In fact, the ethernet is becoming the de facto standard in office, enterprise or business systems and modern CPS. However, the ethernet—as it is known in office or business environments—could not meet the requirements of industrial automation, since it lacks determinism and it has not been designed for real-time applications. Although the introduction of switched ethernet, megabit and gigabit ethernet has notably mitigated such problems, some practical realization of ethernet-based networks in industrial environments still needs special care. The IEC 61784-2 standard introduces terms like "industrial ethernet" or "real-time ethernet" to define a set of ethernet-based technologies able to meet strict industrial and technological requirements.

To address the requirements of industrial level ethernet systems, we follow Kim et al. (2014) by differentiating the types of data transmitted in an IIoT system into *configuration data* and *process data*. Configuration data defines the set of parameters required to configure the system, including remote management and operations work-flow. Process data identifies the process states, thus, depending on the process nature, this type of data may be periodic (when it is constantly generated) or aperiodic (when it is generated sporadically). To illustrate those types of data we can consider a classical robotic pick and place operation, where objects are moved from one place to another and sensors detect the availability of those objects to be placed. Both the measured positions of the joints and the commands to move them are periodic process data. They are generated constantly and describe part of the current state of the process. Also, the information generated by sensors falls into the category of process data. On the other hand, the program controlling the robot during the task execution belongs to a set of configuration data.

Each type of data flow has its own qualitative requirements, update rates, and tolerated latency times. In particular, we can distinguish between the following latency categories:

1. Human-control systems, where humans are involved in the system observation and the characteristic times are of the order of 100 ms.
2. Process control systems, which relate to computer numerical control (CNC) and programmable logic controller (PLC) systems and the characteristic times are of the order of 10 ms.
3. Motion control system, where the timing requirements are less than 1 ms. Motion control latency requirements are also called real-time.

Human and process control systems may be designed using slightly customized ethernet-based technologies. Among others, technologies like Profinet, TCnet, and Powerlink are able to meet process control requirements. Real-time networking solutions (of the order of 1 ms) are too restrictive to provide the level of flexibility that IIoT requires. In fact, CPS internal control mechanism needs to be endowed with enough levels of autonomy to avoid streaming control data flows through the network (goal-oriented control).

3.4.2 A Layered Design for Manufacturing Service Bus

Although it has pertinence in describing a communication system, the OSI Model is inadequate for representing the features of an IIoT system in a convenient way. As we have noticed before, the OSI model consists of many layers that, in general, are defined by dominant communication technologies, as ethernet and derivates. Following Lin et al. (2015), we aggregate the functionalities of the OSI layers into two separate layers. The OSI layers 1 (physical) through 4 (transport) are collected in the *communication transport layer*, to allow the basis for technical interoperability to be addressed along with the ability to reach endpoints along structured networks. On top of the communication transport layer, we define the *connectivity framework layer*. Such a layer spans the functionalities of the OSI layers 5 (session) through 7 (application) and provides the software infrastructure where middlewares are placed. The syntactic interoperability channel (with common and unambiguous data formats) and the ability to create messages without considering the particular endpoint implementations are placed inside this layer. The connectivity framework layer also provides the service discovery functionalities, the resources access handling, the high-level data exchange patterns (peer-to-peer, client-server, publish-subscribe, etc.), the means of security and the interoperability model in different programming languages.

Such subdivision of the connectivity functionalities is of particular benefit for ethernet-based IIoT systems. This comes from the fact that such a subdivision reduces the degrees of freedom of the OSI model and decouples two realms addressing different functional duties inside any ethernet-based systems. We call the collection of the functional components, given by the combination of these two layers, the MSB. This term was coined as an equivalent to enterprise service bus (ESB) (Monostori et al. 2016) in manufacturing systems. However, an MSB is differentiated from an ESB in terms of the quality and complexity of the operation that it can perform over the transit of data it handles. The MSB provides an interface between the low-level fieldbus devices, the CPS, and the business systems, generally based on an ESB and relying on the ISA-95 standard.

This solution allows the IT landscape of the factory floor to be integrated with enterprise-level IT systems. Similar solutions have been addressed by several authors. In MESA (2008), the concepts of SOA are widely explained from the perspective of manufacturing systems and an architecture for MSB is presented. The European Innovation Project SOCRADES (De Souza et al. 2008) presents an MSB system based on web services. The European Innovation project IMC-AESOP (Colombo et al. 2014) proposes an integration approach for CPPS based on cloud technologies. Minguez et al. (2010) explored the concept of MSB from a general perspective. Other proposals may be found in Zhang et al. (2018).

3.4.3 Physical and Logical Network Topologies of the MSB

To achieve coordination and orchestration of a distributed body of CPS, it is mandatory to have a correct structure and organization of the communication functions. For this reason, we address the physical topology issue of the communication hardware through the introduction of a three-tier architecture, sketched in Fig. 3.4.

This architecture collects all nodes of the CPPS's network into three tiers in accordance with its functionalities and requirements. Such tiers are hierarchically structured and each one contains qualitatively different holons or agents. At the lowest tier, almost every agent has a physical part, i.e., a CPS that enables the creation of physical value in implementing OT. At the upper tiers, most of the agent has only a digital part implementing IT for data processing, and retrieving valuable information for high-level decision-making. At the lowest level, we have the edge tier, which implements most of the control-related systems necessary to implement the OT directly in the physical world. In this tier, a proximity network connects sensors, actuators, CPS, and other elements in a bunch of heterogeneous communication technologies which are connected to the same baseline. Such a baseline, also called a *backbone network*, should be based on ethernet and represents the gateway between the shop floor and the office/enterprise networks.

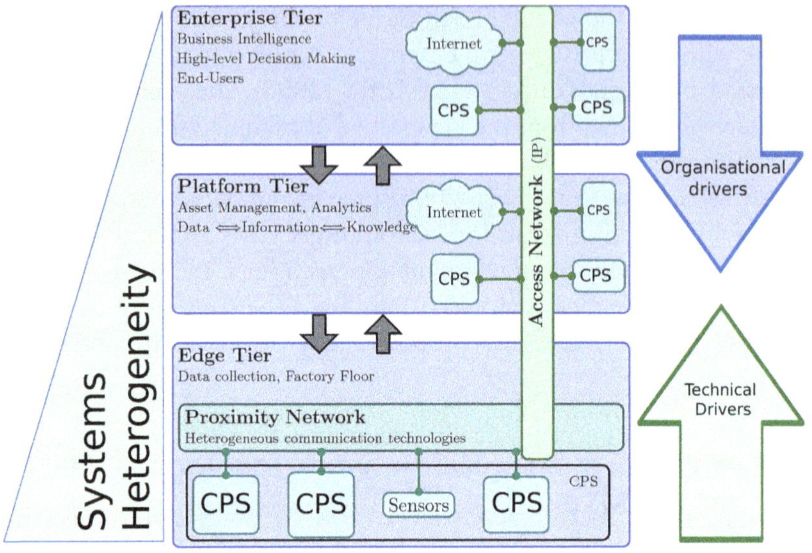

Fig. 3.4 The three-tier architecture

The platform tier implements functional capabilities to enable operability of the hardware and collects information from all instances of the CPPS, providing high-level analysis and intelligence about the overall system. It implements most of the information and operational domains and is composed by digital systems which lie somewhere between desktop computers and customized industrial computers.

The enterprise tier implements most of the business domain functionalities, data analytics and high-level decision-making and is completely composed by desktop-type computers. In our model, the last two tiers are connected to the edge tier through the so-called access network, which, contrary to the proximity network, does not have the same constraints that can be found on the factory floor and connects qualitatively different systems.

It is worth noticing that this architecture not only splits functional components in accordance with the necessary technologies for communication. Indeed, it also splits the decisional task into different levels. At the enterprise tier, strategic and high-level decisions, which are characteristic of the business functional domain, are taken. At the platform tier, the granularity becomes thinner, and more technical aspects are considered to introduce commands into the system. The three-tier architecture proposes a specific kind of network for each tier suitable for the type of digital system lying on the tier. Every actor in a CPPS is represented by a node in some tier and, at the same time, a holon with a digital part associated with such a node.

3.5 Case Study

3.5.1 The Smart Mini Factory

The Smart Mini Factory laboratory of the Free University of Bozen-Bolzano, founded in 2012 and focused on the research area Industrial Engineering and Automation (IEA), aims to reflect the principles of lean and agile production on a small and realistic scale. Inside the Laboratory, three main activities take place: applied research, focused on industry-driven use cases; teaching activities, as part of the bachelor's and master's program for industrial and mechanical engineering; seminars for industry on all aspects of Industry 4.0, to facilitate the transfer of knowledge from research to industry. To support and boost these activities, the Laboratory is equipped with different robotic platforms and devices:

- Adept Cobra i600: four axis manipulator with a SCARA base and one additional wrist joint.
- Adept Quattro s660H: four arm delta robot designed for industrial high-speed applications like packaging.
- Adept FlexBowl: rotary feeder for a wide range of loose parts.
- Two Basler scout giga-ethernet camera.
- Robotiq 85 and 140 adaptive grippers controlled via modbus RTU using RS-485.
- ABB IRB 120: compact anthropomorphic robot able to handle payload of up to 3 kg with a reach of 580 mm.

- KUKA KMR iiwa: combines the strengths of the sensitive LBR iiwa lightweight robot with those of the KMR mobile and autonomous platform.
- Universal Robot UR3: 6-rotating-joint anthropomorphic manipulator suitable for light assembly and high precision tasks.
- Universal Robot UR10: 6-rotating-joint anthropomorphic manipulator suitable for heavier-weight process tasks.
- Ulixes Der Assistent A600: manufacturing assistant system for manual assembly station based on a projector and a visual tracking system.

Since all robots and devices are equipped with an ethernet interface, a proximity network connecting them through a single ethernet baseline has been defined. This network defines the support of the communication transport layer of our MSB (see also Fig. 3.5).

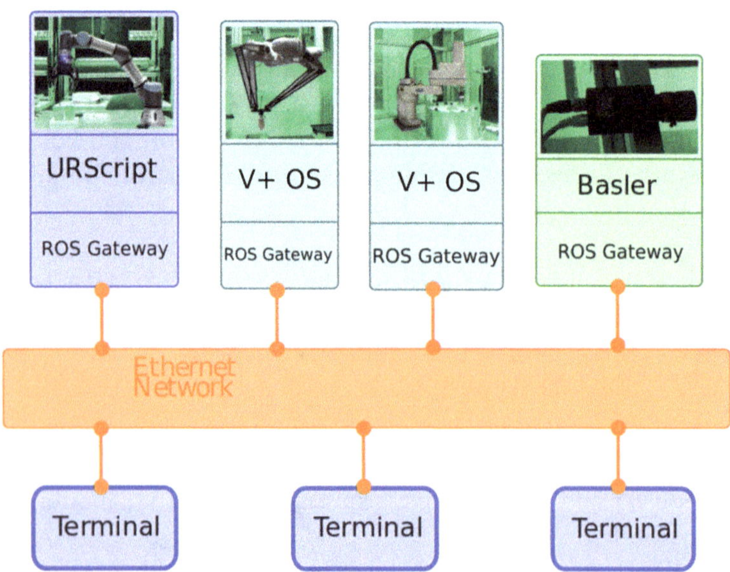

Fig. 3.5 Sketch of the mini-factory network (Reproduced with permission from Smart Mini Factory Lab, unibz)

3.5.2 Design of the Manufacturing Service Bus

Following the layered design guidelines defined in Sect. 3.4.2, we decouple this task in the design of two separate layers: the connectivity framework and transport framework. For the transport network, we choose switched ethernet IP with TCP or UDP protocols at the OSI layer 4. On the other hand, for the connectivity framework layer, we implement the robot operating system (ROS) communication systems. Our MSB should provide the following features common for ESB (Menge 2007):

1. *Invocation*: each software layer will be able to call on the services available in other platforms.
2. *Mediation*: capacity to translate between different data formats (syntactic interoperability).
3. *Adapters*: API wrappers to make each system resource available to the network.
4. *Management*: a logging system for control of process, auditing, etc.
5. *Asynchronous Messaging*: ability to send and receive messages without the need for explicit coordination with the peer.

The invocation and asynchronous messaging features represent the interface between the MSB and the agent application of the CPS. Such an implementation on the native programming language of the CPS is called the MSB API, and should also be available to other end users of the CPS's software layer. To understand the extent to which the MSB may be implemented on the CPS's software layer, it is necessary to introduce the concepts of protocol data unit (PDU) and service access point (SAP) defined in the OSI model. A PDU is a sequence of bits where information is represented using a set of rules to be mutually intelligible by two peer layers. The interface that a lower layer provides to its upper layer for data encapsulation/decapsulation is called SAP.

Each CPS has a vendor-specific software layer that should provide a particular SAP to the programmer. Such an SAP determines the PDU with maximum flexibility available in a CPS. Such PDU should grab all the requirements for the abstraction of the CPS as a virtual identity that can be addressed from the connectivity framework. To provide a connectivity interface, the CPS manufacturer has to implement an SAP on the top of a standard OSI layer (as TCP/IP connectivity) or on the top of a custom protocol. If the PDU is sufficiently flexible, it would be possible to implement the MSB at the CPS as it may be implemented on a desktop computer. If not, it will be necessary to implement and interface between the CPS and the network.

3.5.3 Connectivity Framework Gateways

Normally, ES have a limited software layer for providing a communication interface directly implemented on the OSI layer 7. That means that commands and configuration functions are directly sent through the communication interface. In such cases, it is not possible to implement an MSB API. On the other hand, several CPS do not provide an SAP with such flexibility to allow an MSB API or it is not convenient. To solve this issue, it is necessary to add the required computational features using the concepts of a connectivity framework gateway and administration shield (Adolphs et al. 2015; Ye and Hong 2019).

Unlike a communication transport gateway, a connectivity framework gateway does not convey PDU into wrappers built into other layers as MODBUS over TCP do. Such gateways provide the architectural construct to incorporate connectivity technologies into a device by conveying the semantic meaning of data through different representations. On the other hand, an administration shield gives an object the means to be considered a CPS. Such an administration shield may be implemented using different single board computers available on the market.

3.5.4 The ROS Protocol

The robot operating system is the result of a communized effort to cope with the challenges and characteristics of developing distributed robotic platforms. It is open-source software (OSS) with a strong community of developers and widely available documentation. In spite of this, its name, ROS, is not an operating system but a meta-operative system (O'Kane 2014), in the sense that it is a software platform providing an API and a series of tools for developing and handling a distributed peer-to-peer network of processes called nodes. Nodes are processes at the software layer of a CPS as the agent application. ROS provides two elements of interest for this work: a multi-platform API for developing networking applications and a semantic interoperability protocol. The ROS communication system constitutes a connectivity framework layer to the top of the ethernet TCP/IP protocol which provides services, a publish and subscribe protocol and actions. In this chapter, we will only consider publish-subscribe messaging systems and services. The channels of the ROS publish-subscribe systems are called topics and each topic defines the specific data type that it conveys.

Our MSB is a middleware for IIoT where process data is conveyed using messages in a publish-subscribe pattern and configuration data is conveyed using services. The specific data representations are taken from the ROS specification. The ROS network relies on a *master* which acts as the broker in publish-subscribe systems. The master is also a node that provides node registration and lookup for services and topics to the other nodes. The ROS master is basically an XML-RPC server maintaining a list of topics and services with their corresponding linked nodes: publishers and subscribers in the case of topics; providers and clients in the case of services. As a rule, ROS only allows one node to provide one single service. The master node is also responsible for providing a centralized data storage mechanism available to all other nodes of the network. Peer-to-peer data connections between nodes are also negotiated through XML-RPC. Such data connections can be established through TCP/IP or UDP, depending on the particular applicative context. Inside the ROS implementation, publishers act as servers and subscribers as clients. When a node registers a publisher to a desired topic, it sends the XML-RPC call with the URL of its own XML-RPC server together with the target topic name. After the call is received, the ROS master pushes the node's URL inside the publisher's list for the given topic and returns a binary status code (failure or success). In the case of success, the publisher node automatically allocates a port to the top of the OSI layer 4 (UDP or TCP) and waits, listening for incoming connections. When a node registers a topic subscriber, it sends the XML-RPC call with the name of the topic and the underlying message type. In response to this call, the ROS master returns a list of all XML-RPC servers already registered as publishers to the given topic. To connect to a specific publisher, the subscriber must interrogate the publisher node through its XML-RPC interface. The publisher replies to this call with the URL of the allocated port and the communication protocols (UDP or TCP). At this point, the subscriber allocates a socket on top of the OSI layer 4 and, as a client, initiates the connection with the publisher. Services are implemented in ROS in a simple client-server architecture. Service providers are registered through the ROS master's XML-RPC call specifying its own URI. To access a service, a node interrogates the ROS master using an XML-RPC call where the name of the service is specified. As an answer, the master communicates to the service requester the URI of the service provider. At this point, the client initiates a TCP/IP based connection to retrieve the service.

3.6 Conclusions

We presented a framework for the implementation of an ethernet-based IIoT for CPPS in SME. The framework is rooted in the concepts of CPS connectivity and actor aggregation inside a common MSB as a layer for syntactic interoperability of the production system. The main concern was how to enable interoperability between software layers of different CPS. Thanks to the decomposition of the integration effort into drivers and layers, it was possible to restrain the MSB implementation inside the software layer of the CPS. In the particular case of the Smart Mini Factory laboratory, these features where implemented through a wired ethernet network with a broker-based publish-subscribe system of services.

In spite of its effectiveness, this model has some limitations with respect to industrial-scale IIoT systems. For example, the limited number of devices used in this study does not reflect the characteristic of an IIoT system which has thousands of devices. Such a characteristic requires a more detailed analysis of the transport framework and a larger effort in system integration. On the other hand, thanks to our mild time-delay constraints we could avoid the design of a near real-time channel for data streaming. However, as we are focused on SMEs such a small number of CPS may represent an accurate model. One fundamental limitation of designing an MSB on top of the ROS protocol is given by the need of a master node to route services discovery and resource sharing between nodes. In terms of reliability and resilience, a failure comprising the master node (either in terms of hardware or software) could imply the entire network collapses. Therefore, in industrial environments, a distributed and redundant approach to network resources management and sharing should be preferred. Also, in contrast to broadcast approaches, multicast-based service discovery allows one single data packet to be delivered simultaneously to a group of nodes. Moreover, when wireless networking comes into play, specialized IP mobility solutions are required to handle changes to clients and possibly host locations. Finally, we underline that we did not address our IIoT system from a security perspective (Rehman and Gruhn 2018). Security is a transversal issue in modern IT systems. In regard to CPS, IT-based aggressions may also become physical, implying a coupling between safety and security. Such an important concern requires a dedicated analysis, which is out of the scope of this chapter.

This work presents the first steps for building an IIoT system at the Smart Mini Factory Laboratory. The findings of our research should not only serve as a basis for a further scientific development of CPPS, but also give practitioners an overview of which enabler should be considered in the implementation of Industry 4.0 and especially CPPS in the smart factory of the future.

References

Adolphs, P., H. Bedenbender, M. Ehlich, and U. Epple. 2015. Reference Architecture Model Industrie 4.0 (rami4.0) (Tech. Rep.). VDI/VDE, ZVEI.

Bellman, K.L., and C. Landauer. 2000. Towards an Integration Science. *Journal of Mathematical Analysis and Applications* 249 (1): 3–31. https://doi.org/10.1006/jmaa.2000.6949.

Chappell, D. 2004. *Enterprise Service Bus*. Sebastopol: O'Reilly.

Chen, D., G. Doumeingts, and F. Vernadat. 2008. Architectures for Enterprise Integration and Interoperability: Past, Present and Future. *Computers in Industry* 59 (7): 647–659. https://doi.org/10.1016/j.compind.2007.12.016.

Colombo, A., T. Bangemann, S. Karnouskos, J. Delsing, P. Stluka, R. Harrison, F. Jammes, and J.L. Lastra. 2014. Industrial Cloud-Based Cyber-Physical Systems. The IMC-AESOP Approach.

Cruz-Cunha, M.M. 2009. *Enterprise Information Systems for Business Integration in SMEs: Technological, Organizational, and Social Dimensions: Technological, Organizational, and Social Dimensions*. Hershey, PA: IGI Global. https://doi.org/10.4018/978-1-60566-892-5.

De Souza, L.M.S., P. Spiess, D. Guinard, M. Köhler, S. Karnouskos, and D. Savio. 2008. SOCRADES: A Web Service Based Shop Floor Integration Infrastructure. In *The Internet of Things*, ed. C. Floerkemeier, M. Langheinrich, E. Fleisch, F. Mattern, and S.E. Sarma, 50–67. Berlin and Heidelberg: Springer. https://doi.org/10.1007/978-3-540-78731-0_4.

Felser, M. 2002. The Fieldbus Standards: History and Structures. Technology Leadership Day 2002. MICROSWISS Network.

Gilchrist, A. 2016. *Industry 4.0: The Industrial Internet of Things*. Berkeley, CA: Apress. https://doi.org/10.1007/978-1-4842-2047-4.

Gössler, G., and J. Sifakis. 2005. Composition for Component-Based Modeling. *Science of Computer Programming* 55 (1): 161–183. https://doi.org/10.1016/j.scico.2004.05.014.

Hohpe, G., and B. Woolf. 2004. *Enterprise Integration Patterns: Designing, Building, and Deploying Messaging Solutions*. Boston: Addison-Wesley Professional.

Ji, X., G. He, J. Xu, and Y. Guo. 2016. Study on the Mode of Intelligent Chemical Industry Based on Cyber-Physical System and Its Implementation. *Advances in Engineering Software* 99: 18–26. https://doi.org/10.1016/j.advengsoft.2016.04.010.

Kagermann, H., W. Wahlster, and J. Helbig. 2013. Securing the Future of German Manufacturing Industry: Recommendations for Implementing the Strategic Initiative Industrie 4.0. acatech, Final Report of the Industrie 4.0 Working Group (Tech. Rep.). Acatech.

Kim, J., J. Lee, J. Kim, and J. Yun. 2014. M2M Service Platforms: Survey, Issues, and Enabling Technologies. *IEEE Communications Surveys and Tutorials* 16 (1): 61–76. https://doi.org/10.1109/surv.2013.100713.00203.

Lee, E.A. 2008. Cyber Physical Systems: Design Challenges. In *2008 11th IEEE International Symposium on Object and Component-Oriented Real-Time*

Distributed Computing (ISORC), 363–369. https://doi.org/10.1109/isorc.2008.25.

Lin, S.-W., and B. Miller. 2016. Industrial Internet: Towards Interoperability and Composability (Tech. Rep.). Industrial Internet Consortium.

Lin, S.-W., B. Miller, J. Durand, R. Joshi, and P. Didier. 2015. Industrial Internet Reference Architecture (Tech. Rep.). Industrial Internet Consortium.

Menge, F. 2007. Enterprise Service Bus. In *Free and Open Source Software Conference*, 2, 1–6.

MESA. 2008. Soa in Manufacturing Guidebook (Tech. Rep.). MESA International, IBM Corporation and Capgemini.

Minguez, J., F. Ruthardt, P. Riffelmacher, T. Scheibler, and B. Mitschang. 2010. Service-Based Integration in Event-Driven Manufacturing Environments. In *International Conference on Web Information Systems Engineering*, 295–308. https://doi.org/10.1007/978-3-642-24396-7_23.

Monostori, L., B. Kádár, T. Bauernhansl, S. Kondoh, S. Kumara, G. Reinhart, O. Sauer, G. Schuh, W. Shin, and K. Ueda. 2016. Cyber-Physical Systems in Manufacturing. *CIRP Annals—Manufacturing Technology* 65 (2): 621–641. https://doi.org/10.1016/j.cirp.2016.06.005.

O'Kane, J.M. 2014. *A Gentle Introduction to ROS*. Coleraine: Jason M O'Kane.

Olson, D.L., B. Johansson, and R.A. De Carvalho. 2018. Open Source ERP Business Model Framework. *Robotics and Computer-Integrated Manufacturing* 50: 30–36. https://doi.org/10.1016/j.rcim.2015.09.007.

Rehman, S., and V. Gruhn. 2018. An Effective Security Requirement Engineering Framework for Cyber-Physical Systems. *Technologies* 6 (3): 65. https://doi.org/10.3390/technologies6030065.

Sauter, T. 2014. Fieldbus Systems Fundamentals. In *Industrial Communication Technology Handbook*, ed. R. Zurawski, 1–48. Boca Raton: CRC Press.

Schleipen, M., A. Lüder, O. Sauer, H. Flatt, and J. Jasperneite. 2015. Requirements and Concept for Plug-and-Work. *at-Automatisierungstechnik* 63 (10): 801–820. https://doi.org/10.1515/auto-2015-0015.

Steinberg, D.H., and S. Cheshire. 2005. *Zero Configuration Networking: The Definitive Guide*. O'Reilly Media. https://doi.org/10.1515/auto-2015-0015.

Sztipanovits, J., X. Koutsoukos, G. Karsai, N. Kottenstette, P. Antsaklis, V. Gupta, J. Baras, and S. Wang. 2012. Toward a Science of Cyber-Physical System Integration. *Proceedings of the IEEE* 100 (1): 29–44. https://doi.org/10.1109/JPROC.2011.2161529.

Vernadat, F. 2007. Interoperable Enterprise Systems: Principles, Concepts, and Methods. *Annual Reviews in Control* 31 (1): 137–145. https://doi.org/10.1016/j.arcontrol.2007.03.004.

Vogel-Heuser, B., J. Lee, and P. Leitão. 2015. Agents Enabling Cyber-Physical Production Systems. *at-Automatisierungstechnik* 63 (10): 777–789. https://doi.org/10.1515/auto-2014-1153.

Wagner, H., O. Pankratz, W. Mellis, and D. Basten. 2015. Effort of EAI Projects: A Repertory Grid Investigation of Influencing Factors. *Project Management Journal* 46 (5): 62–80. https://doi.org/10.1002/pmj.21523.

Weber, S. 2004. *The Success of Open Source*. Cambridge: Harvard University Press.

Weiss, G. 1999. *Multiagent Systems: A Modern Approach to Distributed Artificial Intelligence*. Cambridge: MIT Press.

Wooldridge, M., and N.R. Jennings. 1995. Intelligent Agents: Theory and Practice. *The Knowledge Engineering Review* 10 (2): 115–152. https://doi.org/10.1017/S0269888900008122.

Ye, X., and S.H. Hong. 2019. Toward Industry 4.0 Components: Insights into and Implementation of Asset Administration Shells. *IEEE Industrial Electronics Magazine* 13 (1): 13–25. https://doi.org/10.1109/MIE.2019.2893397.

Zhang, Y., Z. Guo, J. Lv, and Y. Liu. 2018. A Framework for Smart Production-Logistics Systems Based on CPS and Industrial IoT. *IEEE Transactions on Industrial Informatics* 14 (9): 4019–4032. https://doi.org/10.1109/TII.2018.2845683.

Zimmermann, H. 1980. OSI Reference Model-the ISO Model of Architecture for Open Systems Interconnection. *IEEE Transactions on Communications* 28 (4): 425–432. https://doi.org/10.1109/TCOM.1980.1094702.

4

The Opportunities and Challenges of SME Manufacturing Automation: Safety and Ergonomics in Human–Robot Collaboration

Luca Gualtieri, Ilaria Palomba, Erich J. Wehrle and Renato Vidoni

4.1 Introduction

This chapter introduces and discusses the main potential and challenges of manufacturing automation in small-and medium-sized enterprises (SMEs) through safety and ergonomics in human–robot collaboration (HRC). Industrial collaborative robotics is a core technology of Industry 4.0 and aims to enhance the operators' work conditions and the efficiency of production systems. It also involves different important

L. Gualtieri (✉) · I. Palomba · E. J. Wehrle · R. Vidoni
Faculty of Science and Technology, Free University of Bozen-Bolzano, Bolzano, Italy
e-mail: luca.gualtieri@unibz.it

I. Palomba
e-mail: ilaria.palomba@unibz.it

E. J. Wehrle
e-mail: erich.wehrle@unibz.it

R. Vidoni
e-mail: renato.vidoni@unibz.it

D. T. Matt et al. (eds.), *Industry 4.0 for SMEs*,
https://doi.org/10.1007/978-3-030-25425-4_4

challenges from an occupational health and safety (OHS) point of view. The methodology chosen for this study is a combination of a literature review and state of the art regarding safety and ergonomics for industrial collaborative robotics and a critical discussion of potentials and challenges identified in the previous analysis.

The structure of this chapter is the following: Sect. 4.1 deals with the introductory concepts of occupational safety regarding industrial robotics. Section 4.2 introduces the main related international standards and deliverables. Section 4.3 explains the ergonomics principles and standards for a human-centered design (HCD) of collaborative workspaces. Section 4.4 discusses the critical challenges for the implementation of collaborative systems from a safety and ergonomics point of view. Finally, Sect. 4.5 concludes and summarizes the concepts illustrated in this chapter. The outcomes of this chapter are not only for the interest of researches, but also for practitioners from SMEs as they give an overview about the potential and challenges of safety and ergonomics in industrial human–robot interaction (HRI).

4.1.1 Introduction to Industrial Collaborative Robotics

A collaborative robot (also known as a cobot or lightweight robot) is a particular kind of industrial robot which is able to physically and safely interact with humans in a shared and fenceless workspace by introducing new paradigms from human–machine interaction (HMI). The International Federation of Robotics defines collaborative industrial robots as those able to perform tasks in collaboration with workers in industrial settings (IFR 2019). The concept of collaborative workspace can be summarized as the "*space within the operating space where the robot system (including the workpiece) and a human can perform tasks concurrently during production operation*" (ISO 2016, p. 8). In general, collaborative robotics is a main cyber-physical enabling technology of Industry 4.0, and aims to improve production performances and operators' work conditions by matching typical machine strengths such as repeatability, accuracy, and payload with human skills such as flexibility, adaptability, and decision-making. Since modern SMEs requires smart process characterized by a scalable degree of automation, collaborative robots can particularly support them in the development of their business by introducing human-centered, lean, adaptable, and reconfigurable manufacturing systems.

When introducing collaborative robotics, a crucial part will be safety. In fact, the main difference between collaborative and traditional industrial robots is that cobots are designed to allow physical HRI in hybrid and fenceless workcells without the necessity of isolating the robot workspace. Traditional industrial robots were introduced to improve production efficiency by replacing human operators in performing heavy, unsafe, and repetitive processes (Huber et al. 2008). Due to safety requirements, a traditional high performance manipulator needs safeguards (physical barriers or optical devices) to isolate the robot activities and therefore to safeguard operators from unexpected and unwanted contacts. Since collaborative applications allow for direct HRI and this is even required, traditional safety solutions for robot isolation are, in general, no longer possible. As a consequence, other systems have to be integrated into the collaborative arm to ensure operators' occupational safety and ergonomics. These systems have to be selected and implemented depending on the robot performance and the level of interaction, and in general, are more demanding and complicated with respect to safety solutions for traditional industrial robotics. For these reasons, the design and integration of OHS aspects will be more challenging in collaborative applications.

There is no doubt that safety and ergonomics are essential in industrial HRI (see Sects. 4.2 and 4.3). Nevertheless, there is a lack in the literature since there are only few scientific documents regarding the application of these topics into SMEs (this is easily verifiable by searching the related keywords in a scientific database like Scopus). For this reason, the proposed identification and discussion about main potential and challenges for safety and ergonomics in industrial HRI aims to support SMEs in the proper consideration and adoption of collaborative systems.

4.1.2 Main Occupational Health and Safety Concepts

In the following, main concepts of OHS are explained. OHS is a multidisciplinary and integrated discipline which deals with aspects of the health and safety of a person during every kind of work. Both health and safety are connected to the concept of work-related risk in terms of work activities and work environments. The concept of risk is strictly related to the definition of hazard: a source with potential for causing harmful consequences (Jensen 2012). Considering the presence of electricity in the workplace, some examples of hazards are accidental contacts with live parts of electric devices or fire, which can originate from electrical malfunctions. Risks are the concrete realization of hazards. According to its widespread definition, a risk is "*the likelihood or possibility that a person may be harmed or suffers adverse health effects if exposed to a hazard*" (Health and Safety Authority 2019, p. 1). Risk is defined as a combination between the probability that harm occurs and the severity of that potential harm:

$$Risk = f(Probability, Severity)$$

The relationship between hazard, risk, and consequence is summarized in Fig. 4.1. Considering the aforementioned example, the relative risk family is an electrical risk and could be realized in the form of electric shocks or burns.

In general, occupational health has a strong focus on primary prevention of occupational disease or infirmity and aims to guarantee healthy work conditions, in terms of mental, physical, and social well-being (World Health Organization 2019). On the other hand, occupational safety is the science which deals with the safeguarding and protection of workers' lives against injuries and accidents. This means to guarantee a condition of physical integrity during work activities and provide a state

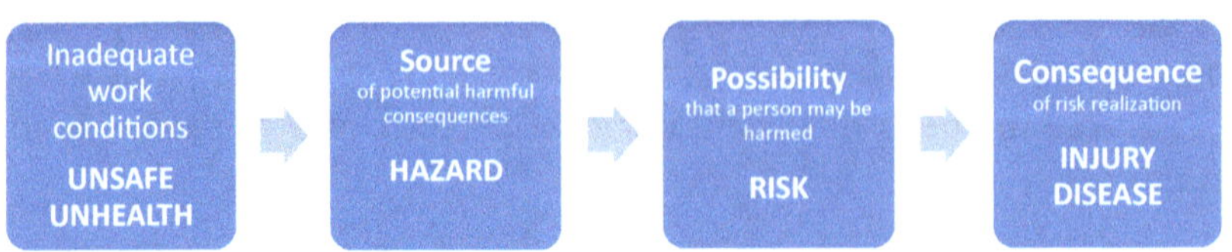

Fig. 4.1 Relationship between hazard, risk, and potential consequence

where the risk has been reduced to a level that is as low as reasonably practicable. Since it is impossible to totally eliminate risks, the remaining risk is generally considered under control and therefore acceptable. As a consequence, safety is a relative condition, which is based on the judgment of the acceptability of risk (Jensen 2012).

Both occupational health and safety deal with the protection of workers by operating on different protection levels. In fact, it is possible to consider safety as an operational measure which safeguards people from unexpected and violent events with potential for causing serious direct physical injuries to human body parts (acute process). Health is the operational measure, which defends a person from possible future occupational diseases caused by long-term exposure to inadequate work conditions (chronic process).

4.1.3 Occupational Health and Safety Standards

OHS requirements are usually worldwide legal requirements (and therefore mandatory), which are interpreted and adopted by industrialized countries in typically similar ways. In order to support technicians in the adoption of complex legal obligations, national, and international standards are developed by recognized and competent organizations. Standards are voluntary reference models, which contain applicable guidelines and technical specifications in the form of documents for the correct implementation of state-of-the-art systems, processes, or products. A formal definition of a standard is provided by the European Committee for Electrotechnical Standardization (CENELEC) as:

> a document that sets out requirements for a specific item, material, component, system or service, or describes in detail a particular method or procedure. Standards facilitate international trade by ensuring compatibility and interoperability of components, products, and services. They bring benefits to businesses and consumers in terms of reducing costs, enhancing performance and improving safety. (CENELEC 2019, p. 1)

In practice, there are standards for diverse fields, i.e., industry, construction, services, informatics, agriculture, telecommunication, etc. Considering the European Union, it is possible to have international, European, and national standards. International standards are developed by international organizations and are used and recognized all over the world. In this context, the main organization is the International Organization for Standardization (ISO). On the other hand, European Standards (ENs) are approved by one of the three European Standardization Organizations, which are recognized as competent in the area of voluntary technical standardization as for the EU Regulations: European Committee for Standardization (CEN), European Committee for Electrotechnical Standardization (CENELEC), and European Telecommunications Standards Institute (ETSI). In addition, recognized European standards will be automatically accepted as national standards in each of the 34 countries which are part of CEN-CENELEC (CENELEC 2019). In general, the standard compliance is not mandatory from the legal point of view. The compliance of standard demonstrates that the proposed solution follows a well-structured and cutting-edge approach, which is a very important advantage especially in the OHS field.

One of the main areas of interest for the OHS in industrial contexts is safety of machinery. This branch aims to adequately reduce the risks related to machines without compromising their ability to perform the planned functions during their life cycle. According to the standard definition, it is possible to define a machine as an "*assembly, fitted with or intended to be fitted with a drive system consisting of linked parts or components, at least one of which moves, and which are joined together for a specific application*" (ISO 2010a, p. 1).

Guidelines on how to realize machines that are safe for their intended use are given in the safety of machinery standards (Jespen 2016). To support these standards, there are also other important and recognizable technical deliverables (such as technical specifications and technical reports) introduced to further integrate the information included. Such standards are divided into the following three main categories, targeted to different levels of details in the design framework for the realization of machines (ISO 2010a, p. 1):

- **Type-A standards** address methodologies and general principles for designing and building machines. They are basic safety standards and can be applied to all machines.

- **Type-B standards** deal with generic safety requirements that are common for designing most of the machines.
- **Type-C standards** deal with detailed safety requirements for a specific machine or group of them. They are machine safety standards providing a presumption of conformity for the essential legal requirements covered in the standard.

4.1.4 Introduction to Industrial Robot Safety

Considering the nature of hazards, there are different kinds of occupational risks related to industrial machines. Recognized hazard categories for industrial machines are the following (ISO 2010a):

- Mechanical hazards
- Electrical hazards
- Thermal hazards
- Noise hazards
- Vibration hazards
- Radiation hazards
- Material/substance hazards
- Work environment-caused hazards
- Combination of hazards

Considering the definition of industrial HRI, collaborative systems allow and require sharing of tasks in a fenceless workspace where the main hazard category will be of a mechanical nature. In fact, due to the combined presence of both humans and robots in a shared workspace, it is possible to have potential non-functional physical interaction between the operator and the mobile parts of the machine, especially with the robot arm and with different types of end-effectors. Unexpected and unwanted contacts can generate different kinds of collisions and crushes if related mechanical risks are not properly identified, predicted, and managed. In particular, a mechanical hazard is a physical hazard which can occur when workers directly or indirectly come into contact with work process related objects. The effects are usually immediate and can cause injuries to human beings. The level of intensity can change according to the physical features of the involved work equipment such as speed, mass, and geometry. Typical mechanical hazard is due to the possibility of being crushed, smashed, cut, trapped, impacted, punctured or stabbed because of machine tools, parts, equipment, or machined/treated objects, production waste, and ejected solid or liquid materials. Other main basic mechanical hazards are due to high-pressure fluid ejection, as well as slipping, or tripping and falling (Koradecka 2010). Besides mechanical risks, there are other risks families (e.g., electrical risk) that must be considered according to the specific applications. The main significant mechanical hazards consequences for traditional and collaborative industrial robots are summarized in Table 4.1.

4.2 Fundamentals of Occupational Safety in Industrial Human–Robot Interaction

4.2.1 Mechanical Risk Analysis in Industrial Robotics: Traditional Versus Collaborative Robotics

Risk assessment is the procedure which combines the machine (limits) specifications with hazard identification and risk estimation (which basically defines risk probability and gravity), in order to judge whether the risk reduction targets have been reached (ISO 2010a). Risk assessment for traditional and collaborative industrial robots is different. In general, due to the standardization of components and diffusion of applications, traditional robotic cells are easier to evaluate from a mechanical risk point of view (Vicentini 2017). In this case, the common main risk features are:

- The robotic cell is isolated (direct physical HRI is usually not allowed).
- The mechanical risk indexes are high and characterized by a high level of gravity and a low level of frequency (high mass and velocity of robot arm involves high kinetic energy levels).
- The safety protection systems are based on prevention, that means a zeroing of mechanical risk probability by using safeguards.

Table 4.1 Main significant mechanical hazards consequences in traditional and collaborative industrial robotics according to ISO 10218-2 and to UR3 robot original instructions (ISO 2011a; Universal Robot 2018)

Traditional robotics	Collaborative robotics
– consequences* due to movements of any part of the robot arm (including back), end-effector or mobile parts of robot cell – consequences* due to movements of external axis (including end-effector tool at servicing position) – consequences* due to movement or rotation of sharp tool on end-effector or on external axes, part being handled, and associated equipment – consequences* due to rotational motion of any robot axes – consequences* due to materials and products falling or ejection – consequences* due to end-effector failure (separation) – consequences* due to the interaction with loose clothing, long hair – consequences* due to the entrapment between robot arm and any fixed object – consequences* due to the entrapment between end-effector and any fixed object (fence, beam, etc.) – consequences* due to the entrapment between fixtures (falling in); between shuttles, utilities; – consequences* due to the impossibility of exiting robot cell (via cell door) for a trapped operator in automatic mode – consequences* due to the unintended movement of jigs or gripper – consequences* due to the unintended release of tool – consequences* due to the unintended movement of machines or robot cell parts during handling operations – consequences* due to the unintended motion or activation of an end-effector or associated equipment (including external axes controlled by the robot, process specific for grinding wheels, etc.) – consequences* due to the unexpected release of potential energy from stored sources	– entrapment of fingers between different parts of robot arm and workspaces equipment/other robot parts – penetration of skin by sharp edges and sharp points on tool/end-effector, tool/end-effector connector and on obstacles near the robot track – bruising due to contact with the robot – sprain or bone fracture due to strokes between a heavy payload and a hard surface – consequences due to loose bolts that hold the robot arm or tool/end-effector – items falling out of tool/end-effector, e.g. due to a poor grip or power interruption – mistakes due to different emergency stop buttons for different machines or due to unauthorized changes to the safety configuration parameters

* crushing, shearing, cutting or severing, entanglement, drawing-in or trapping, impact, stabbing or puncture, friction, abrasion

- There is no possibility of mitigating unexpected contacts between the operators and the mobile parts of the robot due to specific robot design and control.

Therefore, mechanical risk management is more homogeneous and standardized, which means it is less dependent on specific robotic cell applications. Risk management is simplified due to the absence of workers in the robot workspace. On the other hand, in the case of collaborative robots, the situation will be very different and more complex. The main risk features are:

- The robotic cell is collaborative (physical HRI is allowed).
- The mechanical risk indexes are variable in terms of gravity and frequency, depending on the application and on the potential HRI form. In fact, they can vary depending on the single operator tasks during the overall application.
- More probable and less severe mechanical risks related to unexpected human–robot contacts are allowed, which means that safety systems are based on a mixture of risk prevention and attenuation.

Therefore, the risk management is more application-specific, which means heterogeneous, complex, and barely standardized.

4.2.2 Main Safety Standards for Industrial Collaborative Robotics

Considering the mechanical risks that can occur during unwanted contact, there is a short list of general international deliverables regarding the safety of machinery requirements for HRI (see Fig. 4.2). In 2016, a new ISO technical report, ISO TS 15066 (ISO 2016), was published in order to help production technicians and safety experts in the development of safe shared workspaces and in the risk assessment process. This report specifies in greater detail the previous safety requirements for industrial robots included in ISO 10218 part 1 and 2 (ISO 2011a, b). Other useful documents include the EU Machinery Directive (for the European Union) (European Parliament 2006), the standards ISO 12100 (ISO

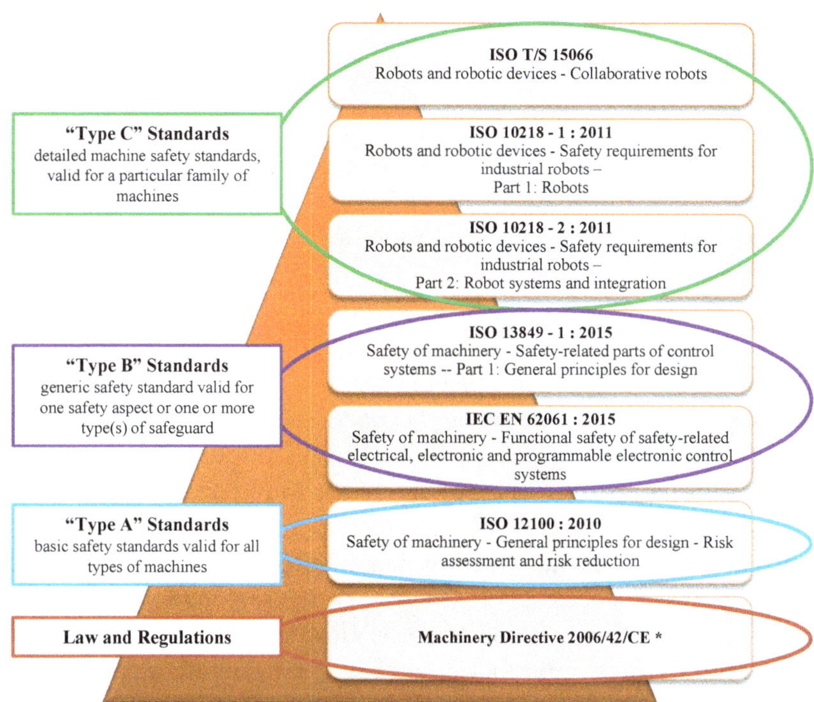

Fig. 4.2 Main standards hierarchy related to industrial collaborative robotics (*Mandatory for European Union nations)

2010a) for risk assessment and ISO 13849 part 1 (ISO 2015a) and IEC EN 62061 (IEC 2015) for the functional safety requirements. In addition, a document which defines main guidelines on safety measures for the design and integration of end-effectors used for robot systems was recently published (ISO 2018). It also includes requirements and suggestions for collaborative applications.

4.2.3 Technical Specification ISO TS 15066 (2016) and Collaborative Operations

The technical specification ISO TS 15066 (ISO 2016) was released in 2016 and explains in more detail the requirements regarding collaborative robots, which were preliminarily introduced in standards

ISO 10218-1 (ISO 2011b) and 10218-2 (ISO 2011a) in 2011. At the moment, this specification represents one of the most detailed document which specifies safety requirements for collaborative industrial robot workcells, especially in terms of mechanical risk. According to specific applications and types of interaction, collaborative robots should be integrated with different kinds of safety devices. The main goal of safe collaboration is to minimize the mechanical risk which could arise from unexpected contacts in terms of gravity and/or probability. ISO TS 15066 (ISO 2016) introduces four methods for safe HRC:

a. **Safety-rated monitored stop**: the robot motion is stopped when an operator is entering the collaborative workspace. The robot enters a controlled standstill mode while the operator is present in the limited workspace. The robotic task can automatically resume when the operator leaves the zone. If there is not an operator in the collaborative workspace, the robot can operate non-collaboratively.

b. **Hand guiding**: the operator can fully control the robot motion by direct physical interaction. In this case, the robotic task is manually guided by the operator at a certain safe velocity by moving the arm through a direct input device at or near the end-effector. Before the operator is allowed to enter the collaborative workspace to conduct the hand-guiding operations, the robot has to achieve a safety-rated monitored stop condition.

c. **Speed and separation monitoring**: the control system of the robot is actively monitoring the relative speed and distance between robot and operator. When the operator is working in the collaborative space, the robot has to dynamically maintain a safe combination of speed and distance in order to stop any hazardous motion before a potential unexpected contact. When the separation distance is below the set protective distance, the robot system stops. This method is designed to prevent unexpected contact between operator and collaborative robot by reducing the probability into safety limits.

d. **Power and force limiting**: the biomechanical risk of unexpected human–robot contacts is sufficiently reduced either through inherently safe means in the robot or through a safety-related control system. In this case, physical unforeseen collisions are allowed, if the pressure (or force) limits for the interaction do not exceed values that are determined during the risk assessment.

The safety-rated monitored stop modality does not represent a real implementation of HRC since the robot just stops its operation if a safety system detects the presence of an object into the limited workspace. The hand-guiding modality represents a marginal form of collaboration since the robot is simply moved by the operator across the workspace manually without obstructing the human intention. This approach is particularly useful for intuitive programming or for guided operation, i.e., the assembly of heavy components by using high-payload robots. Of course, speed and separation monitoring and power and force limiting modalities are more innovative and interesting from a collaborative point of view. The former needs the integration of quite complex and certified vision systems and control algorithms in order to dynamically adapt the robot motion to the operator's behavior. Nevertheless, the level of interaction could be more than satisfactory according to safety device performances, since the robot can continue its works even if there is an operator in the shared workspace. The latter will be particularly useful in case of close-proximity activities between operators and robots, a mandatory condition for a real physical collaboration.

It is important to underline that it could be possible to implement the first three collaborative modalities without the necessity of using an industrial robot which is specifically designed for collaborative applications. The possibility of having that collaborative application depends on the hardware and software features of the safety devices and control systems that will be integrated and regulated in order to properly support the robot during its applications. The "power and force limiting" is the only collaborative operation which requires robot systems specifically designed for this particular type of operation (ISO 2016).

4.2.4 Nature of Human–Robot Contacts

The analysis of human–robot contacts is particularly relevant for the implementation of a "speed and separation" collaborative modality. In general, a contact between a human body part and the robot arm is complex to model, even if there are good approximations. In general, it is assumed that the contact is a partially inelastic collision. The dynamic involves a first rapid part where the two moving objects collide in a more or less intensive way, proceeded by a slight and short physical detachment. After that, depending on the contact conditions, the two objects can proceed together along the same direction or separate. According to the different dynamic conditions of the collision, the kinetic energy exchange can vary. A generic human–robot contact can be classified as shown in Fig. 4.4. The two main contact situations are "transient" type (impulsive) or "quasi-static" type. The main difference refers to the (force or) pressure distribution during the time (Vicentini 2017), which means a different intensity in terms of impact. Another important factor is the presence/absence of constraints in the shared workspace, which means physical objects that can block the human body part during the contact (e.g., a situation where the operator's hand is constrained between the robot gripper and a work table). The common human–robot contact variables and conditions are characterized in Fig. 4.3.

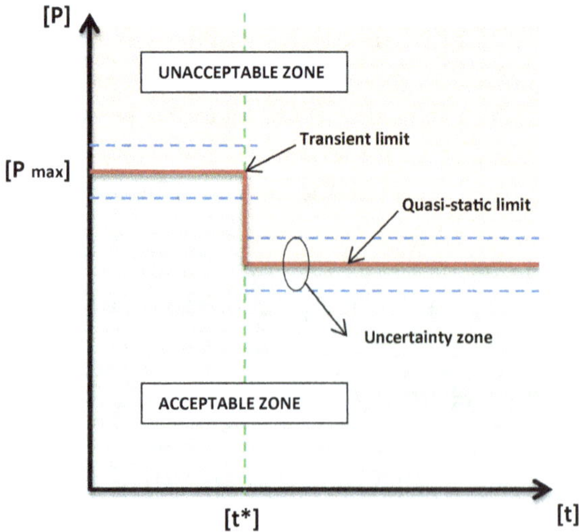

Fig. 4.3 Common human–robot contact variables and condition (*Source* Adapted from Vicentini 2017)

Parameter t* (Fig. 4.3) represents the time limit between the first phase and the second phase of the contact. The first phase is the impact part of the contact, which is a semi-instantaneous phenomenon characterized by the peak of energy transfer between the human body part and the robot arm and it is mainly related to their relative velocity. The second phase is the retention part of the contact, and represents the energy exchange progression after the impact. This part is mainly related to the mechanical characteristics of the contact parts (mass and stiffness) and to the presence or absence of physical constraints. The parameter P_{max} represents the upper limit of pressure for transient and quasi-static situations. This limit divides the unacceptable region for pressure (or force) from the acceptable region for pressure (or force) in the first and second contact phase. In general, an unacceptable value of pressure means that the contact is unsafe because the human pain that is theoretically associated with the involved body part is not admissible from a biomechanical point of view. The uncertainty zone represents the region in which the limit values of pressure (or force) are not exactly defined. It is necessary to consider this uncertainty during the risk assessment procedure. In the following Fig. 4.4, the four cases of human–robot contact are explained.

According to ISO TS 15066 (ISO 2016), there are different risk reduction measures for the management of quasi-static or transient contacts. It is possible to classify these measures into passive or active types. The main design measures are summarized in Table 4.2.

4.3 Human-Centric Design and Ergonomics

The fourth industrial revolution has not only introduced new manufacturing paradigms, but is also changing the role played by humans. Humans play a key role that cannot easily be replaced by advanced technologies. Conversely, the introduction of new technologies complicates manufacturing systems and increases the need for highly skilled, well-trained workers (Tan et al. 2019). Therefore, the fourth industrial revolution renews an anthropocentric approach to conceiving new technologies, changing the question from how to replace humans to how

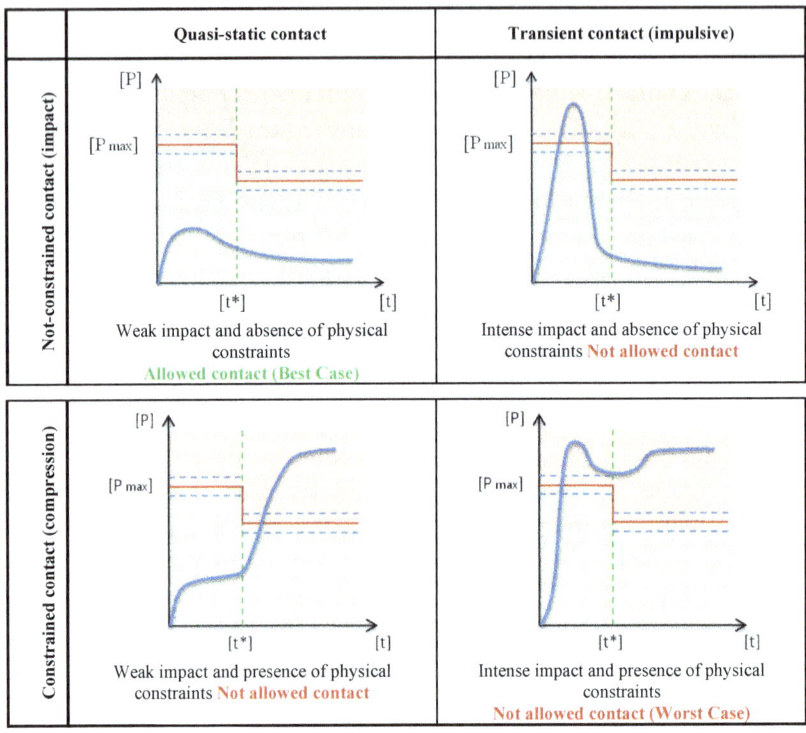

Fig. 4.4 Human–robot contact classification (*Source* Adapted from Vicentini 2017)

to better complement and assist humans. A human-centric approach to design manufacturing systems is introduced. ISO 9241-210 (ISO 2010b) describes HCD as "*an approach to systems design and development that aims to make interactive systems more usable by focusing on the use of the system and applying human factors/ergonomics and usability knowledge and techniques.*" Therefore, such an approach aims to improve human well-being, together with user satisfaction, sustainability, and accessibility, while preventing the potential side effects to human health, safety, and performance due to the use. ISO 9241 part 210 provides guidelines on how to redesign processes to identify and plan effective and timely HCD activities, defining six key principles:

Table 4.2 Main passive and active risk reduction measures for the management of quasi-static or transient contacts according to ISO TS 15066 (ISO 2016)

Passive safety design measures	Active safety design measures
Address the mechanical design of the robot system (a) **increasing the contact surface area**: (1) rounded edges and corners; (2) smooth surfaces; (3) compliant surfaces. (b) **absorbing energy, extending energy transfer time, or reducing impact forces**: (1) padding, cushioning; (2) deformable components; (3) compliant joints or links. (c) **limiting moving masses**	Address the control design of the robot system (1) limiting forces or torques; (2) limiting velocities of moving parts; (3) limiting momentum, mechanical power or energy as a function of masses, and velocities; (4) use of safety-rated soft axis and space limiting function; (5) use of safety-rated monitored stop function; (6) use of sensing to anticipate or detect contact (e.g., proximity or contact detection to reduce quasi-static forces

a. Design based on clear comprehension of users, tasks, and environments
b. User involvement throughout design and development
c. User-centerd evaluation to drive/refine design
d. Iterative process
e. Based on the whole user experience
f. Adoption of multidisciplinary skills and perspectives.

The HCD approach can be incorporated into every kind of design approach (e.g., object-oriented, waterfall and rapid application development). One of the crucial aspects for correct application of the HCD is the respect of the ergonomic principles. ISO 26800 (ISO 2011c) defines ergonomics as being like a "*scientific discipline concerned with the understanding of interactions among human and other elements of a system, and the profession that applies theory, principles, data and methods to design in order to optimize human well-being and overall system performance.*" Such a discipline studies how to best design a workplace and the related equipment or products, in general, in order to optimize them for human use. A design based on ergonomic principles takes into account several human characteristics, e.g., weight, height, age, hearing, and sight. Ergonomics is therefore often referred to as human factors engineering.

The International Ergonomics Association has identified three domains of specialization within this discipline in accordance with peculiar characteristics and attributes of human interactions. The three main fields of research, briefly described in the following, are physical, cognitive, and organizational ergonomics.

a. **Physical ergonomics** deals with human physical activities and therefore, it investigates human characteristics related to anatomy, anthropometry, physiology, and biomechanics. The most relevant topics addressed in this field are related to working postures, handling of materials, repetition of movements, musculoskeletal disorders caused by work, and layout, safety and health of workplaces. The principles of physical ergonomics are widely used and useful not only for the design of workplaces and industrial products but also for the design of consumer products.

b. **Cognitive ergonomics** focuses on the mental interactions between humans and any elements of a system. It considers several human cognitive abilities and mental processes, among others, motor response, reasoning, perception, and memory. Cognitive ergonomics should be taken into account mainly in the design of automated, high-tech, or complex systems. Indeed, as an example, a non-user-friendly interface of an automated industrial equipment, not only may decrease production and quality, but also may result in a life-threatening accident. The main topics tackled by such a discipline include work stress, mental workload, human reliability, decision-making, skilled performance, human–computer interaction.

c. **Organizational ergonomics** addresses the optimization of socio-technical systems in terms of efficiency maximization of their structures, policies, and processes. It aims to achieve a totally harmonized work system able to ensure both the job satisfaction and the commitment of the employees. To this end, organizational ergonomics deals with teamwork, participatory design, community ergonomics, cooperative work, new work paradigms, virtual organizations, telework, and quality management, communication, crew resource management, work design, and design of working times.

The next Sects. 4.3.1 and 4.3.2, will focus on physical ergonomics, by reviewing the main risk factors and the methods to assess them (Sect. 4.3.1) as well as the main related standards (Sect. 4.3.2). An overview on the state of art of ergonomics for designing hybrid (human–robot) workspaces will be provided in Sect. 4.3.3.

4.3.1 Risk Factors and Musculoskeletal Disorders

The risk factors related to physical ergonomics are those related to a job or a task that can lead to biomechanical stress on workers, resulting in musculoskeletal disorders (MSD). An MSD is a health problem of the locomotor apparatus that can affect tendons, muscles, joints, ligaments, nerves, blood vessels, and so on. Among the most common MSDs are carpal tunnel syndrome, tendinitis, trigger finger, muscle strains, and low back injuries. MSD can be aggravated or even induced by work and circumstances of its performance. Additionally, an MSD from a light, transitory disorder can become an irreversible, disabling injury. Therefore, it is very important to identify the risk factors that can cause or contribute to an MSD. Among the most likely risk factors for MSD, the following ones have been identified through reviews of scientific evidence and laboratory studies (da Costa and Vieria 2010):

- Forceful exertions
- Load
- Awkward postures
- Static postures
- Duration
- Frequency
- Repetition
- Cold temperatures
- Contact stress
- Vibration

Typically, the hazard is created by a combination of several risk factors related to physical ergonomics, even if the exposure to just one of them can be enough to contribute to or cause an MSD. Therefore, the potential risk factors related to physical ergonomics should be analyzed also in sight of their combined effect.

The assessment and prevention of the risks are among the main issues in physical ergonomics. Several methods, often complementary, have been developed in the literature for performing ergonomics risk assessment of different regions of the human body considering specific risk factors. The most common assessment methods include the following.

a. **Ovako Working Posture Analysis System (OWAS)** (Scott and Lambe 1996) assesses the risk of the working posture. It describes the full-body posture by identifying the most common working postures for back, arms, and legs, and the weight of the load handled. Each posture is attributed a four-digit code, which is then compared with reference values. The 252 postures contemplated by the method are classified into four action categories indicating needs for ergonomic changes.

b. **Manual handling Assessment Chart (MAC)** (Monnington et al. 2003) assesses the risks related to handling, lifting, and carrying activities. The high-risk manual handling activities in the workplace are identified by means of the known related risk factors associated with such activities already categorized for risk level. The risk level of manual handling tasks is denoted by both numerical and a color-coding score.

c. **Job Strain Index (JSI)** (Kuta et al. 2015) provides a quick and systematic assessment of the hand, wrist, forearm and elbow postural risks to a worker. It estimates the risks of injury to the aforementioned parts starting from the assessments of the following task variables: force, repetition, speed, posture, and duration. The product of the scores given to each task variable is the Strain Index score, which is compared with a gradient that identifies level of task risk.

d. **Rapid Entire Body Assessment (REBA)** (Stanton et al. 2004) evaluates the full-body postural risk and MSD associated with job task. Overall, six hundred postural examples have been coded taking into account static and dynamic postural loading factors as well as the coupling between the human and the load. The data on body postures, forceful exertions, type of movement or action, repetition, and coupling are collected into a single page worksheet. After that, a single score is generated denoting the level of MSD risk.

e. **Occupational Repetitive Actions (OCRA)** (Colombini 2002) assesses the risk of upper limb repetitive actions. Such an assessment includes time-based exposure variables such as recovery and frequency. It is more comprehensive than most other methods; indeed, it is included as a reference method for risk assessment in ISO 11228-3 (ISO 2007a) and EN 1005-5 (CEN 2007). Moreover, the final risk score that predicts the risk of developing MSD is based on epidemiological research.

f. **Composite Ergonomics Risk Assessment (CERA)** (Szabó and Németh 2018) is an easy-to-use paper-and-pencil method, which gives a simple evaluation after a separate determination of the different ergonomic risks. The method is based on the observations of real activities through images. According to EN 1005, this method allows risks related to posture, manual handling, effort, repetitive movements, subjective discomfort, workplace history and improvement ideas to be assessed appropriately.

g. **Ergonomic Assessment WorkSheet (EAWS)** (Schaub et al. 2012) assesses every biomechanical risk to which workers can be exposed during work. It provides detailed results in the following four sections: body postures, action forces, manual materials handling, and upper limbs. This method has been developed by an international team of experts on ergonomics and it is constantly improved.

h. **Rapid Upper Limb Assessment (RULA)** (McAtamney and Corlett 1993) assesses the biomechanical and postural load on the neck, trunk, and upper extremities associated with tasks. The required body posture, force, and repetition are evaluated using a one page worksheet. After the evaluation, the level of MSD risk is represented by means of a single score.

Risk assessment is the preliminary step to guaranteeing safe working conditions. After the assessment has been performed and evaluated, if necessary, the workplace must be redesigned and reorganized according to ergonomic principles. Job activities, tasks, and work environment should be designed to limit the exposure to ergonomic risk factors by taking preventative measures. Such preventive measures as well as the ergonomic principles are defined in the main standards ruling ergonomics in the next section.

4.3.2 Main Standards on Physical Ergonomics

Physical ergonomics deals with the physical interactions between humans and the other elements of a system. However, human beings have very different attributes (i.e., gender, age, height, weight, etc.). Good equipment design should satisfy healthy work conditions for all the operators, according to basic ergonomic principles, regardless of their attributes. To this end, ISO 7250-1 (ISO 2017a) provides a description of anthropometric measurements, which can be used as a basis for the creation of databases for the anthropometry. The basic list of the specified measurements can be used as a guide for ergonomists for defining population groups, whose specifications are, in turn, used for designing the places where people work and live. Anthropometric data are included also in EN 547-3+A1 (CEN 2008a). In particular, these data come from an anthropometric survey which includes at least three million European men and women. All ranges of human abilities and characteristics are taken into account by EN 614-1+A1 (CEN 2009). Based on these, it defines the ergonomic principles to guarantee the health, safety, and well-being of humans as well as the overall system performance.

According to standards, an ergonomic design is to consider at least the 5th–95th percentiles. However, for safety aspects, the 1st to 99th percentiles shall be used. The anthropometric data defined in the previous standards are used by EN 547-1+A1 (CEN 2008b) and EN 547-1+A2 (CEN 2008c) to specify the dimensions of openings for full-body access applied to machinery. These standards also show how to define suitable allowances for the anthropometric data in order to account for factors neglected during their measurements, such as clothing, body movements, equipment, machinery operating conditions, or environmental ones. Body space requirements for equipment for performing normal operations in both sitting and standing positions are defined in ISO 14738 (ISO 2008a), always on the basis of the anthropometric measurements. EN 1005 parts 2–5 deal with manual handling of materials (CEN 2008d), recommended force limits (CEN 2008e), postures and movements during work (CEN 2008f), and load repetition of the upper limbs in machinery operation (CEN 2007). ISO 11226 (ISO 2000) focuses on the evaluation procedure of static working postures. The same standard also defines recommended limits for static working postures on the basis of external force exertion, body angles, and time aspects. Manual lifting and carrying are addressed in ISO 11228-1 (ISO 2003), where guidance for their assessment is provided and the recommended limits defined considering the intensity, the frequency and the duration of the task. Manual pushing and pulling tasks are tackled by ISO 11228-2 (ISO 2007b) which provides a procedure for assessing the risks related to such tasks as well as the full-body suggested limits. Finally, ergonomic recommendations for manual handling of low loads at high frequency in repetitive work tasks are given in ISO 11228-3 (ISO 2007a). These standards on ergonomics provide procedures and design considerations which can be applied in a wide range of situations and ensure the safety and health of both consumers and workers as well as improved work efficiency.

4.3.3 Ergonomics in Human–Robot Collaboration

The underlying idea of collaborative robotics is to have advanced technologies able to help and support humans. An example of this interaction is when cobots lift components for a worker. Although cobots can improve the physical ergonomics of the workplace and hence reduce the worker exposure to MSD, they could also cause workers mental stress and psychological discomfort, if cognitive ergonomics principles are not considered. Indeed, cobots should behave in accordance with the operator's expectations (Mayer and Schlick 2012); their presence has not to be a source of stress for humans or even perceived as a hazard. Human acceptability of the cobots can be improved e.g., by implementing anthropomorphic trajectories for the robot (Kuz et al. 2014; Rojas et al. 2019). Although we are still far from an industrial implementation of hybrid workspaces based on physical and cognitive ergonomic principles that are compliant with standards, some academic results are available. In Faber et al. (2015), a hybrid workspace with improved ergonomics features and flexibility has been designed. It is based on the robot operating system (ROS) and a cognitively automated assembly planner. An anthropocentric design for the workspace was presented in D'Addona et al. (2018), where the process tasks have been classified according to their cognitive complexity. In Michalos et al. (2018) it is presented as a multi-criteria approach and an algorithm to task assignment able to plan the human–robot hybrid cell layout and tasks at the same time. In Müller et al. (2016), a skill-based task assignment approach is proposed. It is based on an assembly task description model and assigns the tasks between human and robot by comparing their skills according to the requirements.

Virtual reality can be an aid in the design of an ergonomic workplace, which allows the simulation of assembly tasks for HCD (Peruzzini et al. 2019). In fact, it is possible to improve human posture and stress by assessing different setups in digital manufacturing tools and adding digital human models and other virtual resource models (Caputo et al. 2018). Although such an approach is able to simulate assembly tasks in workplaces, taking into account ergonomic aspects such as posture, workload, and stress, it neglects higher anthropocentric aspects like human satisfaction (Romero et al. 2015) and emotion. Although progress has been made in the design of an ergonomic workstation, a strategy to design a hybrid workstation which is completely human-centered and which satisfies all the physical and cognitive ergonomic principles is not yet possible.

4.4 Discussion About Potential and Challenges in Safety and Ergonomics in Human–Robot Collaboration

The following section identifies the potential and challenges in safety and ergonomics in HRC by introducing the main organizational and technological future research areas. Figure 4.5 summarizes the overall classification.

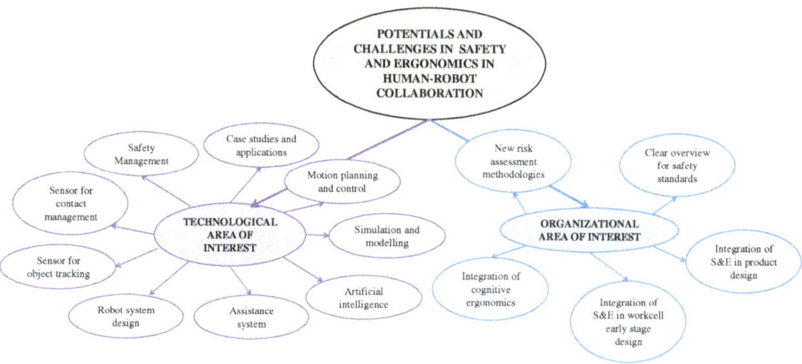

Fig. 4.5 Potential and challenges in safety and ergonomics (S&E) in HRC: research areas

There are two main research areas of interest for safety and ergonomics in HRC: technological and organizational. The former is related to the development of techniques for the improvement of the safety of HRI, while optimizing robot performances. The latter is related to management tools for better design and evaluation of safety and ergonomics solutions in collaborative systems.

4.4.1 Main Technological Research Areas of Interest

After a preliminary systematic literature review (SLR) about safety in modern industrial HRC, it is possible to classify nine dominating research areas of interest (see Fig. 4.5). The SLR workflow is summarized in Fig. 4.6. The review was performed using Scopus as a database by applying different filters to identify only English language and recent documents (period 2015–2018) related to safety and ergonomics in industrial HRC. In order to obtain highly relevant results, only journals, reviews, and articles in press documents related to engineering and computer science were selected. The starting number of identified papers was 93. After a detailed content check, the number of relevant papers was reduced to 42. The selected papers were successively subdivided into two main categories: contact avoidance and contact detection

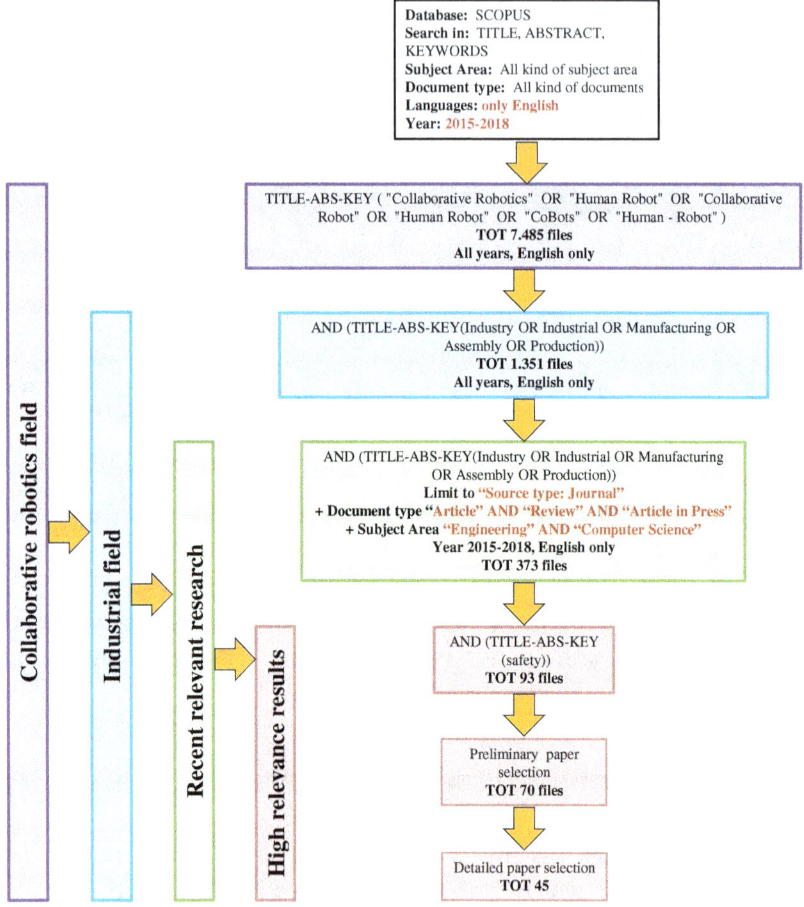

Fig. 4.6 SLR workflow

and mitigation. As shown in Fig. 4.7, the contact detection and mitigation category is less discussed (with a percentage of 35.7% of the total number of identified papers) in comparison with the contact avoidance category (which represents 64.3% of the total number of identified papers). Table 4.3 shows the data classification about the percentage of papers which contain a specific topic (note that a generic paper can be addressed to more than one topic).

Table 4.3 Classification data of the identified technological leading research areas of interest: percentage of relevant papers which contain a specific topic

Category	Motion planning and control	Simulation and modelling	Artificial intelligence	Assistance system	Robot system design	Sensor system for object tracking	Sensor system for contact management	Safety management	Case studies and applications
Contacts avoidance	44.4	11.1	11.1	11.1	0.0	33.3	0.0	33.3	3.7
Contacts detection and mitigation	40.0	6.7	0.0	0.0	33.3	0.0	33.3	20.0	6.7

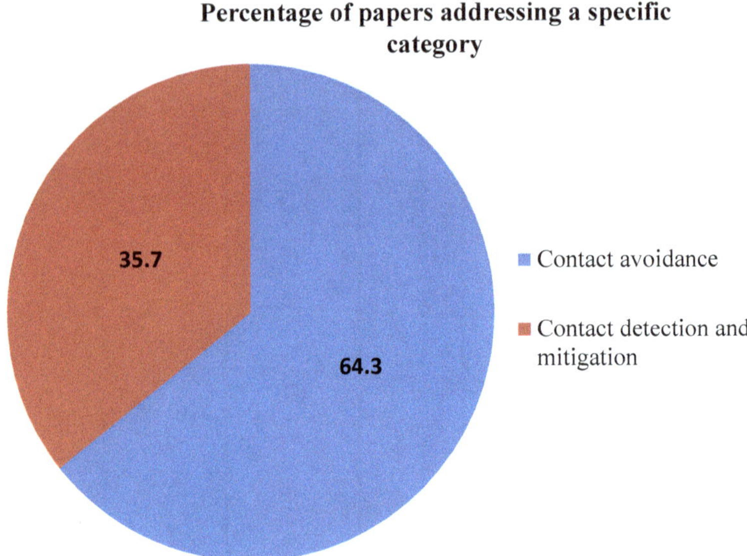

Fig. 4.7 Comparison between contact avoidance and contact detection and mitigation categories

According to the data illustrated in Fig. 4.8, it is clear that some research topics and interests are more structured than others. In addition, not all the leading technological research areas of interest are parts of both categories. The reason is that these topics are strictly related to specific technologies which depend on the methodology used to ensure safety during different levels of interaction between humans and robots. For example, artificial intelligence, assistance system and sensor system for object tracking are only present in the contact avoidance category. On the other hand, robot system design and sensor system for contact management are, of course, present only in the contact detection and mitigation category. For both the categories, motion planning and control, sensor system and safety management are mature topics and found in more than 30% of the works (in exception of safety management for contact detection and mitigation wich is equal to 20%). On the contrary, simulation and modelling, artificial intelligence, assistance systems, and case studies and applications are emerging research topics

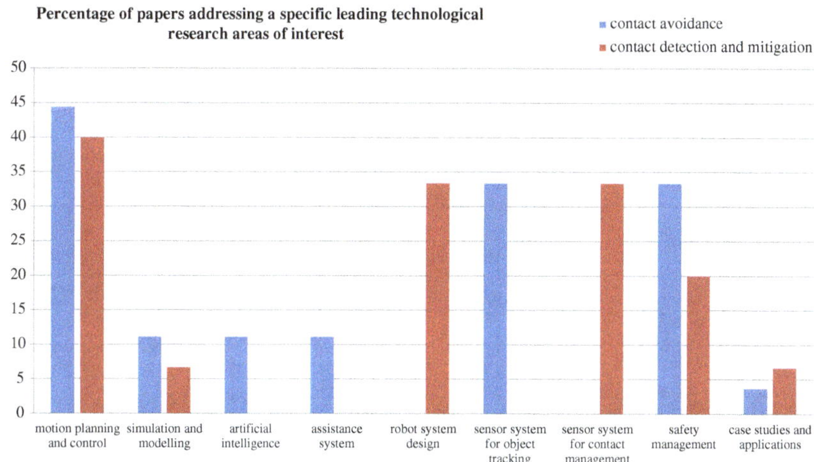

Fig. 4.8 Comparison of leading technological research areas of interest according to a preliminary SLR

with unknown future developments and a percentage lower than 12%. The reason for such limited diffusion could be related to the novelty of certain kinds of research topics (especially for artificial intelligence) or to a less efficient use with respect to other types of equivalent technologies (i.e., for assistance systems) (Fig. 4.8).

4.4.2 Main Organizational Research Areas of Interest

In this section, the main organizational research areas of interest are proposed in order to comment on and discuss future developments and the main possibilities and innovations for safety and ergonomics in industrial collaborative robotics. The following list was developed according to different research results achieved by the authors in the Smart Mini-Factory (SMF) laboratory of the Free University of Bolzano-Bozen (Gualtieri et al. 2018).

a. **Creation of a clear overview of safety standards for industrial collaborative robotics**: in order to correctly implement a collaborative system, the OHS requirements must be fulfilled and hence

Table 4.4 Relationships between main OHS standards and other deliverables related to industrial HRC

Title	Code	ISO 10218-1	ISO 10218-2	ISO 12100	ISO 13849-1	ISO 13854	ISO 13855	ISO 13857	ISO 14118	ISO 14120	ISO TS 15066	IEC 62046
Safety requirements for industrial robots—Part 1: Robots (ISO 2011b)	ISO 10218-1		x	x	x							
Safety requirements for industrial robots—Part 2: Robot systems and integration (ISO 2011a)	ISO 10218-2			x	x	x	x	x		x		x
General principles for design—Risk assessment and risk reduction (ISO 2010 a)	ISO 12100											
Safety-related parts of control systems—Part 1: General principles for design (ISO 2015a)	ISO 13849-1			x								
Minimum gaps to avoid crushing of parts of the human body (ISO 2017b)	ISO 13854			x				x				
Positioning of safeguards with respect to the approach speeds of parts of the human body (ISO 2010c)	ISO 13855			x				x				
Safety distances to prevent hazard zones being reached by upper and lower limbs (ISO 2008b)	ISO 13857			x								
Prevention of unexpected start-up (ISO 2017c)	ISO 14118			x	x							
General requirements for the design and construction of fixed and movable guards (ISO 2015b)	ISO 14120			x			x	x				
Collaborative robots (ISO 2016)	ISO TS 15066	x	x	x			x					
Application of protective equipment to detect the presence of persons (IEC 2018)	IEC 62046			x	x		x					

the standards (and other deliverables) on safety of machinery applied. However, the use of standards and other deliverables for the development of safe HRCs is usually difficult (Gualtieri et al. 2018), due to the unavoidable evolution of technologies increasing the complexity of requirements and the importance of risk-related topics. Additionally, such indications are often very coupled, i.e., the information content is often distributed and linked across different technical documents. This condition makes their consultation and their proper implementation complex, time-consuming, and demanding in terms of technical skills. Table 4.4 shows the mutual relationships between main OHS standards and other deliverables related to industrial HRC.

A clear overview of the main industrial standards and other deliverables for collaborative robotics is needed. A structured framework and a well-defined set of technical guidelines or reports based on related standards will support the designers and safety technicians in an easy and rapid fulfillment of safety requirements and as a consequence, improve the adoption of safe and efficient solutions for HRC.

b. **Development of new methodologies for risk assessment**: the possibility of direct physical interaction between humans and industrial robots represents a new paradigm in the field of OHS. Of course, the introduction of collaborative systems allows new possibilities from a manufacturing point of view but also new challenges in terms of operator safety and ergonomics. For this reason, new and robust risk assessment methods should be developed by including the possibility of interaction between humans and robots. There are various methodologies for the assessment of occupational risks. At the moment, one of the more complete methods for machinery risk assessment is the hybrid method (ISO 2012). This method identifies three different risk conditions by colors: red meaning safety measures are required, yellow meaning safety measures are recommended, green meaning no safety measures are required. These risk conditions are calculated by using a risk matrix through five qualitative variables: Severity (Se), Frequency (Fr), Probability (Pr), Avoidance (Av), and Class (*CI*, which is the sum of the previous indices). These indices were estimated according to the guidelines found in (ISO 2012) and (ISO 2016). Considering only mechanical risks, in the case of collaborative robotics, the main potential harms are mild or moderate in intensity. In general, the Probability and Frequency indices are the more relevant values. The Avoidance index can be medium since there is a real possibility of avoiding unintentional contacts due to the limited robot speed range, which is required for collaborative applications. The final goal of the risk assessment will be risk reduction through the adoption of protective measures implemented by the designer and the user in order to reach an acceptable value of residual risk. The risk reduction could be evaluated through the different values of risk class before (*CI*) and after (*CI**) the introduction of safety solutions. Due to the complexity of the safety requirements related to collaborative applications, the use of an extended method like to one described here is suggested for proper risks assessment.

c. **Integration of cognitive ergonomics considerations**: the sharing of workspaces and the physical interaction between humans and industrial robots could affect the cognitive ergonomics of the collaborative work (see Sect. 4.3.1). In this context, it will be mandatory to minimize the mental stress and psychological discomfort conditions which could be established during hybrid operations. Even if safety measures are well designed and implemented, the mere presence of the robot should not to be perceived as a hazard or as a source of stress for humans. Designers should consider these kinds of cognitive ergonomics problems in order to develop anthropocentric and human-friendly collaborative workstations also from a psychological point of view.

d. **Integration of safety and ergonomics of collaborative workspaces in the early-stage product design by Design for Collaborative Assembly (DFCA)**: the application of concurrent engineering (CE) methods for the integration of safety and ergonomics in the early-stage design of shared spaces will be crucial for the optimization of collaborative systems. CE is a "*comprehensive, systematic approach to the integrated, concurrent design and development of complex products, and their related processes*" (i.e., manufacturing, logistics, disposal, etc.) (Verhagen 2015). The aim is to improve productivity and to reduce production costs by decreasing product development and time-to-market. This methodology requires the consideration of all the early-stage features of the product life cycle, starting from conception and concluding with disposal (Verhagen 2015). Due to the nature of risks, which are potentially related to industrial HRI, concurrent design of a collaborative system necessarily involves the early consideration of OHS requirements for integrated products and process development. In order to develop safe and efficient solutions, the general design of an industrial machine (and related process) should consider the integration and the optimization of safety systems with functional systems in the early design stages. Nevertheless, it is a common procedure to design and develop the machine without considering the safety requirements and then to add them after the fact. This condition makes the design results unavoidably more inefficient and time-consuming, especially for collaborative systems.

From the engineering design point of view, product design is another fundamental part of CE. In order to satisfy customer requirements, engineering design allows the creation and transformation of ideas and concepts into functional products and processes. At the moment, large parts of optimized products, which are involved in manual operations, are designed for manual manufacturing and assembly. This means they present technical features and production processes notably designed for manual picking, handling, assembly, and/or manufacturing. Obviously, these components do not have suitable features for robotic or automated processing (Boothroyd et al. 2001; Boothroyd 2005). Considering that the industrial collaborative robot market is continually growing (Djuric et al. 2016), it is reasonable to suppose that collaborative operations will be an interesting challenge in the near future. For this reason, it will be particularly useful to create new product design approaches, which consider requirements for human–robot physical interaction during collaborative tasks. Therefore, a new research field for product design could be to enrich commonly known "Design For X" (DFX) techniques by adding a new "Design For Collaborative Assembly" (DFCA) method. Some possibilities should include the design of product features according to the minimization of mechanical risk which could arise during physical HRI.

4.5 Conclusions

Collaborative robotics is a key enabling technology of the fourth industrial revolution. The possibility of having side-by-side and fenceless HRI allows the combination of both operators' and robots' strengths and advantages, but it also involves the necessity to solve challenges in regard to OHS requirements. This chapter aims to provide an introduction about safety and ergonomics in industrial HRI by discussing the main potential and challenges in SME manufacturing automation. In particular, a brief introduction about main principles and definitions of OHS is presented. The core concepts and the classification of international OHS standards and deliverables regarding the safety of the machinery context are explained, also providing a summary about main industrial robotics technical documents. Particular attention is paid to mechanical risk, since HRI allows and requires a sharing of tasks and a more or less direct physical collaboration in a fenceless workspace. The four recognized methodologies for safe HRC according to ISO TS 15066: 2016 are introduced, and the nature of human–robot non-functional contacts is discussed. In addition, a general overview about main ergonomics standards in relation to human-centered workplace design is presented in order to have a clear vision about the integration of safety requirements with physical ergonomics considerations. Finally, a detailed discussion about leading potentials and challenges in safety and ergonomics in industrial HRC is discussed. This section introduces main future research areas of interest by dividing the results into technological and organizational types. The data obtained from a preliminary SLR and from the research results achieved by the authors in the SMF laboratory supported the discussion about the identified potential and challenges. Of course, the defined issues are demanding for SMEs both from a technological and organizational point of view, but also concretely achievable if companies are properly supported by investment and research.

References

Boothroyd, G. 2005. *Assembly Automation and Product Design*. CRC Press. http://dx.doi.org/10.1201/9781420027358.

Boothroyd, G., P. Dewhurst, and W.A. Knight. 2001. *Product Design for Manufacture and Assembly, Revised and Expanded*. Boca Raton: CRC Press.

Caputo, F., A. Greco, E. D'Amato, I. Notaro, and S. Spada. 2018. On the Use of Virtual Reality for a Human-Centered Workplace Design. *Procedia Structural Integrity* 8: 297–308. http://dx.doi.org/10.1016/j.prostr.2017.12.031.

CENELEC. 2019. What Is an European Standard? https://www.cencenelec.eu/standards/DefEN/. Accessed on Mar 2019.

Colombini, D. 2002. *Risk Assessment and Management of Repetitive Movements and Exertions of Upper Limbs: Job Analysis, Ocra Risk Indicies, Prevention Strategies and Design Principles* 2. Amsterdam: Elsevier.

da Costa, B.R., and E.R. Vieira. 2010. Risk Factors for Work-Related Musculoskeletal Disorders: A Systematic Review of Recent Longitudinal Studies. *American Journal of Industrial Medicine* 53 (3): 285–323. http://dx.doi.org/10.1002/ajim.20750.

D'Addona, D.M., F. Bracco, A. Bettoni, N. Nishino, E. Carpanzano, and A.A. Bruzzone. 2018. Adaptive Automation and Human Factors in Manufacturing: An Experimental Assessment for a Cognitive Approach. *CIRP Annals* 67 (1): 455–458. https://doi.org/10.1016/j.cirp.2018.04.123.

Directive, E.M. 2006. 42/EC of the European Parliament and the Council of 17 May 2006 on machinery, and amending Directive 95/16/EC (recast). *Official Journal of the European Union L* 157: 24–86.

Djuric, A.M., R.J. Urbanic, and J.L. Rickli. 2016. A Framework for Collaborative Robot (CoBot) Integration in Advanced Manufacturing Systems. *SAE International Journal of Materials and Manufacturing* 9 (2): 457–464. https://doi.org/10.4271/2016-01-0337.

European Committee for Standardization. 2007. Safety of Machinery—Human Physical Performance—Part 5: Risk Assessment for Repetitive Handling at High Frequency (CEN Standard No. 1005-5). http://store.uni.com/catalogo/index.php/en-1005-5-2007.html.

European Committee for Standardization. 2008a. Safety of Machinery—Human Body Measurements—Part 3: Anthropometric Data (CEN Standard No. 547-3 + A1). http://store.uni.com/catalogo/index.php/en-547-3-1996-a1-2008.html.

European Committee for Standardization. 2008b. Safety of Machinery—Human Body Measurements—Part 1: Principles for Determining the Dimensions Required for Openings for Whole Body Access into Machinery (CEN Standard No. 547-1 + A1). http://store.uni.com/catalogo/index.php/en-547-1-1996-a1-2008.html.

European Committee for Standardization. 2008c. Safety of Machinery—Human Body Measurements—Part 2: Principles for Determining the Dimensions Required for Access Openings (CEN Standard No. 547-2 + A1). http://store.uni.com/catalogo/index.php/en-547-2-1996-a1-2008.html.

European Committee for Standardization. 2008d. Safety of Machinery—Human Physical Performance—Part 2: Manual Handling of Machinery and Component Parts of Machinery (CEN Standard No. 1005-2:2003 + A1). http://store.uni.com/catalogo/index.php/en-1005-2-2003-a1-2008.html.

European Committee for Standardization. 2008e. Safety of Machinery—Human Physical Performance—Part 3: Recommended Force Limits for

Machinery Operation (CEN Standard No. 1005-3 + A1). http://store.uni.com/catalogo/index.php/en-1005-3-2002-a1-2008.html.

European Committee for Standardization. 2008f. Safety of Machinery—Human Physical Performance—Part 4: Evaluation of Working Postures and Movements in Relation to Machinery (CEN Standard No. 1005-4 + A1). http://store.uni.com/catalogo/index.php/en-1005-4-2005-a1-2008.html?josso_back_to=http://store.uni.com/josso-security-check.php&josso_cmd=login_optional&josso_partnerapp_host=store.uni.com.

European Committee for Standardization. 2009. Safety of Machinery—Ergonomic Design Principles—Part 1: Terminology and General (CEN Standard No. 614-1 + A1: 2009). http://store.uni.com/catalogo/index.php/en-614-1-2006-a1-2009.html.

Faber, M., J. Bützler, and C.M. Schlick. 2015. Human-Robot Cooperation in Future Production Systems: Analysis of Requirements for Designing an Ergonomic Work System. *Procedia Manufacturing* 3: 510–517. https://doi.org/10.1016/j.promfg.2015.07.215.

Gualtieri, L., E. Rauch, R. Rojas, R. Vidoni, and D.T. Matt. 2018. Application of Axiomatic Design for the Design of a Safe Collaborative Human-Robot Assembly Workplace. *MATEC Web of Conferences*, 223, 1003. EDP Sciences. http://dx.doi.org/10.1051/matecconf/201822301003.

Gualtieri, L., R. Rojas, G. Carabin, I. Palomba, E. Rauch, R. Vidoni, and D.T. Matt. 2018. Advanced Automation for SMEs in the I4.0 Revolution: Engineering Education and Employees Training in the Smart Mini Factory Laboratory. In *IEEE International Conference on Industrial Engineering and Engineering Management (IEEM)*, 1111–1115.

Health and Safety Authority. 2019. Hazard and Risk. https://www.hsa.ie/eng/Topics/Hazards/. Accessed on Mar 2019.

Huber, M., M. Rickert, A. Knoll, T. Brandt, and S. Glasauer. 2008. Human-Robot Interaction in Handing-Over Tasks. In *RO-MAN 2008—The 17th IEEE International Symposium on Robot and Human Interactive Communication*, 107–112. http://dx.doi.org/10.1109/ROMAN.2008.4600651.

International Electrotechnical Commission. 2015. Safety of Machinery—Functional Safety of Safety-Related Electrical, Electronic and Programmable Electronic Control Systems (IEC Standard No. 62061:2005 + AMD1:2012 + AMD2:2015 CSV). https://webstore.iec.ch/publication/22797.

International Electrotechnical Commission. 2018. Safety of Machinery—Application of Protective Equipment to Detect the Presence of Persons (IEC Standard No. 62046). https://webstore.iec.ch/publication/27263.

International Federation of Robotics. 2019. IFR Publishes Collaborative Industrial Robot Definition and Estimates Supply.

https://ifr.org/post/international-federation-of-robotics-publishes-collaborative-industrial-rob. Accessed on Mar 2019.

International Organization for Standardization. 2000. Ergonomics—Evaluation of Static Working Postures (ISO Standard No. 11226). https://www.iso.org/standard/25573.html.

International Organization for Standardization. 2003. Ergonomics—Manual Handling—Part 1: Lifting and Carrying (ISO Standard No. 11228-1). https://www.iso.org/standard/26520.html.

International Organization for Standardization. 2007a. Ergonomics—Manual Handling of Low Loads at High Frequency (ISO Standard No. 11228-3). https://www.iso.org/standard/26522.html.

International Organization for Standardization. 2007b. Ergonomics—Manual Handling—Part 2: Pushing and Pulling (ISO Standard No. 11228-2). https://www.iso.org/standard/26521.html.

International Organization for Standardization. 2008a. Safety of Machinery—Anthropometric Requirements for the Design of Workstations at Machinery (ISO Standard No. 14738, Including Cor 1:2003 and Cor 2:2005). https://www.iso.org/standard/27556.html.

International Organization for Standardization. 2008b. Safety of Machinery—Safety Distances to Prevent Hazard Zones Being Reached by Upper and Lower Limbs (ISO Standard No. 13857). https://www.iso.org/standard/39255.html.

International Organization for Standardization. 2010a. Safety of Machinery-General Principles for Design-Risk Assessment and Risk Reduction (ISO Standard No. 12100). https://www.iso.org/standard/51528.html.

International Organization for Standardization. 2010b. Ergonomics of Human-System Interaction—Part 210: Human-Centred Design for Interactive Systems (ISO Standard No. 9241-210). https://www.iso.org/standard/52075.html.

International Organization for Standardization. 2010c. Safety of Machinery—Positioning of Safeguards with Respect to the Approach Speeds of Parts of the Human Body (ISO Standard No. 13855). https://www.iso.org/standard/42845.html.

International Organization for Standardization. 2011a. Robots and Robotic Devices—Safety Requirements for Industrial Robots—Part 2: Robot Systems and Integration (ISO Standard No. 10218-2). https://www.iso.org/standard/41571.html.

International Organization for Standardization. 2011b. Robots and Robotic Devices—Safety Requirements for Industrial Robots—Part 1: Robots (ISO Standard No. 10218-1). https://www.iso.org/standard/51330.html.

International Organization for Standardization. 2011c. Ergonomics—General Approach, Principles and Concepts (ISO Standard No. 26800). https://www.iso.org/standard/42885.html.

International Organization for Standardization. 2012. Safety of Machinery—Risk Assessment—Part 2: Practical Guidance and Examples of Methods (ISO/TR Standard No. 14121-2). https://www.iso.org/standard/57180.html.

International Organization for Standardization. 2015a. Safety of Machinery—Safety-Related Parts of Control Systems—Part 1: General Principles for Design (ISO Standard No. 13849-1). https://www.iso.org/standard/69883.html.

International Organization for Standardization. 2015b. Safety of Machinery—Guards-General Requirements for the Design and Construction of Fixed and Movable Guards (ISO Standard No. 14120). https://www.iso.org/standard/59545.html.

International Organization for Standardization. 2016. Robots and Robotic Devices—Collaborative Robots (ISO/TS Standard No. 15066). https://www.iso.org/standard/62996.html.

International Organization for Standardization. 2017a. Basic Human Body Measurements for Technological Design—Part 1: Body Measurement Definitions and Landmarks (ISO Standard No. 7250-1). https://www.iso.org/standard/65246.html.

International Organization for Standardization. 2017b. Safety of Machinery—Minimum Gaps to Avoid Crushing of Parts of the Human Body (ISO Standard No. 13854). https://www.iso.org/standard/66459.html.

International Organization for Standardization. 2017c. Safety of Machinery—Prevention of Unexpected Start-Up (ISO Standard No. 14118). https://www.iso.org/standard/66460.html.

International Organization for Standardization. 2018. Robotics—Safety Design for Industrial Robot Systems—Part 1: End-Effectors (ISO/TR No. 20218). https://www.iso.org/standard/69488.html

Jensen, R.C. 2012. *Risk-Reduction Methods for Occupational Safety and Health.* Wiley. http://dx.doi.org/10.1002/9781118229439.

Jespen, T. 2016. *Risk Assessments and Safe Machinery: Ensuring Compliance with the EU Directives.* Springer. http://dx.doi.org/10.1002/9781118229439.

Koradecka, D. 2010. *Handbook of Occupational Safety and Health.* Boca Raton: CRC Press.

Kuta, Ł., J. Ciez, and I. Golab. 2015. Assessment of Workload on Musculoskeletal System of Milkers in Mechanical Milking Through the Use of Job Strain Index Method. *Scientific Papers: Management, Economic Engineering in Agriculture & Rural Development* 15 (1). http://dx.doi.org/10.1201/EBK1439806845.

Kuz, S., M. Faber, J. Bützler, M.P. Mayer, and C.M. Schlick. 2014. Anthropomorphic Design of Human-Robot Interaction in Assembly Cells. In *Advances in the Ergonomics in Manufacturing: Managing the Enterprise of the Future*, 265–272. Boca Raton: CRC Press.

Mayer, M.P., and C.M. Schlick. 2012. Improving Operator's Conformity with Expectations in a Cognitively Automated Assembly Cell Using Human Heuristics. In *Advances in Ergonomics in Manufacturing*, 302–311. http://dx.doi.org/10.1201/b12322-36.

McAtamney, L., and E.N. Corlett. 1993. RULA: A Survey Method for the Investigation of Work-Related Upper Limb Disorders. *Applied Ergonomics* 24 (2): 91–99. https://doi.org/10.1016/0003-6870(93)90080-S.

Michalos, G., J. Spiliotopoulos, S. Makris, and G. Chryssolouris. 2018. A Method for Planning Human Robot Shared Tasks. *CIRP Journal of Manufacturing Science and Technology* 22: 76–90. http://dx.doi.org/10.1016/j.cirpj.2018.05.003.

Monnington, S.C., C.J. Quarrie, A.D. Pinder, and L.A. Morris. 2003. Development of Manual Handling Assessment Charts (MAC) for Health and Safety Inspectors. In *Contemporary Ergonomics*, 3–8. London: Taylor & Francis.

Müller, R., M. Vette, and O. Mailahn. 2016. Process-Oriented Task Assignment for Assembly Processes with Human-Robot Interaction. *Procedia CIRP* 44: 210–215.

Peruzzini, M., M. Pellicciari, and M. Gadaleta. 2019. A Comparative Study on Computer-Integrated Set-Ups to Design Human-Centred Manufacturing Systems. *Robotics and Computer-Integrated Manufacturing* 55: 265–278. https://doi.org/10.1016/j.rcim.2018.03.009.

Rojas, R.A., M.A.R. Garcia, E. Wehrle, and R. Vidoni. 2019. A Variational Approach to Minimum-Jerk Trajectories for Psychological Safety in Collaborative Assembly Stations. *IEEE Robotics and Automation Letters* 4 (2): 823–829. https://doi.org/10.1109/LRA.2019.2893018.

Romero, D., O. Noran, J. Stahre, P. Bernus, and Å. Fast-Berglund. 2015. Towards a Human-Centred Reference Architecture for Next Generation Balanced Automation Systems: Human-Automation Symbiosis. In *IFIP International Conference on Advances in Production Management Systems*, 556–566. Springer, Cham. http://dx.doi.org/10.1007/978-3-319-22759-7_64.

Schaub, K.G., J. Mühlstedt, B. Illmann, S. Bauer, L. Fritzsche, T. Wagner, A.C. Bullinger-Hoffmann, and R. Bruder. 2012. Ergonomic Assessment of Automotive Assembly Tasks with Digital Human Modelling and the 'Ergonomics Assessment Worksheet' (EAWS). *International Journal of Human Factors Modelling and Simulation* 3 (3–4): 398–426. https://doi.org/10.1504/IJHFMS.2012.051581.

Scott, G.B., and N.R. Lambe. 1996. Working Practices in a Perchery System, Using the OVAKO Working Posture Analysing System (OWAS). *Applied Ergonomics* 27 (4): 281–284. https://doi.org/10.1016/0003-6870(96)00009-9.

Stanton, N.A., A. Hedge, K. Brookhuis, E. Salas, and H.W. Hendrick (eds.). 2004. *Handbook of Human Factors and Ergonomics Methods*. CRC Press. http://dx.doi.org/10.1201/9780203489925.

Szabó, G., and E. Németh. 2018. Development an Office Ergonomic Risk Checklist: Composite Office Ergonomic Risk Assessment (CERA Office). In *Congress of the International Ergonomics Association*, 590–597. Cham: Springer. http://dx.doi.org/10.1007/978-3-319-96089-0_64.

Tan, Q., Y. Tong, S. Wu, and D. Li. 2019. Anthropocentric Approach for Smart Assembly: Integration and Collaboration. *Journal of Robotics*. https://doi.org/10.1155/2019/3146782.

Universal Robot. 2018. UR3/CB3 Version 3.6.0 Original Instructions.

Verhagen, W.J. (ed.). 2015. *Concurrent Engineering in the 21st Century: Foundations, Developments and Challenges*. Cham: Springer.

Vicentini, F. 2017. *La robotica collaborativa - Sicurezza e flessibilità delle nuove forme di collaborazione uomo-robot*. Milan: Tecniche Nuove.

World Health Organization. 2019. Constitution. https://www.who.int/about/who-we-are/constitution. Accessed on Mar 2019.

Part III

Industry 4.0 Concepts for Smart Logistics in SMEs

5

Requirement Analysis for the Design of Smart Logistics in SMEs

Patrick Dallasega, Manuel Woschank, Helmut Zsifkovits, Korrakot Tippayawong and Christopher A. Brown

5.1 Introduction

The fourth industrial revolution, also called “Industry 4.0,” is currently transforming the manufacturing and connected supply chain industry. After the advent of mechanization, electrification, and computerization, it represents the increasing digitization and automation of the manufacturing industry, as well as the establishment of digital value chains to enable communication between products, machines, and human operators (Lasi et al. 2014). The focus of Industry 4.0 is to combine the internet, information, and communication technologies (ICT) with classical industrial processes (Bundesministerium für Bildung und Forschung 2012).

P. Dallasega
Free University of Bozen-Bolzano, Bolzano, Italy
e-mail: patrick.dallasega@unibz.it

M. Woschank · H. Zsifkovits (✉)
Chair of Industrial Logistics, Montanuniversitaet Leoben, Leoben, Austria
e-mail: helmut.zsifkovits@unileoben.ac.at

D. T. Matt et al. (eds.), *Industry 4.0 for SMEs*,
https://doi.org/10.1007/978-3-030-25425-4_5

An important transformation that comes with Industry 4.0 is the shift from centralized to decentralized control to reach a highly flexible production of customized products and services. An increased individualization and personalization of products leads to customer interaction strategies like X-to-order (make-to-order, build-to-order, configure-to-order, and engineer-to-order) and ultimately to the concept of "mass customization" where products can be configured by the customer at costs similar to those of mass production. The increasing fusion of the information technology (IT) environment with production and logistics allows flexible and reconfigurable manufacturing and logistics systems (Spath et al. 2013).

Another important part of Industry 4.0 are cyber-physical systems (CPS) that allow self-organization and self-control of manufacturing and logistics systems (Rauch et al. 2016a, b). CPS are computers, sensors, and actuators that are embedded in materials, equipment, and machine parts and connected via the internet, the so-called Internet of Things (IoT), allowing the physical and digital worlds to be merged (Spath et al. 2013). A further big benefit that comes with the introduction of CPS is the acquisition of a high amount of data that is available in real time. This allows better production planning and control (PPC) and efficient counteraction of unforeseen events in manufacturing and logistics processes (Dallasega et al. 2015a, b). In summary, it can be stated that the potential of a successful implementation of Industry 4.0 is enormous.

So far, Industry 4.0 is mainly brought forward by bigger companies and SMEs are risking not being able to exploit this huge potential. According to the European Commission, SMEs are characterized

M. Woschank
e-mail: manuel.woschank@unileoben.ac.at

K. Tippayawong
Chiang Mai University, Chiang Mai, Thailand
e-mail: korrakot@eng.cmu.ac.th

C. A. Brown
Department of Mechanical Engineering,
Worcester Polytechnic Institute, Worcester, MA, USA
e-mail: brown@wpi.edu

by having not more than 250 employees and an annual turnover of less than €50 mio or a balance sheet of no more than €43 mio (Kraemer-Eis and Passaris 2015). In more detail, micro and SMEs provide around 45% of the value added by manufacturing and around 59% of manufacturing employment and therefore, they can be considered as the backbone of the European economy (Vidosav 2014). As the previous economic crisis showed, SMEs proved to be more robust than bigger companies because of their flexibility, their entrepreneurial spirit, and their innovation capabilities (Matt 2007).

As such, the successful implementation of Industry 4.0 has to take place not only in large enterprises but even more important in SMEs. Because of often limited financial and human resources, the implementation of Industry 4.0 represents a special challenge for SMEs. Up to now, SMEs are only partially ready to adapt to Industry 4.0 concepts. In particular, smaller enterprises face a high risk of not being able to benefit from this industrial revolution. As a result, further research and action plans are needed to prepare SMEs for the stepwise introduction of Industry 4.0 (Sommer 2015). According to the authors, SMEs will only benefit from Industry 4.0 by following customized implementation strategies, approaches, concepts, and technological solutions that have been appropriately adopted. Otherwise, the current effort for publication and sensitization of SMEs for Industry 4.0 will not lead to expected results.

This book chapter presents an explorative set of hypotheses of requirements to implement Industry 4.0 concepts in logistics processes in SMEs spread over the world. They were identified in the course of the research project "SME 4.0 – Industry 4.0 for SMEs" by using expert workshops with SMEs from the north-east part of the USA, Central Europe, and Northern Thailand.

5.2 Problem Formulation

Requirements of larger companies regarding the successful adoption of Industry 4.0 concepts and technologies may not be suitable for SMEs. Compared to bigger companies, SMEs have at their disposal fewer resources in terms of budget and qualified workforces for doing research and innovation actions. Up to now, there are only a few studies available that propose requirements for the adoption of Industry 4.0, especially considering companies in specific geographical areas. However, little to almost no studies are available that deal with the requirements for SMEs to support the logistics part. Therefore, the authors conducted workshops with SMEs from South-East Asia, central Europe, and Northern USA to identify the first hypothesis of requirements to support logistics management with Industry 4.0 concepts.

5.3 Related Work

In order to build a profound theoretical foundation for the explorative investigation of the specific requirements of SMEs to use Industry 4.0 concepts in their logistics management, the authors have conducted a systematic literature review. Thereby, the main results will be briefly outlined within the next paragraphs (Dallasega et al. 2019).

In summary, Glass et al. (2018) identified barriers for implementing Industry 4.0 in SMEs by investigating the literature and using a survey approach in German companies. They emphasized specific barriers to Industry 4.0 implementation such as missing standardization and an inappropriate company strategy. Maasouman and Demirli (2015) developed a lean maturity model and designed a framework for assessment of concepts like just-in-time in manufacturing cells. Schumacher et al. (2016) developed models for assessing the readiness and maturity of Austrian companies regarding Industry 4.0. The model considers elements like the strategy of the organization, customer, people, and technology. Qin et al. (2016) developed a framework to show the gap between the state of the art in UK companies and the requirements for Industry 4.0 readiness. Similarly, Benešová and Tupa (2017) analyzed the Industry 4.0 requirements of Czech Republic companies where the digital representation of a factory in real time, the horizontal and vertical data integration, and the self-controlling of manufacturing and logistics processes emerged. Furthermore, the results showed that education and qualification of employees is one of the main requirements for the implementation of Industry 4.0. Kamble et al. (2018) used interpretive structural modeling (ISM) and fuzzy technology to analyze the barriers to implementing Industry 4.0 in Indian companies. Barriers like the lack of clear comprehension about IoT benefits, employment disruptions, organizational, and process changes (needed to implement Industry 4.0) emerged. Luthra and Mangla (2018) evaluated key concepts and challenges of implementing Industry 4.0 in Indian manufacturing companies with a special focus on sustainable supply chain management. Here, technological as well as an appropriate strategic orientation of the companies emerged as the main challenges to Industry 4.0-implementation.

5.4 Research Design/Methodology

In accordance with the theoretical foundation, the authors have conducted expert workshops, which followed pre-defined methodological guidelines to systematically evaluate the requirements of smart logistics in SMEs. In total, the research team conducted six workshops with 37 participating SMEs and 67 participating experts in Italy, Austria, USA, and Thailand in the timeframe between 9 June 2017 and 22 March 2018. The quantitative content analysis resulted in 548 statements for further investigation. Moreover, the statements for smart logistics in SMEs were divided into the sub-sections "smart and lean x-to-order Supply Chains," "intelligent logistics through ICT and CPS," "automation in logistics systems and vehicles," and "main barriers and difficulties for SMEs." Thereby, the results are briefly summarized in Table 5.1.

Axiomatic design theory states that the best design solution first, maintains the independence of the functional elements (axiom one), and second, minimizes the information content (axiom two).

Table 5.1 Facts and figures of SME workshops

	Italy	Austria	Thailand	USA	Total
Number of workshops	1	2	1	1	5
Participating SMEs	10	7	10	10	37
Participants	13	13	25	16	67
Total statements	213	97	98	140	548
Logistics-related statements	93	41	36	33	203

The axiomatic design method applies the theory during top-down, parallel decompositions of the functional requirements (FRs), and the design solutions, or parameters (DPs) and possibly process variables (PVs) which are how the DPs are produced or realized.

The decompositions start with the abstract concepts and develop detailed functions, FRs, the design problems, and solutions, DPs the design solutions, with more and more detail, level by level for each functional branch. At each level of detail, for each functional branch, the candidate design solutions are tested against the axioms. The individual DPs are finally integrated into a complete physical solution to the design problem (Suh 1990).

The value of the design is established by the customer needs (CNs) in the customer domain. In the functional domain, the FRs satisfy the CNs, and FRs are the next link in the value chain. In the physical domain, the next link in the value chain, the DPs, fulfills the FRs. In the process domain, PVs produce the DPs. The elements of the decomposition, FRs and DPs, at each level should be complete, or collectively exhaustive, with respect to the higher levels of abstraction in their domain. At each level, they should be mutually exclusive with respect to each other in their domains to satisfy axiom one.

Functional metrics can be selected for the FRs. Physical metrics should be implicit in the DPs. The relations between the parent and children FR, DP, and PVs, adjacent decomposition levels, in each of their domains, are described with decomposition equations. The relations between the FRs and DPs are described with design equations, which form the design matrix. The relations between the DPs and the PVs are described in the process equations that form the process matrix.

Design thinking demands the development of the functions first. The FRs should be stated to foster creativity in the design solution. FRs should create a large solution space. The FRs should be developed carefully, because no design solution can be better than the FRs. The FRs are the independent functions that define criteria for the success of the design solution. To maintain independence, each FR needs a different DP. Customer needs, like "low cost" or "ease of use," often should be non-FRs, which are represented in selection criteria (SCs), or optimization criteria (OCs), which are used to select between candidate DPs (Thompson 2013).

There is a tendency, especially for engineers, to seek and select physical design solutions before the functions have been adequately defined. This is not good design thinking.

If FRs contain physical attributes, this can inappropriately limit the solution space and inhibit creativity. Generally, it should be possible to generate several candidate DPs for each FR. If this is not possible, then perhaps the FR is too limiting and it should be reformulated. The solution space can be enlarged so it is possible to find more candidate DPs to fulfill the FRs by appropriate reformulation of the FRs. This can be done by asking why an FR is required. In essence, the FR should be moved closer to the CN and further from the DP, which makes a larger space for more creative, potential solutions.

5.5 Hypothesis of Requirements for Smart Logistics in SMEs

This section presents the identified statements as the hypothesis of requirements, which were recorded in the course of the SME workshops as an explorative approach based on a systematic literature analysis. Consequently, the statements were assigned to nine thematic clusters.

5.5.1 Lean and Agility

This cluster contains the requirements of the usage of advanced planning techniques that, for example, allow a production on-demand and delivery just-in-time. Workshop participants mentioned the requirement to ensure flexible supply chains. The identification and avoidance of material flow breaks and the timing of orders, to minimize transportation costs, were recorded. The reduction of buffer stocks, raw material, WIP, and finished parts was mentioned.

Strategies to increase the material efficiency in automated logistics systems such as the optimization of material yields at the vendor and the grouping of trucking routes of complimentary suppliers were recorded. Additionally, the reduction of buffer stocks at the workplace, preventive "rhythms" (delivery, preparation, etc.), efficient storage and removal systems for the holding of raw material, WIP, finished parts, parts produced and packaged at machines, and moved to shipping were mentioned.

5.5.2 Real-Time Status

This cluster includes the requirements for an infrastructure and digital feedback system, which monitors the status of production, storage, and shipping in real time. In particular, the short-term availability of information about the shipment/delivery status of material is very important for proper supply chain management. Moreover, the visibility of the supplier's status in real time for quick access to information for improved supplier risk management was mentioned. Even further, participants mentioned that a system, which enables real-time status information, could also assist in the predictive maintenance process.

5.5.3 Digitization, Connectivity, and Network

This cluster entails the necessity for an improved customer–supplier connection to gain the ability to communicate and/or share capacity, materials, infrastructure, and information with internal and external customers and suppliers. Even more, information should be provided and visualized everywhere and every time to reduce waiting times and unnecessary delays. Thereby, the requirements included the automated tracking of prices, the automation of processes (e.g., the generation of bill of materials), and the automated communication between multiple systems. The material flow should be visualized from upstream to downstream companies. This includes the visualization of tools and parts used throughout the supply chain processes. The requirement to increase transparency by visualizing stock and delivery times throughout the supply chain through the interconnection of suppliers with manufacturers and customers over the internet was recorded. Here, aggregator websites to determine the short-term availability of material or capacity in the supplier network, following the example of Skyscanner for the search of flights, were mentioned.

Moreover, the interconnection of customers with suppliers to avoid causes of missing parts/materials and to increase the reliability of supplies was recorded. Here, the following practical example was named by the experts: "when an order is received, the system should generate the bill of materials and automatically send the purchase order to the suppliers." In more detail, the need for an automatic on-site measurement and electronic submission of order data to the fabrication shop was collected. Digitalization should be implemented especially in the order receiving and procurement processes. Some of the participating experts stated that digitalization should limit the accessibility of related stakeholders to obtain optimal data. Moreover, the workshop participants mentioned the sharing of transport capacities. As a practical example, the geographical visualization of transport routes for the analysis of losses and inefficiencies in delivery routes was recorded. Another mentioned requirement was flexibility regarding the scalability of logistics systems and the predictive maintenance of logistics systems. Systems should be synchronized throughout the supply chain to avoid re-work and communication interruptions. Data are required to be integrated in order to support a uniform database system.

5.5.4 Tracking, PPC, and WMS

This cluster includes the requirements of digitally tracking and localizing (tracing) products throughout supply chains. Tracking systems should provide better information about the status of inventory, the tracking of multiple parts through multiple processes being able to monitor the status of production in real time.

Advanced PPC methodologies and tools should forecast demand changes quickly by interacting with internal and external systems for planning, control, and logistics. As a specific requirement, the automatic triggering of orders for tools and materials when processed orders come in was recorded. Moreover, the need for automatic "Pull" systems that allow a synchronized workflow across networked machines to minimize downtime, tool changes, and predictive maintenance was listed.

Furthermore, a better knowledge of the state of the art in warehouse management systems (WMS) was listed. Here, as a specific requirement, an automatic adjustment of inventory levels through low inventory levels that automatically trigger stock runs was collected. Furthermore, according to the participants, warehouse management systems should be implemented in a way that allows for easy exchange and storage of all needed information concerning command control and logistics. Moreover, an automated and permanent inventory control by comparing planned vs. actual data and the intuitive visualization of where the material is stored in the warehouse were mentioned.

Other requirements like automated assistance in order and distribution processes based on historical assumptions were mentioned. The provision of data for inventory decision making, such as inventory turns and reorder point arrangements to support economic order quantities (EOQ), were recorded. WMS should also be able to allocate and optimize storage locations and display accurate locations for product pick up.

5.5.5 Culture, People, and Implementation

In this cluster, the SME's needs to access the financial, informational, digital, physical, and educational resources to ensure that Industry 4.0 is fully implemented rather than passed by are summarized. The requirement to increase the visibility of Industry 4.0 among professionals who might not have been exposed to it otherwise was collected. Moreover, top management should be aware and support Industry 4.0 to avoid missing acceptance throughout the company. The need for qualified and trained employees to implement and handle Industry 4.0 concepts in daily business was recorded. Here, the participants stated that employees should be specifically trained in software and data collection. For successful implementation of Industry 4.0 into SMEs, the necessity of having an overview of existing Industry 4.0 concepts and tools for logistics and their suitability for SMEs for specific industry sectors was mentioned. Here, the need for a specific distinction of SMEs in countries with high-labor cost and countries with low-labor cost was specified.

5.5.6 Security and Safety

According to the workshop participants, security and safety issues should not be forgotten while implementing smart logistics in SMEs. Here, specifically, the internal traffic optimization for safety and efficiency in the workplace and required ICT to monitor and control safety in driverless transport systems were mentioned. Moreover, the ensuring of data security and intellectual property protection were recorded.

5.5.7 Ease of Use

According to the workshop results, the implementation of smart logistics concepts should be easily understandable and easy to use. The requirement for intuitive and role-related user-interfaces for software or machine control was mentioned. Different views for different roles (e.g., operator, supervisor) should be provided. Digital assistant systems should facilitate the work for operators, the communication of needs to R&D, and the communication of work metrics to supervisors.

5.5.8 Transportation

This cluster contains the automated material transport by using automated guided vehicles (AGVs) including all related activities (e.g., loading, transport, unloading, safety issues) aiming at a fast and cost-efficient distribution of materials.

5.5.9 Automation

This includes requirements for decreasing the manual workload in logistics systems. Thereby the experts mainly focused on the automated labeling of products, automatic picking and delivery, automated storage systems for materials and transport containers, and the automated removal of scrap in the course of the production process. Moreover, the participants were interested in cause–effect analyses aimed at the impact of automation approaches on business success.

5.6 Creativity and Viability Through Axiomatic Design

Industry 4.0 is often defined by the physical solutions, DPs, like a vision system and algorithms for managing the supply chain. It is supposed that these are fulfilling FRs, which satisfy CNs, which are common to many enterprises. The responses to the workshops often reflect the desire to use these physical solutions.

These Industry 4.0 solutions generally fit into the segmentation somewhere in an intermediate level of abstraction. Higher levels might result in maximizing return on investment (ROI), an appropriate upper level CN for an enterprise. In a manufacturing enterprise, maximizing ROI can be decomposed to add appropriate value and minimize cost.

Rather than starting with the highest level CNs and developing FRs first, Industry 4.0, like the previous industrial revolutions, tends to begin in the middle of the domains and the decomposition. To discover if implementing some aspect of a design solution like this is appropriate for a particular enterprise, and consistent with good design thinking, the higher levels of the decomposition and the functional domain need to be considered.

5.7 Conclusions and Outlook

Industry 4.0 has been mainly brought forward by bigger companies and SMEs are risking not being able to exploit this huge potential. However, micro and SMEs provide around 45% of the value added by manufacturing and around 59% of manufacturing employment and therefore, they can be considered as the backbone of the European economy (Vidosav 2014). Therefore, Industry 4.0 concepts should not only be conceived and implemented in bigger companies but even more importantly, customized implementation strategies, approaches, concepts, and technological solutions should be proposed for efficient implementation in SMEs.

The book chapter presents an explorative set of hypotheses of requirements for the implementation of Industry 4.0 concepts in logistics processes for SMEs. The hypotheses were identified by using expert workshops with SMEs, conducted by the Free University of Bolzano (Italy), the University of Leoben (Austria), the Worcester Polytechnic Institute (USA), and the Chiang Mai University (Thailand). Axiomatic Design was used as a methodology to structure the identified hypothesis of requirements. From the customer needs, functional requirements are to be derived. Following the Axiomatic Design approach, we can define design parameters, in order to specify the properties of the system to implement.

The recorded statements were clustered into nine thematic groups and different elements emerged as being important for the definition of requirements like "Lean and agility," "Real-time status," "Digitization, connectivity, and network," "Tracking, PPC, and WMS," "Culture, people, and implementation," "Security and safety," "Ease of use," "Transportation," and "Automation."

When applying the axioms and domains of Axiomatic Design, it can clearly be seen that a high percentage of the requirements defined by the respondents are not solution-neutral. This applies mainly to the categories "Digitization, connectivity, and network," "Tracking, PPC, and WMS," "Transportation," and "Automation" which clearly include not FRs but rather DPs or PVs. This, in turn, limits the solution space, and it is not possible to ensure the best solution is reached. Again, the inputs from the workshops need refinement to derive the "true FRs" behind them. The FR derivation technique seems a good methodology to derive solution-neutral requirements.

Future research will consist of validating the identified hypotheses by using a structured survey. So far, the survey has been launched in Austria, Italy, and Slovakia and in the near future, it will be launched in the USA and Thailand. Furthermore, after having validated the hypotheses of requirements, an assessment of the maturity and level of implementation/application of different Industry 4.0 concepts to satisfy the requirements will be undertaken. This will take place during the second phase of the research project, "SME4.0 – Industry 4.0 for SMEs."

References

Benešová, A., and J. Tupa. 2017. Requirements for Education and Qualification of People in Industry 4.0. *Procedia Manufacturing* 11: 2195–2202. https://doi.org/10.1016/j.promfg.2017.07.366.

Bundesministerium für Bildung und Forschung. 2012. *Zukunftsbild „Industrie 4.0"*. Berlin.

Dallasega, P., E. Rauch, and D.T. Matt. 2015a. Sustainability in the Supply Chain Through Synchronization of Demand and Supply in ETO-Companies. *Procedia CIRP* 29: 215–220. https://doi.org/10.1016/j.procir.2015.02.057.

Dallasega, P., E. Rauch, D.T. Matt, and A. Fronk. 2015b. Increasing Productivity in ETO Construction Projects Through a Lean Methodology for Demand Predictability. In *2015 International Conference on Industrial Engineering and Operations Management (IEOM)*, 1–11. IEEE. http://doi.org/10.1109/IEOM.2015.7093734.

Dallasega, P., M. Woschank, S. Ramingwong, K.Y. Tippayawong, and N. Chonsawat. 2019. Field Study to Identify Requirements for Smart Logistics of European, US and Asian SMEs. In *Proceedings of the International Conference on Industrial Engineering and Operations Management*.

Glass, R., A. Meissner, C. Gebauer, S. Stürmer, and J. Metternich. 2018. Identifying the Barriers to Industrie 4.0. *Procedia CIRP* 72: 985–988. https://doi.org/10.1016/j.procir.2018.03.187.

Kamble, S.S., A. Gunasekaran, and R. Sharma. 2018. Analysis of the Driving and Dependence Power of Barriers to Adopt Industry 4.0 in Indian Manufacturing Industry. *Computers in Industry* 101: 107–119. https://doi.org/10.1016/j.compind.2018.06.004.

Kraemer-Eis, H., and G. Passaris. 2015. SME Securitization in Europe. *Journal of Structured Finance* 20 (4): 97–106. https://doi.org/10.3905/jsf.2015.20.4.097.

Lasi, H., P. Fettke, H.-G. Kemper, T. Feld, and M. Hoffmann. 2014. „Industry 4.0". *Business & Information Systems Engineering* 6 (4): 239–242. https://doi.org/10.1007/s12599-014-0334-4.

Luthra, S., and S.K. Mangla. 2018. Evaluating Challenges to Industry 4.0 Initiatives for Supply Chain Sustainability in Emerging Economies. *Process Safety and Environmental Protection* 117: 168–179. https://doi.org/10.1016/j.psep.2018.04.018.

Maasouman, M.A., and K. Demirli. 2015. Assessment of Lean Maturity Level in Manufacturing Cells. *IFAC-PapersOnLine* 48 (3): 1876–1881. https://doi.org/10.1016/j.ifacol.2015.06.360.

Matt, D.T. 2007. Reducing the Structural Complexity of Growing Organizational Systems by Means of Axiomatic Designed Networks of Core Competence Cells. *Journal of Manufacturing Systems* 26 (3–4): 178–187. https://doi.org/10.1016/j.jmsy.2008.02.001.

Qin, J., Y. Liu, and R. Grosvenor. 2016. A Categorical Framework of Manufacturing for Industry 4.0 and Beyond. *Procedia CIRP* 52: 173–178. https://doi.org/10.1016/j.procir.2016.08.005.

Rauch, E., P. Dallasega, and D.T. Matt. 2016a. The Way from Lean Product Development (LPD) to Smart Product Development (SPD). *Procedia CIRP* 50: 26–31. https://doi.org/10.1016/j.procir.2016.05.081.

Rauch, E., S. Seidenstricker, P. Dallasega, and R. Hämmerl. 2016b. Collaborative Cloud Manufacturing: Design of Business Model Innovations Enabled by Cyberphysical Systems in Distributed Manufacturing Systems. *Journal of Engineering* 2016 (3): 1–12. https://doi.org/10.1155/2016/1308639.

Schumacher, A., S. Erol, and W. Sihn. 2016. A Maturity Model for Assessing Industry 4.0 Readiness and Maturity of Manufacturing Enterprises. *Procedia CIRP* 52: 161–166.

Sommer, L. 2015. Industrial Revolution—Industry 4.0: Are German Manufacturing SMEs the First Victims of This Revolution? *Journal of Industrial Engineering and Management* 8 (5): 1512–1532.

Spath, D., O. Ganschar, S. Gerlach, T.K. Hämmerle, and S. Schlund. 2013. *Produktionsarbeit der Zukunft – Industrie 4.0*, 2–133. Stuttgart: Fraunhofer Verlag.

Suh, N.P. 1990. *The Principles of Design*. New York: Oxford Press.

Thompson, M.K. 2013. A Classification of Procedural Errors in the Definition of Functional Requirements in Axiomatic Design Theory.

Vidosav, D.M. 2014. Manufacturing Innovation and Horizon 2020-Developing and Implement „New Manufacturing" (9. edit.): 3–8.

BY

6

Consistent Identification and Traceability of Objects as an Enabler for Automation in the Steel Processing Industry

Helmut Zsifkovits, Johannes Kapeller, Hermann Reiter, Christian Weichbold and Manuel Woschank

6.1 Introduction

An intelligent product in manufacturing systems is a physical product that can itself provide data for its own, virtual image in the manufacturing process (Kagermann et al. 2013). Thus, a basis is provided at all times

H. Zsifkovits (✉) · M. Woschank
Chair of Industrial Logistics, Montanuniversitaet Leoben, Leoben, Austria
e-mail: helmut.zsifkovits@unileoben.ac.at

M. Woschank
e-mail: manuel.woschank@unileoben.ac.at

J. Kapeller
Boston Consulting Group, Vienna, Austria

H. Reiter
ZKW Group, Wieselburg, Austria

C. Weichbold
Voestalpine Stahl Donawitz, Leoben, Austria

D. T. Matt et al. (eds.), *Industry 4.0 for SMEs*,
https://doi.org/10.1007/978-3-030-25425-4_6

for reacting to the product-specific parameters during production, to initiate decentralized decisions and to identify areas for process optimization more easily. From the point of view of production plants, it is equally advantageous if the digital connection between the products and the actual plant enables intelligent automation of the processing steps. This requires permanent and real-time traceability of objects and their states.

Traceability is essential to gain knowledge of causes of deviations in product attributes, since it makes it possible to trace product deviations back to root causes within the production process. Further benefits from traceability include lot uniformity in production, and the reduction of the extent to which products are affected by product recalls. Traceability supports fact-based decision-making and continuous improvement (Kvarnström 2008).

Moreover, the issue of traceability is by no means only for the benefit of companies. There are a number of industries where there is a pronounced demand in this regard. In particular, in the aerospace, chemical, pharmaceutical, food, and automotive industries, the requirements are regulated in great detail. From the point of view of the end customer, safety must be guaranteed, all demands on quality must be complied with, and there should be a minimum of risk potential. From the point of view of the company, it is above all, a question of liability or, in the B2B area, the possibility of recourse between companies, i.e., the avoidance of economic business risk.

In the current ISO 9001, the international standard that sets out the criteria for a quality management system (QMS), the following requirements around identification and traceability are defined: "*Use suitable means to identify outputs when it is necessary to ensure the conformity of products and services. Identify the status of outputs with respect to monitoring and measuring requirements throughout production and service provision. Control the unique identification of the outputs when traceability is a requirement, and retain documented information to enable traceability*" (ISO 9001, 2015). Thus, these requirements can be broken down into three distinct elements: (i) output identification, (ii) process stage, and (iii) traceable identification.

Depending on the required level of traceability, various methods may be employed, from individual component unique stamping or bar coding to whole lot/batch identification, in combination with documenting information manually or electronically gathered from suppliers and during processing. This can be through part labels, job travelers, work orders, production plans, route sheets, process validation worksheets, lot/batch control, test certificates, "inspected" labels, or any means to identify outputs and their status.

Object identification and traceability provide the information required to enable five functional areas which, depending on the existing basic architecture, will be more or less developed or further developed in the company (Bischoff 2015).

- **Data collection and processing**: This includes the collection and evaluation of data on processes, quality, products, production facilities, employees, and their environment. For the virtual image of reality, IT-based data collection of customer, product, production, and usage data is essential. Data evaluations include the analysis of overall equipment effectiveness and big data analyses, with the focus on improving process and quality. By capturing and analytically analyzing the data, considerable efficiency gains can be tapped, which have not been fully utilized until now.
- **Assistance systems**: These aim to provide the employee with the necessary information as quickly and easily as possible, anytime, anywhere. They summarize all the technologies that support employees in carrying out their work so that they can concentrate on their core tasks. These are, in particular, technologies for providing information (e.g., visualization systems, mobile devices, tablets, data glasses). Especially with regard to the ever-increasing individualization of products with decreasing quantities, the companies that use them have a great opportunity to design value-added processes efficiently.
- **Networking and integration**: This includes integration between divisions and departments within a company (vertical integration) and also between different companies (horizontal integration). The goal of digital networking is to improve collaboration, coordination, and transparency across business units and along the supply chain.
- **Decentralization and service orientation**: Decentralization necessitates the modularization of products and processes, decentralized

control and the change to service orientation. The goal is to make increasing complexity manageable, along with coordination, steering, ultimately, so that the decision can be decentralized.
- **Self-organization and autonomy**: Here technologies and processes that carry out an automatic data evaluation are summarized. Based on these results, the systems in the process, such as products and machines, are then to react independently. This should create a closed loop, which leads to a self-configuration and self-optimization of systems. The most important requirement is data analysis and data exchange in real time.

These functional areas provide the informational basis for advanced systems in the sense of Industry 4.0. We will further investigate approaches and technologies which enable identification and traceability in the sense described above. In applying these, there are major differences with regard to product types, production flow, environmental conditions, and industry.

In process industries, non-discrete products, continuous flow with no natural batches, reflux flows, mixing, and intermediate storages make it even more difficult to achieve a high level of traceability in continuous processes (Kvarnström 2008). We will investigate some of these challenges and possible solutions in a case study from the steel industry, specifically focusing on the requirements and restrictions of SMEs. It has to be kept in mind that SMEs are usually able to control only minor segments of the supply chain, so they are not in a position to make their approaches and solutions an obligation for their suppliers and/or customers. Thus, they have to go for pragmatic, easy-to-implement solutions which do not require immense investments of time and finance.

6.2 Background and Literature Review of Identification and Traceability

Kvarnström (2008) distinguishes the concepts of traceability, traceability systems and traceability methods. Traceability is managed by traceability systems (Moe 1998). A traceability system enables traceability in a process by combining process information with data covering the product flow throughout the process. The product flow data can be continuously recorded or modeled with different traceability methods, e.g., individual component stamping to whole lot/batch identification.

In this chapter, we will use identification as the wider, more general term encompassing these concepts. As the traceability method is very often a physical or electronic label or marker, we will refer to this as labeling. The labeling method is the process of applying labels of a certain kind, e.g., by printing, engraving, or laser marking.

6.2.1 Labeling Type and Content

For better understanding, the most widely used types of labeling are briefly explained below, even simple manual, non-electronic labels create a significant basis for the entire system in terms of the virtual representation of reality (Zsifkovits 2013). Normally, the tag contents are different and depending on the code type, more or less information may be included directly in the code. However, the information on a label on the product—no matter what type of code—can be enriched with a unique identification digitally with a variety of other information that results from the real processes. For example, using a standard bar code with limited content, the virtual image of the product could be enriched with additional, digital data, up to mapping the entire life cycle. In every process that manipulates or uses it, more data are captured and stored. Several of these types of labels were developed decades ago in the 1960s; they are still widely used, though.

Plain Text (OCR code): This is the use of a plain text, which can be read by human operators without technical aids. The information content is limited to the available area on the product. The OCR code (Optical Character Recognition) is meant to be easily readable by both machines and humans. There are two types of monospaced (fixed-width) fonts, OCR-A and OCR-B. The OCR-A uses only the uppercase letters of the alphabet, numbers from 0 to 9, and some special characters. The OCR-B may also contain lowercase letters and other special characters (Schulte 2013). With the progress made in character recognition technology and software, virtually any typographic text can be converted, so the standard is not required and not widely used anymore.

1D Code: The 1D code is the widely known "zebra-striped" bar code. This code is the best known and most widely used for goods identification (e.g., the Global Trade Item Number Code (GTIN)). The code is a string of vertical, parallel bars (bars) with different widths and spaces (Koether 2014). This code can be read by scanners or industrial cameras, but the amount of data are quite limited. There are many different types of 1D codes with different information content. Numeric-only bar codes store numbers only, while alphanumeric bar codes contain a combination of numbers and alphabetic characters (letters). The basic architecture of bar codes is shown in Fig. 6.1.

2D Code: According to the defined coding rule, a two-dimensional sequence of dark and light areas within a rectangle is printed. Due to this arrangement, a 2D code can store much more information on the same surface as a 1D code. This code can be read by machine using a scanner or industrial cameras. There are several types of codes, such as QR Code (Quick Response Code) or Data Matrix Code with different information content (Knuchel et al. 2011). These codes can be applied to a component by means of labels, lasers, or embossing methods. The numbers given for storage capacity in Table 6.1 indicate upper limits. As all these codes support variable-length data content, and different error correction levels can be defined, the capacity and symbol size may vary.

Fig. 6.1 Bar code structure (*Source* Adapted from Zsifkovits 2013)

Table 6.1 Comparison of 2D codes (based on Knuchel et al. 2011)

		QR code	Data matrix code	PDF417	Aztec
Type		Matrix	Matrix	Stacked linear bar code	Matrix
Storage Capacity	Numeric	7089	3116	2710	3832
	Alphanumeric	4296	2355	1850	3067
	Binary	2953	1556	1018	1914
	Kanji	1817	778	554	–
Features		High capacity, fast readability, public domain, and free to use	Small footprint, fast readability	Ideally suited for paper prints, number of rows, and columns are configurable	Suited for screens, can still be decoded at low resolution, public domain, and free to use
Application		Marketing, retail, mobile tagging	Logistics, electronics, small items	Transportation, paper-based tickets	Transportation, tickets, boarding passes
Standards		AIM International, ISO/IEC 18004	AIM International, ISO/IEC 16022	AIM International, ISO/IEC 15438	AIM International, ISO/IEC 24778

3D Code: There are some reports in the literature on the development of a 3D code (e.g., Microsoft Research 2007), but there is still no widespread industrial use. Here, by using the third dimension in the form of colors or depth information, the information content can be significantly increased. If a two-dimensional representation (e.g., QR Code) can encode the URL to a picture on the Internet, a 3D code can encode the picture itself. The storage capabilities are approximately 2000 binary bytes, or 3500 alphabetical characters per square inch in its highest density form, using eight colors (Microsoft Research 2007).

RFID: Radio-Frequency Identification tags come in many different shapes; depending on the area of application, they can be different in size, design, and also storage capacity. The storable amount of data depends on the available storage capacity on the RFID tag. RFID systems utilize radio waves and consist of three components: an RFID tag or smart label, an RFID reader, and an antenna. RFID tags contain an integrated circuit and an antenna, which are used to transmit data to the RFID reader which converts the radio waves and transfers data through a communications interface to a host computer system, where the data are stored in a database and analyzed. In contrast to bar codes that require that the scanner to maintain a line-of-sight with each code, RFID is a "near-field" technology, so the scanner only needs to be within the range of the tag to read it. RFID enables "Smart Logistics Zones" as a multiple-use concept of technical systems for identification, localization, and condition monitoring of different object levels in logistics and production processes (Kirch et al. 2017).

6.2.2 Labeling Method

For unique component identification, it is necessary that an identifiable code is used and the component itself is marked with a defined method. The method of labeling depends on the product, application, and the process requirements with regard to the application and environmental conditions.

In practical applications, many different labeling methods or specific marking methods can be found. In particular, in the marking process, it is important that on the selected surface, a mark of sufficient quality and accuracy is applied. This means that the requirements for contrast, dimensional accuracy of the attachment, edge sharpness of the marking, scratch resistance, wipe resistance, and other external influences must be met.

A distinction is made between direct and indirect labeling. In the case of direct marking, the marking is applied directly to the component, while the indirect marking uses another carrier unit for the information (e.g., printed label, RFID tag). This unit is in turn, connected to the component, usually by an adhesive technique. Often direct labeling is slightly worse in terms of quality of readability, especially if various surface treatments occur during production. Nevertheless, there are also a number of advantages (ten Hompel et al. 2008):

- Cost reduction is possible, no additional labels are needed.
- Code generation can be automated.
- The code is inextricably linked to the object.

Direct labeling methods include (Reiter 2017):

Mechanical engraving: By means of electric or pneumatic embossing units (needle embossing or needle scribing), among other things, plain text or 2D codes are applied directly to the surface of the component. Due to the deformation/depression on the surface, this marking method is also applicable to painting processes. However, there are signs of wear in the embossing units.

Laser marking: With the laser, many different characters, symbols, or codes can be applied to a wide variety of surfaces. It is a non-contact marking process in which the material to be inscribed is engraved by the impact of the laser (Müller 2008).

Electrolytic marking: An electrochemical etching process is well suited for applying clean, high-quality light/dark markings to metallic objects. A low-voltage current is passed through a stencil with lettering on the metallic object, thereby the lettering of the template is transferred to the object.

Inkjet printing: This non-contact process uses ink to apply the information. Depending on the field of application, different qualities can be realized.

6.3 Problem Formulation

The labeling of a product makes a connection between the physical material flow and the associated information flow. The focus is on the real processes, the real product flows, and on the improved planning, control, and decision-making along the entire supply chain.

The focus of "digital" labeling is on real-time data generation during production and the continuous analysis of data with regard to a relevant impact on downstream production steps. Digital identification using a unique code and associated data contribute to increase material flow and product quality with a reduced effort to manually capture the state of products and machine tools. At the same time, a defined production sequence can be ensured in order to reduce process costs and avoid errors and defects. An error is an inadvertent mistake caused by a human/an operator, which can result in a defect. Thus, quality is produced the first time and is not "tested" in retrospect (Gerberich 2011). Moreover, this information provides improved traceability and root cause analysis in the event of an internal or external customer complaint.

Traceability can be defined from the perspective of the customer (upstream traceability) or from the processed product (downstream traceability) (Lichtenberger 2016). For upstream traceability, tracing starts from the customer who has received a product, either a final or intermediate product. In the event of a complaint, the product is used to check where the defective component is coming from or where this component was potentially installed. Downstream traceability applies a different view. A product is traced from a defective component through all stages of production. This also includes the customers who were supplied with the final product. The different views are shown in Fig. 6.2.

In most cases, current supply and production processes in industrial companies do not allow tracing of the product and process parameters at the level of the individual components and products. The parameters are only recorded either in isolated or highly aggregated form and are usually only available in batches. In case of serious customer complaints, the violation of security or quality of health standards, often

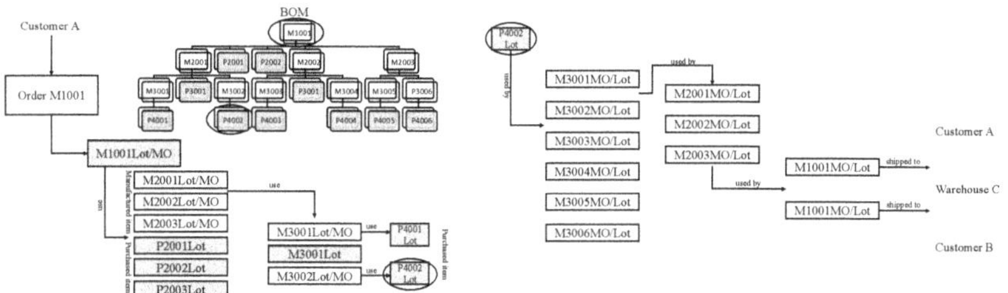

Fig. 6.2 Upstream and downstream traceability (*Source* Adapted from Yuan et al. 2011)

several thousand final products have to be recalled, as the affected production batch might contain a few defective or hazardous products. This in turn, could lead to high costs for repair, penalties, and compensation for enterprises throughout the entire supply chain.

Ultimately, noncompliance with quality requirements can lead to business-critical consequences, such as loss of competitiveness, reduced order intake, litigation due to quality problems, negative media coverage, and resulting loss in shareholder value.

The identification and the understanding of causal relationships in the production process, as well as the future reduction or avoidance of sources of error within industrial production processes thus represent an increasingly important factor for industrial companies as a basis for adaptive quality control.

6.4 Methods/Methodology

6.4.1 Developing a Traceability Model

To achieve digital transparency in a real production system and implement a generic concept, the following domains of the production system have to be taken into account and designed in detail (Reiter 2017):

1. **Material flow**: An analysis of the manufacturing process from the first delivery of the raw materials to the delivery of the finished products is required. In this context, it must be precisely defined when and where

processes are performed, manually, by humans or fully automated by a plant. This applies in particular to transport, conveyance, and goods manipulation.

2. **Production concept**: Within the framework of production, it is important to understand whether entire batches or individual products "flow" through production. This allows the conditions for the generation of the digital image to be worked out. The delivery of goods may be a batch-oriented manipulation and mapping of the batch properties, whereby the properties apply to the entire quantity delivered. The actual unit (e.g., piece, liter, kg) is of less importance. In the manufacturing steps, there may well be different requirements both on batch production and single-part production. Overall, it must be ensured that the reality of production with the associated production steps and the transport containers used is displayed digitally. This can be reflected in the form of a complete batch, a single container, or individual product.

3. **Identification concept**: The labeling method and content must be defined here. At the same time, the infrastructure must be provided for the generation of identification means (e.g., labels, RFID tags), as well as for the collection of static or dynamic information. In this context, it is very important that, in particular, direct types of marking, such as embossed data, are not damaged by the production steps or that their quality is not so impaired that automated identification during production is no longer possible. This decision must be made very early, as the marking should be made on a predefined part of the component. It is therefore advisable to consider this topic within the framework of the production and plant design.

4. **Data model**: When creating the data model, the requirements of the industry, the specific requirements of the customer, and the associated standards must be taken into account. Basically, two levels are relevant, static and dynamic information. Some examples for these are presented in Fig. 6.3 (based on Lichtenberger 2016):

- *Static information*: Data that form the general framework for production, such as the material number of a component, quantity, unit, parts lists and routings, drawings, inspection plans, measuring equipment, and their respective revision levels.

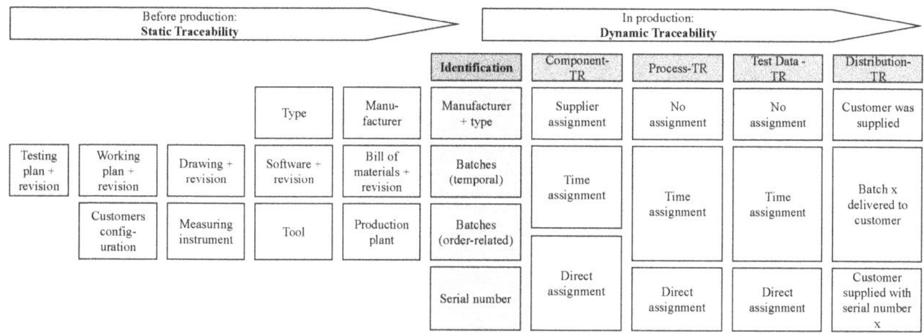

Fig. 6.3 Traceability before and within the production process (Lichtenberger 2016)

- *Dynamic information*: Data that are generated during series production or delivery to the customer and that can be assigned directly to a batch or a component, both in terms of time and content, such as the process data of a plant, results data of an inspection, or delivery information to a customer.

5. **Identification and data generation**: The interaction between static and dynamic information, as mentioned above, must be ensured. When generating data, the existing systems (e.g., ERP—Enterprise Resource Planning, MES—Manufacturing Execution Systems, PLM—Product Lifecycle Management) must be linked to each other, in accordance with the data model. This includes the use of markings on transport containers or individual products. Any manipulation of goods, any changes in product characteristics, or features must be recorded digitally. Depending on the manufacturing step, it can be a manual, partly, or fully automated process.

- *Manual data generation*: Product identification and data collection is done by humans, assisted by technical systems/devices such as scanners, to enable faster and error-free data capture.
- *Fully automated data generation*: In the case of technical production steps in a plant, a large number of process data are generated. These must be digitally connected in real time to the component. Therefore, an automated component identification has to be realized at the plant, and a connection to external systems is required.

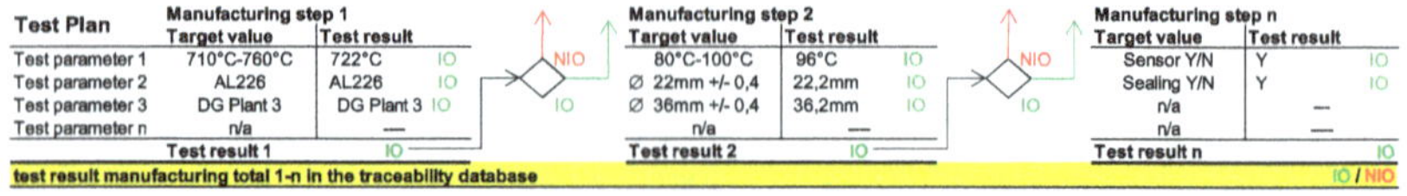

Fig. 6.4 Compressed view on traceability database (Reiter 2017)

6. **Data analysis and real-time control**: There are advanced and efficient technologies for data generation and storage, but the presence of data without context-oriented evaluation and interpretation does not represent any company-related added value. Therefore, it is necessary to process a high volume and large variety of data using models for filtering, aggregating, clustering, analyzing, and visualizing the results and, above all, providing them in real time at the place where they are needed. For this purpose, a traceability database with high availability can be used (see Fig. 6.4). Control can either be handled manually or by autonomous systems that react accordingly.

- *Manual control*: Care must be taken that the results for humans are presented in a simple, comprehensible, and transparent manner at the time of the demand. On this basis, the person either makes decisions or implements the proposed decisions (e.g., the check result is NIO/reject—parts are removed from the process, check result is IO/accept—parts are processed further).
- *Automated control*: The basic principle is the same as explained above; a system must react autonomously, though. In this context, it is important that the component is identified before the next value-adding manufacturing step begins to enable the review of the result of the upstream manufacturing step, so the system can react to it in real time. An example of the basic logic is shown below.

From the domains described above, a generic, hierarchical model for the implementation of digital product labeling is defined. The traceability model describes the relations between process steps, labeling type and method, information systems, and data. It aims to create a virtual image of business-related reality and permanent knowledge of quality-related parameters and results per batch or product to increase quality and reduce error costs.

6.4.2 Traceability Issues in Process Industries

Research literature on traceability is dominated by descriptions of traceability issues in parts production. In the flow of discrete components and products, various kinds of identification markers can be attached to a product or batch and followed, so traceability is usually high.

Process industries very often—at least in some stages of their supply chain—handle bulk material in continuous processes. In transport, a distinction is made between the two classes of goods—unit loads and bulk goods. Unit loads do not change shape during the transport process and can be handled individually. They are identifiable in terms of unit numbers and can be broken down into containers, bales, boxes, machine parts, etc., and their characteristics (dimensions, shape, mass, etc.). Bulk materials are defined as a variety of granular or dusty individual goods with relatively small dimensions, which have a low viscosity and change shape during the transport process. They cannot be formed into a single unit (unitized) without using containers. Bulk density, density, angle of repose, grain size, moisture content, etc., are physical properties of bulk materials (Martin 2014).

Continuous processes are processes where the products are refined gradually and with minimal interruptions through a series of operations (Fransoo and Rutten 1994). In industries using continuous processes, as are commonly found in the paper, food, mining, and steel industries, creating traceability implies major challenges that are rarely addressed in the literature (Kvarnström 2008). In some of these industries, such as mining, the traceability throughout the process from handling the raw materials to the final product is very limited.

Process flows in these industries can be serial, parallel, convergent, divergent, and reflux, often mixed with batch flows. The products are usually non-discrete; they change state and structure in the process, e.g., through chemical treatment or grinding (Fransoo and Rutten 1994).

With an uninterrupted flow, there are no natural batches. In order to achieve a high level of traceability, there has to be a way to divide the product flow into traceable units, like lots or batches.

Reflux flows, mixing operations, and intermediate storages make traceability based on the order of appearance (first-in-first-out principle) often inappropriate.

The methods of material identification commonly used in production and transport chains are only applicable to unitized goods and partly to discretizable and unambiguously determinable bulk goods (Martin 1999).

In recent years, attempts have been made in various industries to track the inhomogeneous, continuous flow of goods through various technological approaches. For example, attempts have been made to detect the bulk material by introducing markers which have identical flow characteristics as the bulk material but at least one easily distinguishable feature (Hötger 2005). In order to achieve a reliable identification of the material, physical properties (color, magnetism, radioactivity), chemical parameters (tracers), or also auto-ID methods (codes, RFID) can be used.

We will address some of these issues in the following part, using the case of an international manufacturer of high-quality steel products.

6.5 A Case—Tracing Continuous Flow in Process Industries

The company focuses on the production and machining of steel bar products that are in the alloyed quality range, rolled or bright, pre-fabricated, surface-treated, or ready-to-install components. They are the basis for high-tech components, camshafts, steering gears, construction and agricultural machinery, diesel injection units, piston rods, and chain pins.

The products are used wherever components are subject to high levels of strain or it is essential that components function safely. They are machined with maximum precision and meet the tight tolerance requirements of extremely demanding sectors, like automotive, engine and plant construction, and specialist applications.

6.5.1 Initial Situation and Project Steps

In the steel processing industry, initiatives toward digitalization and Industry 4.0 are commonly in quite a premature state. Research is limited and the potential applications still need to be researched. There is an awareness of potential gains in productivity, efficiency, flexibility, and competitiveness, though. A lack of competence in these areas, and the harsh conditions prevailing in the industrial environment, such as high dust and dirt levels, noise, and high temperatures, makes the situation more difficult.

In an industrial environment, each individual product or assembly, such as a threaded rod in an automobile, is identified by a specific serial number. In principle, this makes it easy for the end customer to reorder spare parts. However, the serial number is also used in the area of quality assurance in order to be able to initiate specific recall actions in the event of faulty products. Furthermore, by identifying a defective product, targeted process improvement initiatives can be launched within the company or throughout the supply chain. In most cases, the general conditions prevailing in the industry prevent a direct assignment of product defects to the associated product and process parameters. Serial numbers are only used for batch identification, so that traceability at the level of the individual products cannot be carried out.

Improved identification and traceability at the individual product level would be the basis for adaptive quality control and thus generate considerable potential for product and process improvement. In addition to significant cost savings within the supply chain, improved quality can also be achieved.

The project outlined here deals with the development of adaptive quality control based on the continuous identification and traceability of individual products within industrial production processes. This will enable an early detection and avoidance of production errors and defects by a continuous analysis of product and production parameters. We will give a brief overview of the basic processes relevant to the project.

Continuous casting (also called strand casting) is a process whereby molten steel is solidified into a billet (length of metal with a round or square cross section, 30–150 mm square), bloom (produced by a first pass of rolling, cross section 150–400 mm square), or slab (rectangular in cross section) for subsequent rolling in the finishing mills steel processing. These semi-finished casting products are first cut into sections, then are transferred to the rolling mill. Rolling is the process in which metal is passed through one or more pairs of rolls to reduce the thickness and make the thickness uniform. This results in elongation of the workpiece which is consequently cut into rods. These are then bundled, according to customer orders, to enable efficient loading and transport. The process investigated in the project starts with the receipt and storage of billets from various suppliers, before the rolling operation is performed.

In the first project phase, an evaluation of the state of the art of material identification systems was carried out, analyzing the prevailing conditions in the steel processing industry. In a material flow analysis, existing production processes and their interfaces were collected, visualized, and evaluated. Then, based on the knowledge gained, the model of adaptive quality control was developed. The product and process data generated at individual product level will be integrated into a centralized data management system, which still has to be designed and implemented.

The production data transmitted by the suppliers can only be partially used for quality assurance, due to the missing link with the products or with the production process of the processing company. There is also a break in the usage of media within the supply chain (customer, supplier). Data formats are inhomogeneous and incompatible; some information is passed on by paper or fax (see Fig. 6.5).

A synchronized information flow based on data standards would make it possible to retrieve product/production process data from faulty parts afterwards, e.g., temperature, composition, cycle times. The development of an adaptive, learning quality management system (see Fig. 6.6), becoming increasingly effective by constantly recording the data and feeding it into an analysis database, opens up much potential for improvement.

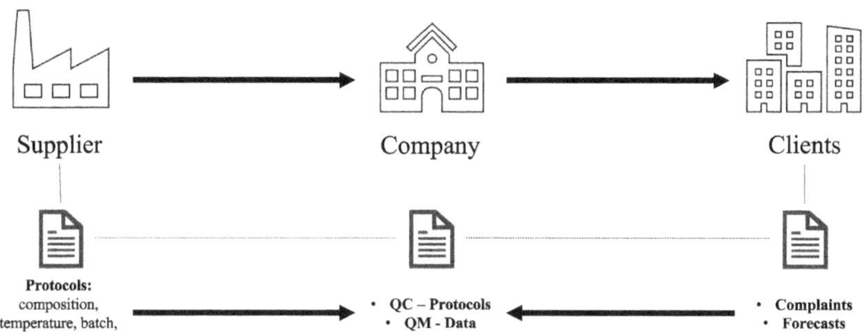

Fig. 6.5 Data flow in the supply chain

As a first step in the project, a process survey for rolling mill/heat treatment was carried out. The physical separation of the input material for the areas of rolling, continuous annealing (conventional heat treatment), and inductive annealing (inductive heat treatment) was documented in a process diagram.

Next, the process survey of the bright steel division was carried out. The analysis concentrated on the areas of peeling, straightening, and testing. In the course of this process survey, it became obvious that there is a risk of mixing products classified as "accept/IO" and "reject/NIO" during separation, sampling, and post-processing (e.g., the hardening process). The findings of this survey were also illustrated in a process diagram.

All the critical transition points and interfaces in the rolling mill process were identified. For full traceability of the manufactured products, a continuous identification in the rolling mill process was required. For this reason, special attention was paid to these processes during the concept generation. The bright steel, with a more linear sequence of process steps, appeared to be less challenging.

Consequently, alternative solutions for identification and traceability were conceived (concepts I to III, and eventually concept IV). These proposals are the result of the process survey, the restrictions identified through this survey, and the discussion with the people involved in the process.

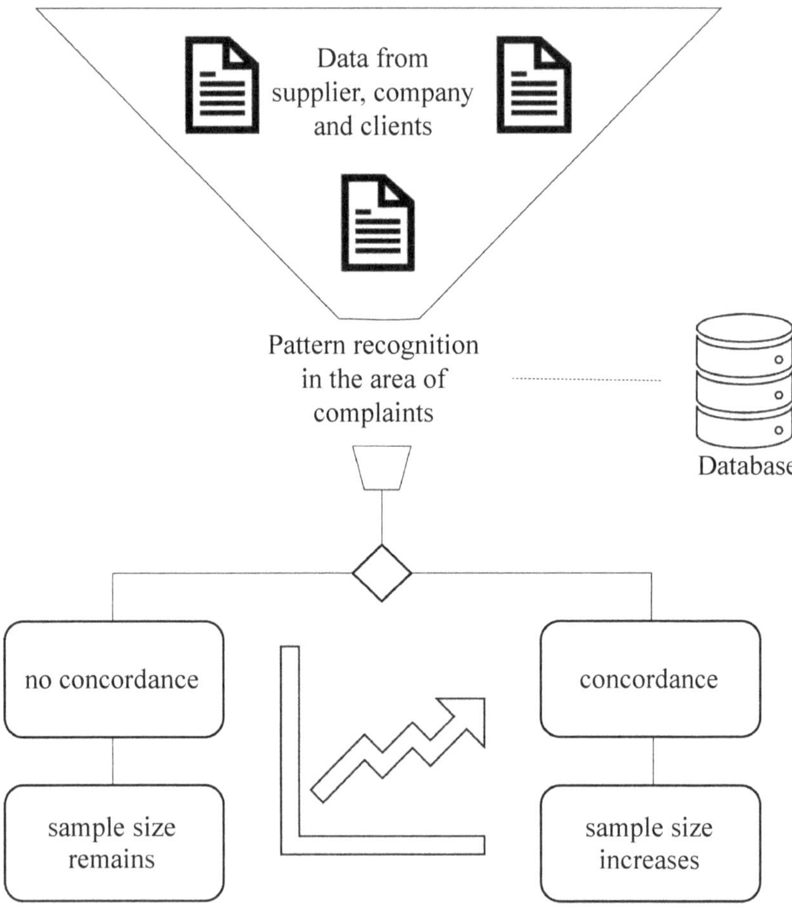

Fig. 6.6 Adaptive quality control

6.5.2 Evaluation of Proposed Solutions

Concept I comprises identification at rod level and uses a combination of bar code, laser marking, and inkjet print on paper.

One way of identifying individual rods is to mark them with heat-resistant bar codes. Seven codes per section are applied or sprayed longitudinally during transport of the rolled wire sections. Matching the conveyor speed with spray unit operations is required for the precise spacing of the codes, so that after another separation, each rod is provided with a bar code. The marking by a spray head is done between the separation of the rolling strands and the cooling process.

As it is possible, and probable, that several billets will be mixed during the formation of a bundle after the separation of the rolling strand sections, it is necessary to apply the markings at this stage. The presence of a spray head, which can apply bar codes generated by a system using heat-resistant paint, is necessary for the implementation of this variant. When the rods are combined to form bundles of approx. five tons of weight, the rods can come from up to ten different strands and, as a result, from up to four different billets. Therefore, it would be possible to assign the individual rods of a bundle to the source billets with only four different bar codes. However, in order to obtain a unique identification at rod level, each rod is marked with a unique code.

Due to the heat-resistant marking, this can be maintained until before the peeling process. After the inductive heat treatment, the code must be scanned, and converted into a QR code which points to the source billet and is applied frontally at both ends of each rod by a laser (face marker). As shown in the following graph, this takes place between the inductive heat treatment and the peeling process:

Before the next cutting process, in which the ends of the rods are removed, the face marker is scanned. All information on the individual rods can now be stored on a standard bar code which is applied to the finished product using an inkjet process. Concept I is illustrated in Fig. 6.7.

Concept II comprises identification at billet level and uses a combination of color coding, laser marking, and inkjet print on paper.

Another option for providing traceability in this process is a combination of color and laser marking, or a combination of color marking and embossing. Color codes are applied after strand separation ("flying shears") and a laser engraving (bar code, QR code) is applied on the front side in the material storage area.

From the survey carried out in the area of the flying shears, a color coding using two colors came out as a possible solution. This can be justified by the fact that for the product under consideration five-ton bundles are produced after the hot separation line. Based on the data, these

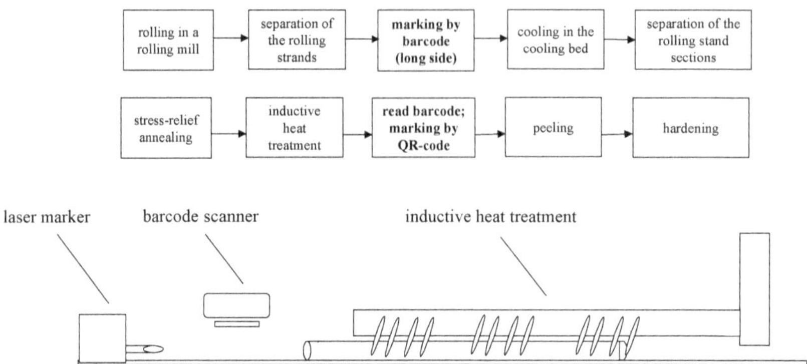

Fig. 6.7 Concept I—bar code and laser marker

bundles consisted of about 400 rods, which corresponds to an approximate number of ten strands or four billets. Therefore, a positive identification of source billets can be achieved by the use of two colors and the permutations possible.

For the further and/or continuous identification in the course of this concept, a separation of the bundles at the new interface (material storage location) is planned. The bundles are opened after arrival and fed to a separating machine.

This machine is equipped with optical recognition which identifies and systemically assigns the respective billets on the basis of the color coding. After the identification has been carried out, a laser engraver or embossing tool is used to apply a unique identification to the front side of the material. In addition to storing relevant data, this allows the sections to be numbered consecutively, so their traceability is clearly guaranteed. The advantage of this procedure is that it ensures identification despite any environmental influences (surface scale, heat). In order to implement this concept, it is not necessary to make any major changes to the existing process, since the frontal marking is retained until the last step of the bright steel process, the separation. After this step, the finished individual rods can be marked for the end customer using ink-jet or paper labeling (see Fig. 6.8).

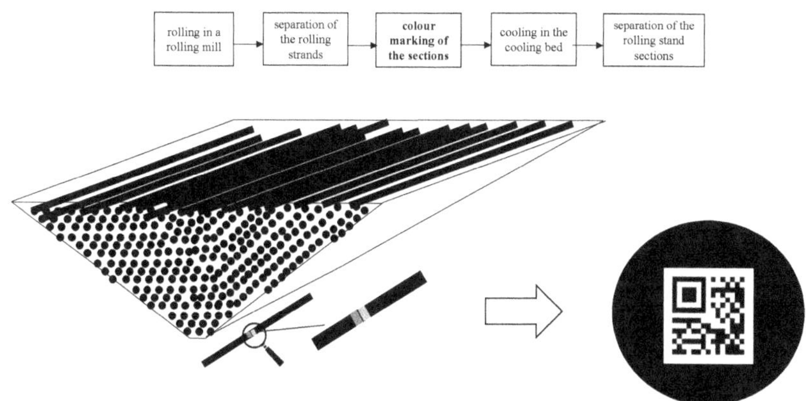

Fig. 6.8 Concept II—color marking

Concept III comprises identification at rod level. This requires a structural modification of the cooling bed and uses laser marking.

This solution aims to prevent the mixing of consecutive rolling strand sections, and requires a physical modification in the production line, between the flying shears and the hot cutting line. A low barrier will prevent the mixing of several rolling strands directly after cutting them to length with the flying shears.

This structural modification, would prevent the mixing of several rolling strand sections. The process-related sequence would ensure a clear allocation of the sections up to the area of the hot separation line. Another modification has to be made in the cooling bed itself. A part of the cooling bed is separated from the rest of the cooling bed by means of a movable cut-off plate, thus ensuring an equal supply to the hot separation line. Hydraulically/pneumatically retractable spacers also ensure the distance between the respective rolling strand sections. The precise positioning of rods also serves the further process step of frontal marking. During the cutting process, the products can be marked on the front by means of laser markers or embossing. By preventing sequence mixing, an unambiguous marking for each individual rod can be guaranteed and assigned to the sequence stored in the work plan.

In order to implement this concept, it is not necessary to make any major changes to the existing process, since the frontal marking is retained until the last step of the bright steel process, the cutting of the rod ends. After this step, the finished individual rods can be marked for the end customer using inkjet or paper labeling. Concept III is illustrated in Fig. 6.9.

The proposed alternative solutions were further evaluated in terms of costs, reliability, and their impact on processes and cycle times. Potential problems were considered, due to environmental influences (surface scaling, mechanical stress, temperature, etc.). The major advantages and disadvantages of the three concepts and some remarks from the discussion are summarized in the following Table 6.2.

The possible problems and barriers to implementation of the previously described concepts resulted in the development of yet another concept, concept IV. The critical interface of the hot separation line resulting from the process evaluation formed the starting point of this concept.

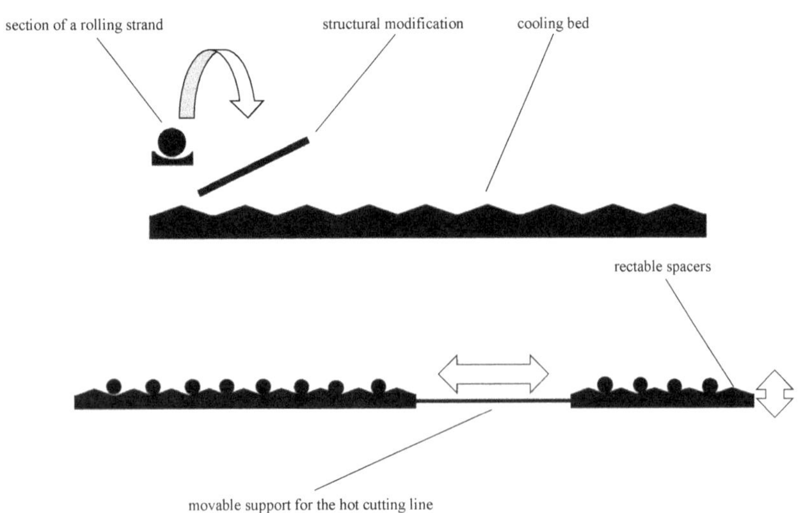

Fig. 6.9 Concept III—modification of the cooling bed

Table 6.2 Comparison of the proposed concepts

	Concept I	Concept II	Concept III
Level of identification	Rod	Billet	Rod
Advantages	• no demand for additional space • low impact on cycle times • flexible change of the production sequence between the processes	• no additional, large area demand • application is not specific for one product only • no change in the lead time, as the existing material flow can be guaranteed by using the buffer times in the material warehouse	• only one marking necessary up to the last step of cross-cutting • application is not specific for one product only • modification of the cycle time can be controlled via the adaptation of the cooling bed
Disadvantages	• Ink/lacquers as additional costs • Possible scaling of the color marking areas • investment costs	• investment costs of separating machine, optical recognition systems, laser coding, and color marking • possible scaling of the color marking area • cost of color/paint • longer cycle times	• investment costs of construction measures for the cooling bed and the installation of the laser marking/print stamping
Notes	• Cooling bed impurities due to color • Color scaling	• Paint will be scraped off in the cooling bed • Separation as a separate workplace • Face of rods is not bare and right-angled • Squeezing through straightening process	• Difficult because of modifications

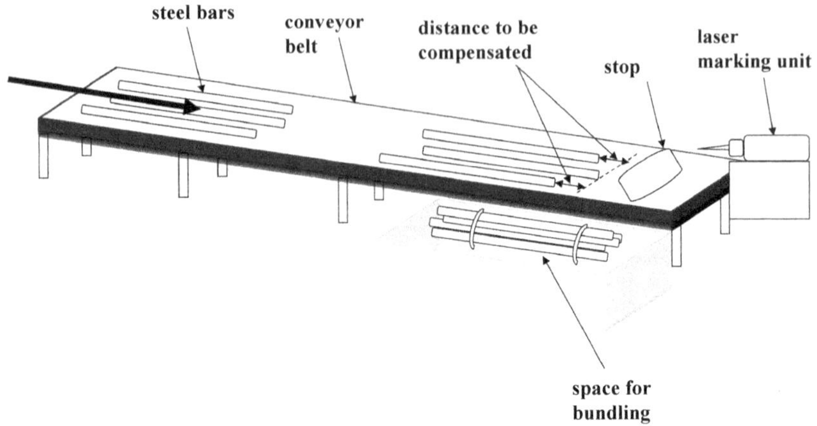

Fig. 6.10 Concept IV—Schematic illustration

This concept is based on the idea of an unambiguous identification of the individual rods through the production sequences stored in the work plan by means of a laser marking device. A mixing of the bars at the cooling bed could only occur in case of a malfunction, so a structural modification was not considered necessary. The workplace for laser marking and bundling has to be redesigned, though. Some of the advantages and limitations of the other concepts are avoided, like scaling. There is no negative impact on cycle times. Investment costs for the marking and bundling workplace are higher, though.

The project partners agreed to test this identification solution under real conditions. This pilot testing was carried out after a market analysis and led to positive results, which further underlined the feasibility of the project. It became clear that several marking devices or laser markers had to be purchased for successful implementation. It also became obvious that a structural modification of the hot cut-off line outlet in the form of a mechanical barrier (stop) was necessary, which, however, did not cause any reduction in cycle time. Concept IV is illustrated in Fig. 6.10.

Further feasibility checks will be carried out, and it will also be clearly established that there is a need for Level II automation for the successful implementation of this project. Level II automation is considered a further step toward Industry 4.0, with an increased ability to react on product-specific parameters during the manufacturing process, to enable autonomous, decentralized decisions at the level of machines and production lines, and to identify areas for process optimization more easily.

6.6 Discussion and Conclusions

The preceding chapter discussed features of intelligent products, machines, and facilities in manufacturing and logistics systems, as objects that are able to provide data for their own, virtual image. Thus, "digital twins" can be established, of products and process flows, to be used for planning and control of the plant. The digital connection between the products and the actual plant enables intelligent automation of the processing steps, and smart. This requires permanent and real-time traceability of objects and their states.

Traceability makes it possible to better control processes and trace product deviations back to root causes within the production process. Improved uniformity in production, and the reduction of product recalls can be named as further benefits from traceability of lots. In a number of industries, such as the aerospace, chemical, pharmaceutical, food, and automotive industries, there is a pronounced demand for traceability. The requirements are regulated in legislation, standards, guidelines, and common practices, to hold paramount the safety and wellbeing of consumers and users of the product, to ensure high levels of quality, and to reduce risk.

To ensure traceability, a structured approach should be used, taking into regard the systems and processes in material flow, production, and identification, and establishing systems and technologies required for data storage, identification and data generation, and data analysis and real-time control.

For object identification, labels and markers of various types (e.g., bar codes, 2D codes) are used, applied with different methods (e.g., print, electronic tags). Digital labeling provides real-time data generation during production and the continuous analysis of data, using unique codes for unambiguous identification of products and their states.

The methods of identification widely used are mostly only feasible for discrete, unitized goods. Many industrial processes, in food processing, paper mills, mining, and steel industries, in contrast, handle bulk material in continuous processes. These materials cannot be unitized, and made identifiable without using containers. Attempts have been made in various industries to track the inhomogeneous, continuous flow of goods through various technological approaches, e.g., introducing markers into the material, or using physical properties, chemical tracers, or auto-ID methods (codes, RFID).

We discussed the problems in product identification and traceability in these processes, and the potential application of methods and standards established in discrete processes. The case of a steel processing company served to illustrate the challenges involved, and how to develop alternative solutions.

Further efforts, both in research and application areas, will be needed to develop effective identification and traceability in these critical sectors of industry and supply chains. Also, standardization and legislation have to keep up with new technologies, their opportunities, and risks. Non-discrete products represent a considerable share of all the products created, especially in the early stages of their value chain. Being able to better control their flow and quality will open up opportunities for improved effectiveness and efficiency in the way we produce and handle products, and for more sustainable operations and material flows.

References

Bischoff, J. 2015. *Erschließen der Potenziale der Anwendung von ‚Industrie 4.0' im Mittelstand.* Studie im Auftrag des Bundesministeriums für Wirtschaft und Energie (BMWi). Mühlheim an der Ruhr: agiplan GmbH.

Fransoo, J.C., and Rutten, W.G.M.M. 1994. A Typology of Production Control Situations in Process Industries. *International Journal of Operations & Production Management* 14 (12): 47–57. https://doi.org/10.1108/01443579410072382.

Gerberich, T. 2011. *Lean oder MES in der Automobilzulieferindustrie: Ein Vorgehensmodell zur fallspezifischen Auswahl.* Wiesbaden: Gabler. https://doi.org/10.1007/978-3-8349-6754-1.

Hötger, G. 2005. *Verfahren zur Kennzeichnung von Schüttgut, derart gekennzeichnetes Schüttgut, sowie Verwendung von Lebens- oder Futtermittel für ein derartiges Schüttgut.* Münster: Habbel & Habbel.

ISO. *ISO 9001:2015 Quality Management Systems—Requirements.*

Kagermann, H., W. Wahlster, and J. Helbig. 2013. *Umsetzungsempfehlungen für das Zukunftsprojekt Industrie 4.0: Abschlussbericht des Arbeitskreises Industrie 4.0.* München: acatech – Deutsche Akademie der Technikwissenschaften e.V.

Kirch, M., O. Poenicke, and K. Richter. 2017. FID in Logistics and Production—Applications, Research and Visions for Smart Logistics Zones. *Procedia Engineering* 178: 526–533. https://doi.org/10.1016/j.proeng.2017.01.101.

Knuchel, T., T. Kuntner, E.C. Pataki, and A. Back. 2011. 2D-Codes: Technology and Application. *Business & Information Systems Engineering* 3 (1): 45–48. https://doi.org/10.1007/s12599-010-0139-z.

Koether, R. 2014. *Distributionslogistik: Effiziente Absicherung der Lieferfähigkeit* (2. Aufl.). Wiesbaden: Springer Gabler.

Kvarnström, B. 2008. *Traceability Methods for Continuous Processes.* Luleå: Luleå University of Technology.

Lichtenberger, S. 2016. *Entwicklung eines Modells zur Darstellung von Traceabilitydaten in Abhängigkeit von Traceabilityanwendungsfällen.*

Martin, H. 1999. *Praxiswissen Materialflußplanung.* Wiesbaden: Vieweg + Teubner Verlag. https://doi.org/10.1007/978-3-322-96885-2.

Martin, H. 2014. *Transport- und Lagerlogistik.* Wiesbaden: Springer Fachmedien Wiesbaden. https://doi.org/10.1007/978-3-658-03143-5.

Microsoft Research. 2007. High Capacity Color Barcode Technology. https://www.microsoft.com/en-us/research/project/high-capacity-color-barcodes-hccb/. Accessed on April 15, 2019.

Moe, T. 1998. Perspectives on Traceability in Food Manufacture. *Trends in Food Science & Technology* 9 (5): 211–214. https://doi.org/10.1016/S0924-2244(98)00037-5.

Müller, H. 2008. Markieren mit dem Laser: Ein schnelles, automatisierbares und robustes Verfahren. *Laser Technik Journal* 5 (2): 45–47. https://doi.org/10.1002/latj.200890008.

Reiter, H. 2017. *Anwendungsmöglichkeit digitaler Produktkennzeichnung zur Erhöhung der Qualität am Beispiel der Automobilzuliefererindustrie.*

Schulte, C. 2013. *Logistik: Wege zur Optimierung der Supply chain* (6. Aufl.). München: Vahlen.
ten Hompel, M., H. Büchter, and U. Franzke. 2008. *Identifikationssysteme und Automatisierung*. Berlin, Heidelberg: Springer.
Yuan, M., H. Yeh, and G. Lu. 2011. The Development of Products Traceability for Enterprise Resource Planning System. 2018 *IEEE 18th International Conference on Industrial Engineering and Engineering Management*, 475–479. https://doi.org/10.1109/ICIEEM.2011.6035203.
Zsifkovits, H.E. 2013. *Logistik*. Konstanz: UVK.

7

State-of-the-Art Analysis of the Usage and Potential of Automation in Logistics

Helmut Zsifkovits, Manuel Woschank, Sakgasem Ramingwong and Warisa Wisittipanich

7.1 Introduction—Automation in Production Logistics

In general, logistics management can be seen as one of the major success factors increasing the competitive advantage of small- and medium-sized enterprises (SME) and industrial enterprises. During recent years, logistics-related technologies have fundamentally changed. They have become more affordable and therefore within reach of SMEs. These technologies assist them in improving their efficiency by

H. Zsifkovits (✉) · M. Woschank
Chair of Industrial Logistics, Montanuniversitaet Leoben, Leoben, Austria
e-mail: helmut.zsifkovits@unileoben.ac.at

M. Woschank
e-mail: manuel.woschank@unileoben.ac.at

S. Ramingwong
Department of Industrial Engineering, Chiang Mai University, Chiang Mai, Thailand
e-mail: sakgasem.ramingwong@cmu.ac.th

D. T. Matt et al. (eds.), *Industry 4.0 for SMEs*,
https://doi.org/10.1007/978-3-030-25425-4_7

using transport management systems (TMS), warehouse management systems (WMS), enterprise resource planning systems (ERP), product lifecycle management solutions, inventory management software, etc. (Inboundlogistics 2018). In this context, a multitude of empirical studies were able to demonstrate significant positive effects of improvement initiatives in logistics on various performance measures (e.g., costs, delivery times, quality, and/or flexibility).

Significant positive effects were investigated by Agus and Hajinoor (2012) on product quality performance and business performance, by Birou et al. (2011) on financial performance, by Chen et al. (2007) on firm performance, by Danese and Kalchschmidt (2011) on operational performance, by Juga et al. (2010) on service quality and loyalty, and by Spillan et al. (2013) on firm competitiveness.

Thereby, logistics management research and practical applications have isolated a multitude of success factors, which can be used to design and improve a logistics system. Besides basic principles, e.g., organizational measures, improved planning heuristics, flow orientation, process alignment, and product design, automation can be identified as one of the major opportunities for logistics to improve overall performance and competitiveness.

In this context, automation and robotics in logistics systems was ranked as one of the most important megatrends for logistics, among others, e.g., data analytics and artificial intelligence, autonomous trucks, unmanned aerial vehicles, cloud computing, and blockchain technology (SCI Verkehr 2018). Moreover, Tractica (2012) has forecast that the potential sales of service robots in logistics will rise to US$31,910 billion in 2020. This will also increase the necessity for a redesign of jobs in production and logistics (LivePerson 2018) and a realignment of the professional working environment.

W. Wisittipanich
Center of Excellence in Logistics and Supply Chain Management, Chiang Mai University, Chiang Mai, Thailand
e-mail: warisa.w@cmu.ac.th

In conclusion, automation in logistics holds considerable potential and great opportunities for performance-enhancing incentives. According to Groover (2008), the reasons for implementing automated processes can be summarized as follows:

- Increased labor productivity
- Lower labor costs
- Mitigation of the effects of labor shortage
- Reduction and/or elimination of routine manual and clerical tasks
- Improved workplace safety
- Improved product quality
- Reduced lead time
- The accomplishment of processes that cannot be done manually, and
- The avoidance of high costs in comparison to manual processes.

Thereby, the automation of material flow processes in logistics systems depends on the integration of information and communication technologies, the compatibility of hardware and software, standardized interfaces, modular designed systems, consistent storage of information, and interoperable hardware and software (Krämer 2002). Moreover, modern automation is highly dependent on state-of-the-art identification technologies and technological concepts, which will be briefly outlined in the next section.

7.2 Problem Formulation and Methods/ Methodology

The importance of automation approaches in logistics systems is mostly recognized by larger companies, mainly in the industrial environment. Unfortunately, small- and medium-sized companies still lack knowledge regarding the effects, state-of-the-art technologies, and the implementation of automation concepts.

Therefore, this chapter systematically discusses studies that investigate the effects of automation in logistics systems on various performance measures. The authors present a structured analysis of enablers of automation in logistics systems by focusing on both, identification technologies and technological concepts for automation. Moreover, the authors present recent developments in automation (e.g., agent-based automation through enhanced process control, automated guided vehicles and robots in logistics systems, conveyor belts and sorting systems, automation through augmented reality [AR]) and introduce a case study of automation by using conveyor belts and sorting systems in an SME in Thailand.

7.3 Enablers of Automation in Logistics

The successful implementation of automation in logistics is dependent on a variety of organizational, procedural, technological, and socio-economic success factors. In this context, recent literature has developed a multitude of partial and divergent frameworks, models, and conceptualizations which should be used to support the efficient implementation, continuous operation, and further development of digitalization strategies of Industry 4.0 initiatives. However, existing conceptualizations can still be regarded as unspecific, because of a missing holistic approach and/or a missing unambiguous classification (Zsifkovits and Woschank 2019).

In this context, the authors will further discuss identification technologies and technological concepts for automation as one of the main prerequisites for enhanced material and information flow processes in logistics.

7.3.1 Identification Technologies for Automation

The successful implementation of automated logistics processes depends on the consistent identification, tracking and tracing of raw materials, semi-finished components, and finalized goods.

One prerequisite of automation is that the products in logistics systems constantly contain all necessary information. Therefore, state-of-the-art identification technologies should be used to ensure clear identification and constant tracking within the entire supply, production, and distribution process.

Moreover, modern enterprises try to implement smarter products which are based on product-enabled information devices (PEID), such as RFID, sensors, actors which allow the interoperability of systems by dynamically exchanging product data and additional in-depth information regarding the lifecycle management (Kiritsis 2011) and further process-relevant, real-time information.

Another often-recognized trend is the shifting from identification on batch level to identification on product level. This is forced by market requirements (e.g., product liability laws), efficiency initiatives (e.g., lower number of recalls in case of possible product errors), and enhanced planning strategies (e.g., lower quality management incentives based on a better understanding and control of the production process).

Furthermore, environmental conditions in production facilities (e.g., changes in temperature, heat, dust, changes due to surface treatments) lead to enhanced requirements for new identification technologies. In most cases, the identification on product level will require a combination of different identification technologies because the surface condition of the products will constantly change during the production process.

In the next section, the authors outline the most important identification technologies as enabling factors for automation in logistics. Direct labeling technology identifies the material directly without any additional tools. Indirect labeling uses additional code carriers, e.g., various labeling technologies or RFID tags (ten Hompel et al. 2008).

Direct labeling can be realized by using lasers (McKee 2004), lasers on painted ground layers (InfoSight Corporation 2017), direct printing with inkjet technology (ten Hompel and Schmidt 2005), marking by needle printing (Seegert 2011), and labeling by stamping devices (Henning and Müller 2001).

Indirect labeling is established by using thermal-printed labels (Drews 2008; ten Hompel et al. 2008), sheet metal labels (Henning and Müller 2001), and/or radio-frequency identification (RFID) devices (Finkenzeller 2015).

It is important to note that research into automation tries to further develop both direct and indirect labeling technologies, in order to ensure better identification, particularly in harsh industrial environments. For example, printed labels nowadays are able to resist temperatures up to 1100 °C (Alpine Metal Tech GmbH 2019) and RFID tags can now be used on metallic materials without interferences (Feinbier et al. 2011).

An RFID tag for metals can be used without interference of the RFID signal. Furthermore, research is constantly developing new technologies for the identification and the continuous tracking and tracing of bulk material (Weichbold and Schuster 2017). For further elaboration of identification and traceability, we refer to Sect. 7.3.

7.3.2 Technological Concepts for Automation

Moreover, automation depends on the implementation of new technological concepts. In this section, the most advanced concepts of CPS, IoT, and PI will be briefly outlined and their potential application in automation will be discussed (Zsifkovits and Woschank 2019).

Cyber-Physical Systems (CPS) are physical objects or structures, such as products, devices, buildings, means of transport, production facilities, and/or logistics components, that include embedded systems in order to ensure interactive communication (Bauernhansl et al. 2014). The systems are connected through local and global digital networks (Broy 2010) by using sensors and actors in closed control loops (Lee 2010). CPS detect, analyze, and capture their surrounding environment by using sensors combined with available information and services. Moreover, actors are used to interact with physical objects. CPS act autonomously, in a decentralized way, can easily build up networks among themselves, and can independently optimize themselves according to the principles of self-similar fractal production systems. The Smart Factory interacts with human resources and/or machines and is able to organize itself in a decentralized, real-time way (Bauernhansl et al. 2014). A virtual image of the real production environment is permanently analyzed and updated with real-time information. Therefore, the virtual environment, often entitled the "digital twin" is always synchronized with information from the real environment. This can be seen as the starting point in order to connect the Internet of Mankind to the IoT and to the Internet of Services (IoS) (Padovano et al. 2018).

Internet of Things (IoT) can be seen as an essential and important part of the CPS which is often associated with RFID technologies. Thereby, the IoT is used to identify and track objects (e.g., products, container, machines, vehicles) in logistics systems and supply chains. The objects are constantly processing information from their surrounding environment and can be unambiguously allocated, which increases the effectiveness and efficiency of all related monitoring and control processes (Boyes et al. 2018; Borgmeier 2017).

Physical Internet (PI) is an open, standardized, worldwide freight transport system based on physical, digital, and operative interconnectivity by using protocols, interfaces, and modularization. A provider-free, industry-neutral, and border-free standardization is one of the basic requirements for the usage of the PI which connects and virtualizes material flows, in analogy to the concept of the digital internet. Moreover, standardized containers and carriers are used to ensure maximum utilization of transport vehicles and a better usage of spare capacities. These principles can be applied to internal logistics systems, as well as transportation networks by using self-controlling, autonomous systems in transport and storage processes as one of the central elements of the PI. The usage of shared transport capacities, storage locations, hubs, and delivery points will have a positive effect on both economic (e.g., short transportation times, lower costs of human resources) and ecological (e.g., reduction of traffic and emissions) effects (Montreuil 2011; Pan et al. 2017).

7.4 Discussion of Automation Approaches

In general, there is a multitude of opportunities for implementing automation concepts in logistics systems in order to improve the overall efficiency. In this context, based on a systematic literature review, Granlund

(2014) has clustered and summarized the most commonly occurring applications and types of mechanized automation as follows:

- Automated loading and unloading systems
- Automated guided vehicles (AGVs)
- Automated storage and retrieval systems (AS/RS)
- Automatic fork-lift trucks for mechanized handling
- Various types of carousels, conveyor belts, and conveyor-based sorting systems
- Industrial robots/robotics
- Item-picking devices
- Lift and turntables/aids
- Linear actuators
- Mechanized palletizing
- Moving decks and screening and/or sorting systems.

In the next section, we will discuss the most promising developments for automation, namely, agent-based automation through enhanced process control, automated guided vehicles and robots in logistics systems, conveyor belts and sorting systems, automation through AR, and automation through modularization strategies in SME.

7.4.1 Agent-Based Automation Through Enhanced Process Control

In logistics systems, agent-based automation strategies can be used for the self-organization of the material flow process. Thereby, agents communicate autonomously by constantly transferring information about targets, system conditions, and occurring restrictions (Gudehus 2012). For example, agent-based-controlled AGVs resulted in 25.5% lower processing times, 7.9% reduced traveled distances, and 2.4 2% lower empty runs, which leads to a more robust and more flexible logistics system (Ullman and Sauer 2013).

7.4.2 Automated Guided Vehicles (AGVs) and Robots in Logistics Systems

In general, AGVs are becoming more and more important for industrial companies and for SME. Thereby, Mehami et al. (2018) have identified the ability of reconfiguration, flexibility, and customizability, as the main success factors for effective AGV implementation. In addition, warehousing processes are optimized by the usage of autonomous robots, intelligent carriers, and advanced assistance systems for man–machine interaction (Glock and Grosse 2017).

7.4.3 Conveyor Belts and Sorting Systems

Nowadays, conveyor systems are quite easy to implement. The selection of the conveyor system is dependent on the product that needs to be moved, on the available space, and the space needed for further operations. Conveyors can be located on the ground and/or positioned overhead and can be integrated by using sensors and actors.

Thereby, the machines, devices, systems, and products have the capability to connect with each other without any human intervention (McGuire 2009; Jeschke et al. 2017). Conveyor systems are not as flexible as AGVs, but with frequent transport tasks, the conveyor system is a good solution for automation due to their mechanical simplicity, reliability, and ability to transfer materials very efficiently and flexibility (Greenwood 1988).

7.4.4 Automation Through Augmented Reality (AR)

In the automation of logistics processes, AR is used to enable man–machine interactions through the integration of real and virtual information by using cameras, smartphones, tablets, AR helmets, and data glasses (Jost et al. 2017).

AR is often used to support picking operations. In this context, the pick-by-vision technology can lead to an enhanced picking efficiency

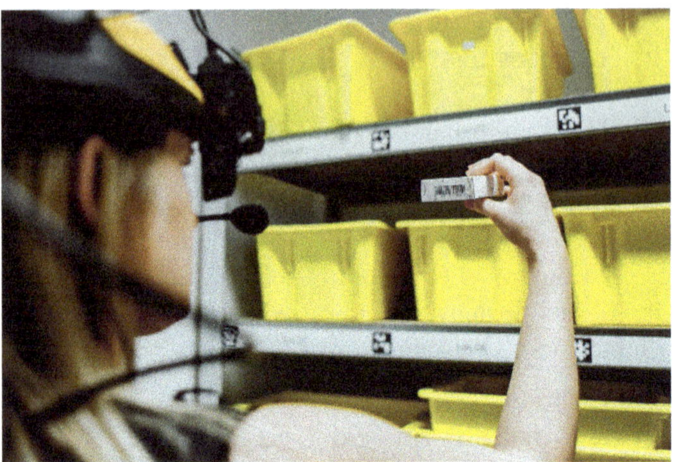

Fig. 7.1 Picking processes by using augmented reality devices (Reproduced with permission from University of Leoben, Chair of Industrial Logistics)

(of up to 25%) by simultaneously eliminating almost all picking errors (Deutsche Post AG 2015). This approach was also transferred to packaging operations leading to higher usage of packaging space (up to 19% more) and lower costs (−30%) (Mättig et al. 2016).

Figure 7.1 displays pilot implementation of a picking process using an AR device (AR helmet) which could also be transferred to the environment of SME.

7.4.5 Automation Through Modularization Strategies

The automation of packaging processes will gain enormously in importance. In this context, researchers are developing modular load carrier strategies to support automation in logistics systems.

In this context, packaging processes are becoming more and more important. Various researchers are developing modular load carrier strategies in order to support automation in logistics systems. Thereby inefficiencies in transport, quality control, administration, and maintenance should be avoided and sustainability should be increased through new pooling and sharing concepts (Zsifkovits and Woschank 2019).

7.5 A Case—Conveyor Belts and Sorting Systems Case Study: Medium-Sized Logistics Service Provider in Thailand

The case study company is a local logistics service provider (LSP) in Northern Thailand. The company provides port-to-door services for customers in Northern Thailand and in the Bangkok metropolitan area. The company uses semi-trailer trucks to deliver goods from their drop-point distribution center (DC) to the destination DC. Then the goods are cross-docked and delivered to the consignees' door by small trucks. Thereby, the range of goods includes construction materials, textile and garment, paper, automotive parts, food and snacks, vegetables, and flowers. Most of the customer's requirements are lesser than a full truckload.

The size, weight, and dimension of goods can vary in a wide range. The packaging also comes in different sizes due to the nature of the products. For examples, snacks come in light boxes. Fabric rolls are often heavy and long. Tyres are individually wrapped up in plastics or packing paper (Fig. 7.2).

The investigated logistics operation is cross-docking where the goods must be unloaded from trucks to the cross-dock area in DC. Today, DC operators manually pick up the goods by hand from the delivery truck and put them on the vertical conveyor belt. Then the goods will

(a) Heavy and Long Fabric Rolls

(b) Light Snack Box

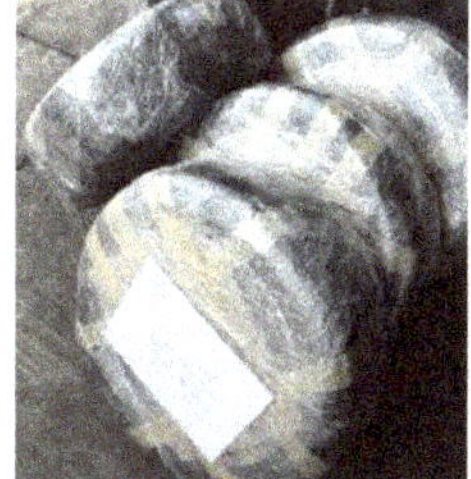

(c) Packed Second handed Tyres

Fig. 7.2 Example of delivered goods (DG) (Reproduced with permission from University of Chiang Mai, Department of Industrial Engineering)

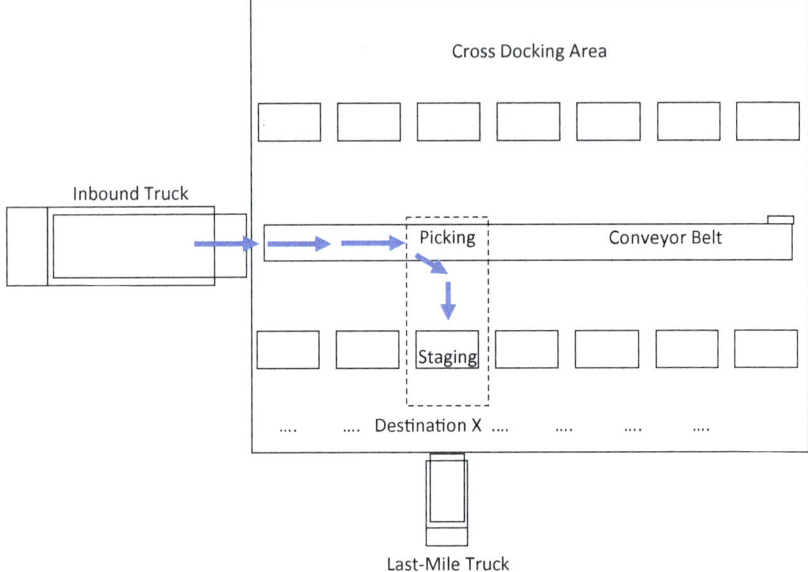

Fig. 7.3 Cross-docking operation

be sorted aligned to their designated destination. The conveyor belts are straight and approximately 20 meters long. There are 14 destinations of last-mile delivery aligned along the belt (Fig. 7.3).

If the good, which belongs to destination X, arrives at the picking area of destination X, the last-mile operator of destination X will pick up the good from the belt and transfer to destination X staging area, where the last-mile delivery will be arranged.

At this picking area, with 14 possible destinations, the goods are identified by the packing label where the picker must recognize the address and determine if it belongs to his/her destination. The identification is manual and requires experiences of picker in order to accurately pick the right goods according to their designated destination.

Today, at this case study DC only, there are more than 30 trucks hauling from 6 Bangkok DCs with more than 10,000 pieces of goods per normal working day. At present, Barcode is embedded with the parcel label. However, it is used only to cross-check the goods with the database once staged. It is not used for AutoID sorting.

The process requires one picker per destination and 2 more DC operators to stage and manage the last-mile delivery per destination. Therefore, at the conveyor belt, there are 42 workers, excluding the DC manager and 3 more unloading workers at the truck. The process is labor-intensive and hence expensive.

The company is now suffering with the labor cost and productivity. Moreover, operator turnover is high due to the hard work and working condition. Staffing becomes more difficult. Thus, the concept of Industry 4.0 of using conveyor belts and sorting systems is considered.

According to expert consultancy under Industry 4.0 scheme, the company was suggested the following potential improvements:

(a) To use inclining conveyor belt to assist truck unloading. This will increase the speed of unloading and reduce the labor cost in loading. The initial investigation suggests that the Payback Period is less than 6 months on the equipment investment (Fig. 7.4).

Fig. 7.4 Truck Loading/unloading conveyor systems (Reproduced with permission from University of Chiang Mai, Department of Industrial Engineering)

Fig. 7.5 Cross belt sorter (Reproduced with permission from University of Chiang Mai, Department of Industrial Engineering)

(b) To use closed-loop conveyor. This will help any bottleneck when picking and unloading from the belt. Often, there are many items to be picked at one destination. Then the conveyor belt must be shut to allow picker to pick all items (Fig. 7.5).
(c) To use cross belt auto sorting system and the declined roller belt. Where barcode is already embedded with the database, the item can be transferred using closed-loop conveyor. They can be auto-sorted. This will reduce picker load. The (b) and (c) proposals yield the Payback Period of 20–24 months. The cross-docking capacity is expected to increase by 20–30%. Labor productivity should increase by 33%. The accuracy should also increase (Fig. 7.6).

The suggestion is only preliminary. Further investigation must be made in terms of item compatibility (extra-large or out-of-shape items), equipment maintenance, facility layout, etc. The company now requests for quotation from the equipment providers for exact equipment specification and cost.

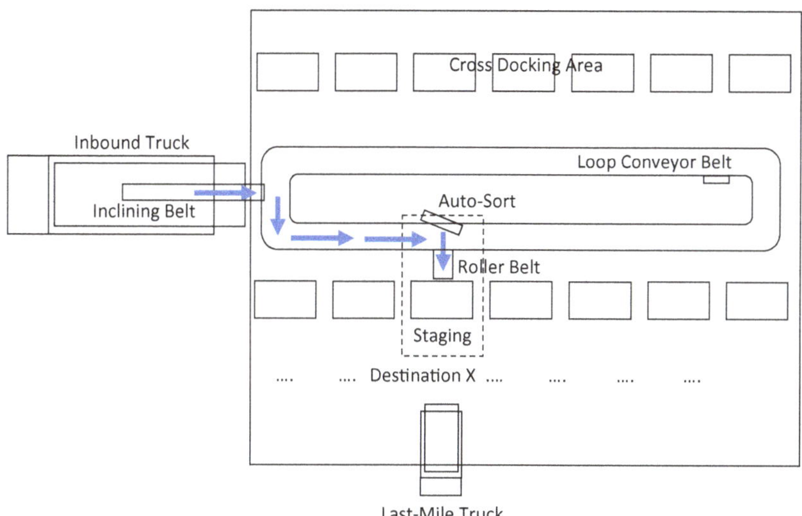

Fig. 7.6 Improved cross-docking operation

7.6 Discussion, Suggestions, and Implications

Logistics is among those areas strongly affected by the upcoming technologies associated with Industry 4.0, both as an opportunity and a risk. Above all, logistics is still the imperative for providing a superior service level at feasible costs.

Automation and IT are not applied for their own sake, to be consistent with market trends, or to satisfy expectations. Automation can help to increase productivity, to lower costs, to gain flexibility, to make routine tasks more efficient, to provide workplace safety, to reduce lead times and time to market, and to improve product and service quality.

In particular, there is a huge challenge for SME to keep up with this development. They often lack the resources, and the competences, to make use of opportunities.

The authors investigated numerous literature sources on automation and information technology in logistics. The approaches and frameworks mostly focus on selected, limited aspects and research is fragmented and inhomogeneous.

A number of basic concepts can be identified which define smart logistics within the framework of Industry 4.0. CPS, IoT, and the PI are high-level, sophisticated concepts that cannot be implemented by just buying some service or software. A step-by-step approach has to be applied, for every enterprise, much more so for SME.

Existing CNC machine tools can be connected and the majority of machines support open communication standards, such as MTConnect or OPC UA, as standards for communication and information modeling in automation. There are constraints to observe, legacy systems, technical, organizational, and financial restrictions.

Further research will be needed, with a focus on SMEs, to further investigate lower-level approaches, and technical as well as organizational solutions to satisfy the functional requirements.

References

Agus, A., and M. Shukri Hajinoor. 2012. Lean Production Supply Chain Management as Driver Towards Enhancing Product Quality and Business Performance. *International Journal of Quality & Reliability Management* 29 (1): 92–121. https://doi.org/10.1108/02656711211190891.

Alpine Metal Tech GmbH. 2019. https://www.alpinemetaltech.com/produkte_services/detail/product/60/.

Bauernhansl, T., M. ten Hompel, and B. Vogel-Heuser (eds.). 2014. *Industrie 4.0 in Produktion, Automatisierung und Logistik: Anwendung, Technologien, Migration*. Wiesbaden: Springer. http://dx.doi.org/10.1007/978-3-658-04682-8.

Birou, L., R.N. Germain, and W.J. Christensen. 2011. Applied Logistics Knowledge Impact on Financial Performance. *International Journal of Operations & Production Management* 31 (8): 816–834. https://doi.org/10.1108/01443571111153058.

Borgmeier, A. 2017. *Smart Services und Internet der Dinge: Geschäftsmodelle, Umsetzung und Best Practices: Industrie 4.0, Internet of Things (IoT), Machine-to-Machine, Big Data, Augmented Reality Technologie*. München: Hanser. https://doi.org/10.3139/9783446452701.

Boyes, H., B. Hallaq, J. Cunningham, and T. Watson. 2018. The Industrial Internet of Things (IIoT): An Analysis Framework. *Computers in Industry* 101: 1–12. https://doi.org/10.1016/j.compind.2018.04.015.

Broy, M. (ed.). 2010. *Cyber-Physical Systems: Innovation Durch Software-Intensive Eingebettete Systeme*. Berlin and Heidelberg: Springer. https://doi.org/10.1007/978-3-642-14901-6.

Chen, H., D.D. Mattioda, and P.J. Daugherty. 2007. Firm-Wide Integration and Firm Performance. *The International Journal of Logistics Management* 18 (1): 5–21. https://doi.org/10.1108/09574090710748144.

Danese, P., and M. Kalchschmidt. 2011. The Role of the Forecasting Process in Improving Forecast Accuracy and Operational Performance. *International Journal of Production Economics* 131 (1): 204–214. http://doi.org/10.1016/j.ijpe.2010.09.006.

Deutsche Post AG. 2015. https://www.dpdhl.com/de/presse/pressemitteilungen/2015/dhl-testet-augmented-reality-anwendung.html.

Drews, K., in: Norbert Bartneck, V. Klaas, H. Schönherr, and M. Weinländer (eds.). 2008. *Prozesse optimieren mit RFID und Auto-ID: Grundlagen, Problemlösungen und Anwendungsbeispiele*. Erlangen: Publicis Corporate Publishing.

Feinbier, L., Y. Yaslar, and H. Niehues. 2011. RFID-Brammenlogistik. techforum.

Finkenzeller, K. 2015. *RFID-Handbuch*, 7th ed. München: Carl Hanser Verlag.

Glock, C., and E. Grosse (eds.). 2017. *Warehousing 4.0: Technische Lösungen und Managementkonzepte für die Lagerlogistik der Zukunft* (1. Aufl.). Lauda-Königshofen: B + G Wissenschaftsverlag.

Granlund, A. 2014. *Facilitating Automation Development in Internal Logistics Systems*. Västerås: School of Innovation, Design and Engineering, Mälardalen University. http://doi.org/10.1504/IJLSM.2014.063984.

Greenwood, N.R. 1988. *Implementing Flexible Manufacturing Systems*. London, UK: Macmillan Education.

Groover, M.P. 2008. *Automation, Production Systems, and Computer-Integrated Manufacturing*, 3rd ed. Upper Saddle River, NJ: Pearson/Prentice Hall.

Gudehus, T. 2012. *Dynamische Disposition: Strategien, Algorithmen und Werkzeuge zur optimalen Auftrags-, Bestands- und Fertigungsdisposition* (3. Aufl.). Berlin: Springer. http://dx.doi.org/10.1007/978-3-642-22983-1.

Henning, A., and U. Müller. 2001. *Entwicklung eines maschinenlesbaren Kennzeichnungssystems für Brammen und Knüppel: Abschlußbericht*. Luxembourg: Off. for Off. Publ. of the Europ. Communities.

Inboundlogistics. 2018. https://www.inboundlogistics.com/cms/article/2018-top-100-logistics-it-market-research-survey/.

InfoSight Corporation. 2015. http://www.infosight.com/support/white-papers/68-bar-code-identification-of-hot-steel.

Jeschke, S., C. Brecher, H. Song, and D.B. Rawat (eds.). 2017. *Industrial Internet of Things*. Cham: Springer.

Jost, J., T. Kirks, B. Mättig, A. Sinsel, and T.U. Trapp. 2017. *Der Mensch in der Industrie – Innovative Unterstützung durch Augmented Reality: Handbuch Industrie 4.0*, Bd. 1. Heidelberg: Springer.

Juga, J., J. Juntunen, and D.B. Grant. 2010. Service Quality and Its Relation to Satisfaction and Loyalty in Logistics Outsourcing Relationships. *Managing Service Quality: An International Journal* 20 (6): 496–510. https://doi.org/10.1108/09604521011092857.

Kiritsis, D. 2011. Closed-Loop PLM for Intelligent Products in the Era of the Internet of Things. *Computer-Aided Design* 43 (5): 479–501. https://doi.org/10.1016/j.cad.2010.03.002.

Krämer, K. 2002. *Automatisierung in Materialfluss und Logistik: Ebenen, Informationslogistik, Identifikationssysteme, intelligente Geräte*. Wiesbaden: Deutscher Universitäts-Verlag.

Lee, E.A. 2010. CPS Foundations. In *Proceedings of the 47th Design Automation Conference*, 737. New York, NY: ACM.

LivePerson. 2018. https://de.statista.com/statistik/daten/studie/863407/umfrage/umfrage-zu-drohenden-jobverlusten-durch-automation-in-verschiedenen-branchen/.

Mätttig, B., T. Kirks, and J. Jost. 2016. Untersuchung des Einsatzes von Augmented Reality im Verpackungsprozess unter Berücksichtigung spezifischer Anforderungen an die Informationsdarstellung sowie die ergonomische Einbindung des Menschen in den Prozess. In *Logistics Journal: Proceedings*.

McGuire, P.M. 2009. *Conveyors: Application, Selection, and Integration*. Boca Raton, FL: CRC Press.

McKee, T., in: Colin E. Webb, and J.D.C. Jones. 2004. *Handbook of Laser Technology and Applications: Applications*. Bristol: Institute of Physics Publishing.

Mehami, J., M. Nawi, and R.Y. Zhong. 2018. Smart Automated Guided Vehicles for Manufacturing in the Context of Industry 4.0. *Procedia Manufacturing* 26: 1077–1086. https://doi.org/10.1016/j.promfg.2018.07.144.

Montreuil, B. 2011. Toward a Physical Internet: Meeting the Global Logistics Sustainability Grand Challenge. *Logistics Research* 3 (2–3): 71–87. https://doi.org/10.1007/s12159-011-0045-x.

Padovano, A., F. Longo, L. Nicoletti, and G. Mirabelli. 2018. A Digital Twin Based Service Oriented Application for a 4.0 Knowledge Navigation in the Smart Factory. *IFAC-PapersOnLine* 51 (11): 631–636. https://doi.org/10.1016/j.ifacol.2018.08.389.

Pan, S., E. Ballot, G.Q. Huang, and B. Montreuil. 2017. Physical Internet and Interconnected Logistics Services: Research and Applications. *International Journal of Production Research* 55 (9): 2603–2609. https://doi.org/10.1080/00207543.2017.1302620.

SCI Verkehr. 2018. https://de.statista.com/statistik/daten/studie/980502/umfrage/megatrends-in-der-logistik-in-deutschland/.

Seegert, S. 2011. Präge- und Stempelschrifterkennung auf metallischen Oberflächen. Kölner Beiträge zur Technischen Informatik.

Spillan, J.E., M.A. McGinnis, A. Kara, and G. Liu Yi. 2013. A Comparison of the Effect of Logistic Strategy and Logistics Integration on Firm Competitiveness in the USA and China. *The International Journal of Logistics Management* 24 (2): 153–179. https://doi.org/10.1108/IJLM-06-2012-0045.

ten Hompel, M., H. Büchter, and U. Franzke. 2008. *Identifikationssysteme und Automatisierung*. Berlin and Heidelberg: Springer.

ten Hompel, M., and T. Schmidt. 2005. *Warehouse Management: Automation and Organisation of Warehouse and Order Picking Systems*. Berlin and New York: Springer.

Tractica. 2012. https://de.statista.com/statistik/daten/studie/870614/umfrage/prognostizierter-umsatz-mit-servicerobotern-weltweit-nach-bereichen/.

Ullmann, G., and J. Sauer. 2013. Dezentrale, agentenbasierte Selbststeuerung von Fahrerlosen Transportsystemen (FTS) (Schlussbericht, IPH – Institut für Integrierte Produktion).

Weichbold, C., E. Schuster, in: H.E. Zsifkovits, and S. Altendorfer-Kaiser (eds.). 2017. *Entwicklung einer Methodik zur Verfolgung von kontinuierlichen, inhomogenen Materialflüssen und deren Eigenschaften (SmartSinter)*. Mering: Rainer Hampp Verlag.

Zsifkovits, H., and M. Woschank. 2019. Smart Logistics—Technologiekonzepte und Potentiale. *BHM Berg- Und Hüttenmännische Monatshefte* 164 (1): 42–45. https://doi.org/10.1007/s00501-018-0806-9.

Part IV

Industry 4.0 Managerial, Organizational and Implementation Issues

8

Development of an Organizational Maturity Model in Terms of Mass Customization

Vladimír Modrák and Zuzana Šoltysová

8.1 Introduction

Production technologies are, these days, mostly affected by dynamic development of information technology and automatic identification technologies. Obviously, technological changes are driven by many factors such as increasing requirements of individual customers, safety and environmental standards, social demands, the diffusion of disruptive innovations, and so on. In general, technology is changing very rapidly and the newest technological developments are reshaping the manufacturing sector in its original form. For example, additive manufacturing, cloud computing, radio frequency identification, fifth-generation wireless systems, and the Internet of Things (IoT) are only a few of the

V. Modrák (✉) · Z. Šoltysová
Department of Manufacturing Management, Technical University of Košice, Prešov, Slovakia
e-mail: vladimir.modrak@tuke.sk

Z. Šoltysová
e-mail: zuzana.soltysova@tuke.sk

D. T. Matt et al. (eds.), *Industry 4.0 for SMEs*,
https://doi.org/10.1007/978-3-030-25425-4_8

new technologies that are driving a paradigm shift in manufacturing. The umbrella term for this new wave of so-called smart manufacturing is Industry 4.0 (Kagermann et al. 2013). The main objectives of Industry 4.0 can be characterized, in a simple way, as the introduction of intelligent applications and smart sensor device in production, logistics and business models. Moreover, new information and communication technology (ICT) and web technologies act as enablers of smart, autonomous, and self-learning factories. According to some authors as Sommer (2015), Rauch et al. (2018), successful implementation of Industry 4.0 has to take place not only in large enterprises but in small- and medium-sized enterprises (SMEs) as well. Therefore, a great challenge for the future lies in the transfer of Industry 4.0 expertise and technologies to this size of manufacturing firms that represent the backbone of regional economies. Although there is high potential from Industry 4.0 in SMEs, the main limit lies in a lack of methodological frameworks for its introduction and wide implementation. In addition, a growing number of factories are facing the challenges of even more individualized and customized products (Modrak 2017). This is also the case among SMEs, which are involved in global business and facing a demand for increased product variety (Brunoe and Nielsen 2016). In this context, this chapter aims to help overcome this gap through proposed approaches and solutions.

The chapter is divided into several sections. After this section, the existing approaches to maturity models for the application of Industry 4.0 (I4.0) concept are presented and analyzed. Next, the problem description gives a short explanation of why managerial and organizational concepts, supporting models, and quantitative indicators can be helpful in the introduction of I4.0 in manufacturing companies. The methodological steps of the presented research are graphically depicted in Sect. 8.4. The development steps of the proposed maturity model (MM) are described in detail in the subsequent Sects. 8.5 and 8.6. This part of the chapter presents its main contribution to the managerial and organizational models for the introduction and implementation of I4.0 in terms of mass customization (MC). The final section offers future directions and summarizes the major results of this chapter.

8.2 Literature Review

In general, advanced technologies, including those related to I4.0, infiltrate permanently into all spheres of human life in developed countries. On the other hand, Cotteleer and Sniderman (2017) argue that "there is little doubt that penetration of Industry 4.0 concept in companies' processes and operations will grow." Moreover, some authors (e.g., Hofmann and Rüsch 2017) were skeptical about companies' efforts in this area since according to them the "concept of Industry 4.0 still lacks a clear understanding." This corresponds with a limited occurrence of literature that deals with the concept of Industry 4.0 or smart manufacturing from methodological viewpoints and clarifies how to successfully implement the main components of I4.0 into manufacturing practice. Smart factories can be characterized by distinctive features that reflect different aspects of the domain of interest. According to Pessl et al. (2017) smart factories represents the connection between digital and physical production networks known also as cyber-physical systems. In particular, the integration of computing, wireless, and Internet technologies makes this connection possible. IoT is the most critical component for connecting devices without wired connection (Avram et al. 2017; Belforte and Eula 2012; Ahuett-Garza and Kurfess 2018). For this reason, some of the biggest challenges for manufacturing companies is to increase the level of digitization, to adapt production lines to new technologies or to define the role of humans within new processes (Fang et al. 2016). Toward these outcomes, different MMs can help to identify where the company currently operates and what needs to change. Moreover, maturity models offer comprehensive guidance and introduce and create a basis for evaluating the progress in the maturity of process or a technology. Most maturity models are dedicated to assessing people, culture, processes, structures, and objects or technology, respectively (Mettler 2011). Tavana (2012) pointed out that "the most important point of critique is the poor theoretical basis of maturity models." Becker et al. (2009) proposed evaluation criteria and a generic methodology for the development of maturity models and applied them to the maturity model for IT management. Several authors adopted their recommended steps for developing maturity models related to I4.0. For instance, Leyh et al. (2017) offered an MM for classifying the enterprise-wide IT and software Landscape from the I4.0 perspective, and recommended activities, which can enable a company to reach individual maturity stages. Similarly, Schumacher et al. (2016), and Sternad et al. (2018) have been inspired by their methodology and applied their recommendations to the development of Industry 4.0 related readiness and maturity models. Kese and Terstegen (2017) categorize four types of MM in terms of I4.0. Another classification concept was proposed by Barata and da Cunha (2017), who recommend classifying MM into two groups, i.e., practical models for specific applications or generic MM for I4.0 and its sub-domains. According to Fraser et al. (2002), it is also useful to distinguish between so-called maturity grids, capability MM models or Likert-like questionnaires MM.

In order to compare and analyze existing readiness maturity models, roadmaps, and conceptual frameworks related to Industry 4.0, it is useful to present them in a structural form by pointing out their relevant attributes. Related works with their characteristics are presented in Table 8.1.

Based on this review, the 20 identified and investigated literature sources were dedicated to one or more of three types of methods namely, I4.0 readiness MMs, roadmaps, and conceptual frameworks. It is necessary to note that specific capability maturity models can also be understood as evolutionary roadmaps for implementing the best practices or methodologies into company processes (Curtis et al. 2001). Here identified systematic roadmaps are presented in explicit form, and they might enable companies to answer questions about what technologies to develop when and how. The quantitative occurrence of each type of method was as follows: I4.0 MMs—17 literature sources, the roadmaps—5 literature sources, and the conceptual frameworks—5 papers.

The literature sources used in Table 8.1 can be classified into two basic categories: academic literature, and nonacademic literature. In general, journal papers and conference proceedings represent scientific research, and the latest developments in a specialized field (Wong and Monaco 1995). Rigorous academic books can be included in the first category, too. Then, the academic literature used in our review consists

Table 8.1 Comparison of literature sources dedicated to MMs, roadmaps, and conceptual frameworks

Author(s) and year of publication	Type of the publication	Nature of the method used	The method description and levels of MMs	Subject(s) of interest
Katsma et al. (2011)	CP	I4.0 readiness MM and architectural framework	It contains four stages that describe the development from ERP to the IoT. The stages are applied to four different dimensions	SCM, In-house logistics
Rockwell Automation (2014)	FP	Roadmap and IT readiness MM	Connection of y to IT. Maturity Model divided into 5 stages	Operation technology, IT, SCM, employees
Lichtblau et al. (2015)	FP	I4.0 readiness MM	MM proposes 6 maturity levels that measure the Industry 4.0 readiness. It contains an action plan to boost the readiness in the context of technology, environment, and organization	Organizational aspects, products, operation technology, employees, data management
Bitkom et al. (2016)	FP	Reference Architecture Model for I4.0	A 3D model reflects smart grid architecture model, which was defined by the European Smart Grid Coordination Group	Business layer, functional layer, information layer, communication layer, integration layer, asset layer

(continued)

Table 8.1 (continued)

Author(s) and year of publication	Type of the publication	Nature of the method used	The method description and levels of MMs	Subject(s) of interest
Geissbauer et al. (2016)	FP	Roadmap and I4.0 readiness MM	Definition of six practical steps that a company needs to take to lead tomorrow's competitive digital landscape. MM divided into 4 levels	Employees, company culture
Ganzarain and Errasti (2016)	JP	Business model framework, Roadmap, and I4.0 readiness MM	Three stage maturity model in SME's toward Industry 4.0	Product, organizational aspects
Chromjakova (2016)	JP	I4.0 readiness MM	Performance management system for Industry 4.0. Three-stage MM	Process, product employees
Leyh et al. (2016)	CP	I4.0 readiness MM	MM enables organizations to evaluate IT capabilities for I4.0. It consists of five levels with four dimensions	IT, software, technological aspects
Schumacher et al. (2016)	CP	I4.0 readiness MM	The model is dedicated to manufacturing companies. It consists of five maturity stages, which are applied to nine dimensions	Products, customers, operation technology, strategy, leadership, data management, culture, employees

(continued)

Table 8.1 (continued)

Author(s) and year of publication	Type of the publication	Nature of the method used	The method description and levels of MMs	Subject(s) of interest
Singh et al. (2017)	CP	Generic enterprise architecture	It describes enterprise architecture for a smart logistics of business processes	Logistics of business processes, big data, production planning
Kumar and Namapuraja (2018)	FP	I4.0 readiness MM	A methodology is based on holistic and multidimensional approach to assess current readiness for Industry 4.0 journey using 6 levels	Organizational aspects, company culture, IS
Hofmann and Rüsch (2017)	JP	I4.0 application model	Focused on logistics management. Divided into two dimensions	SCM, data management
Gökalp et al. (2017)	B	Roadmap and I4.0 readiness MM	Industry 4.0-Maturity Model for manufacturing processes is divided into six levels and five dimensions	Asset management, data management, enterprise management, process transformation, organizational aspects
Jæger and Halse (2017)	B	I4.0 readiness MM	The IoT technological maturity model determines the current IoT implementation level for manufacturing enterprises. It introduces eight maturity stages	IoT-technologies, manufacturing technologies

(continued)

Table 8.1 (continued)

Author(s) and year of publication	Type of the publication	Nature of the method used	The method description and levels of MMs	Subject(s) of interest
De Carolis et al. (2017)	B	Roadmap and I4.0 readiness MM	The digital readiness assessment maturity model (DREAMY) guides manufacturing companies toward digitalization. MM consists of five maturity levels and four dimensions	Processes, monitoring and controlling, manufacturing technology, organizational aspects
Li et al. (2017)	CP	I4.0 readiness MM	Proposed Capability Maturity Model aims to evaluate the maturity and readiness of manufacturing enterprises from the factory operations perspective using 5 levels	Operation technology
Klötzer and Pfaum (2017)	CP	I4.0 readiness MM	Maturity model is dedicated to digital transformation of manufacturing enterprises. MM is divided into 5 levels	Products, IT, organizational aspects, processes, company culture

(continued)

Table 8.1 (continued)

Author(s) and year of publication	Type of the publication	Nature of the method used	The method description and levels of MMs	Subject(s) of interest
Weber et al. (2017)	CP	I4.0 readiness MM	MM analyzes the IT architecture of manufacturing companies to provide a development path toward servitization. MM is divided into 6 levels	IT, IS, data management
Werner-Lewandowska and Kosacka-Olejnik (2018)	CP	I4.0 readiness MM	Maturity model is focused on logistics operations in service companies. MM is divided into 6 levels	Logistics
Sternad et al. (2018)	JP	I4.0 readiness MM	MM enables self-evaluation of the company and covers the development areas. MM is divided into 5 levels for logistics 4.0	Logistics, business models, IT, IS, processes, enterprise management, production planning, production control, human-machine communication

Legend:

Firms' publication (FB)
Conference paper (CP)
Journal paper (JP)
Book or Book chapters (B)

Abbreviations:

ERP—Enterprise resource planning
SCM—Supply chain management
IT—Information technology
IS—Information system

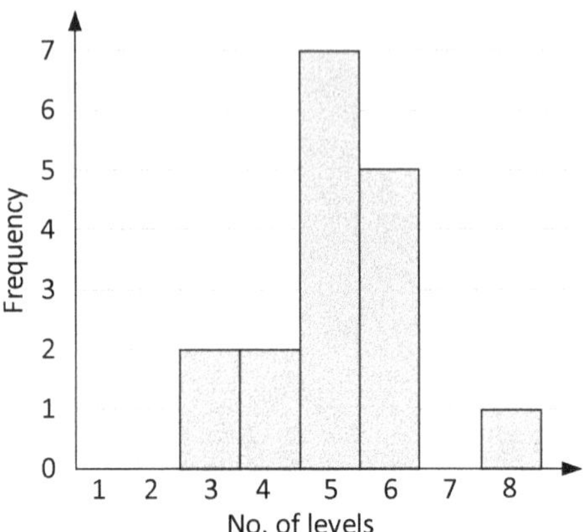

Fig. 8.1 Frequency distribution of the progress levels used in the reviewed MMs

of 15 references, and the nonacademic literature is represented by 5 literature sources.

When analyzing levels of the examined I4.0 MMs, they vary from 3 to 8 stages. Frequency distribution of the levels used in the reviewed MMs is depicted in Fig. 8.1. As can be seen, the most commonly occurring number of levels in the MMs is 5. This is in line, e.g., with the representative generic capability MM for software process program which also uses 5 stages of maturity progress (Paulk et al. 1993).

Our next interest was to learn which domains of MMs dominated in the investigated literature. For this purpose, the following diagram in Fig. 8.2 is provided.

As shown in Fig. 8.2, the domains of interest can be divided into two categories classified as "essential domains" and "recommended domains." Then, the domains with a frequency of 4–6 fall into the first category, and the rest of the subjects of interest belong to the second category.

8.3 Problem Description

As mentioned above, a great challenge for SMEs lies in the transfer of I4.0 expertise and technologies to their environment. Moreover, this challenge also includes transfer toward mass customization. Comparing

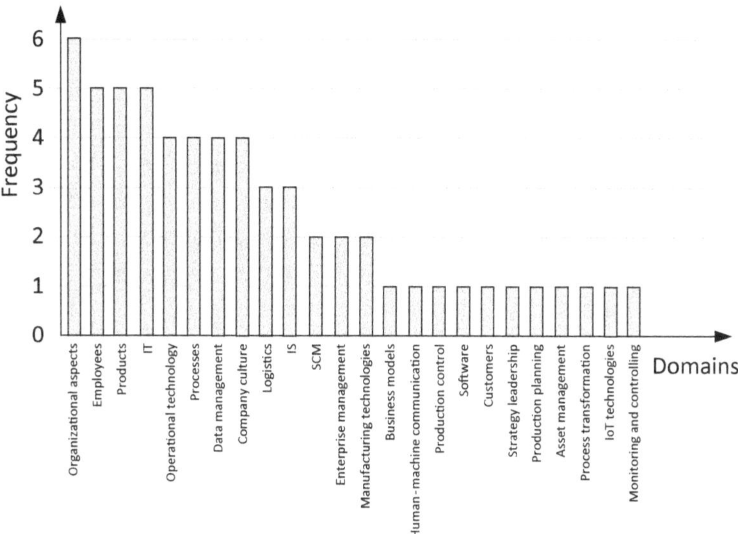

Fig. 8.2 Categorization of subjects of interest based on their appearance in the I4.0 MMs

starting conditions for introduction of the I4.0 concept between large companies and SMEs, it can be stated that larger companies can follow the higher maturity levels in the technological domain for this concept more quickly than SMEs. This is because they can invest more money, time, and expertise into this transfer. On the other hand, an advantage of SMEs against large companies is lower complexity of their business and manufacturing processes. And thus, organizational and cultural changes can be implemented into the whole enterprise much more easily.

Even though several I4.0 MMs which focused on organizational facets were identified in the aforementioned section, none of them can be considered as standardized or universal. Moreover, the described approaches in Table 8.1 were mostly based on self-assessment by using questionnaires which offered answers yes or no, and were oriented toward the identification of a company's current state related to the maturity requirements. Reflecting on the findings of the review from Sect. 8.2, the ambition of our research is to develop a comprehensive I4.0 MM focused on organizational and managerial aspects in terms of mass customization. The proposed I4.0 MM is based on a collaborative approach using a questionnaire method for self-assessment described further in Sects. 8.5.1 and 8.5.2. The outputs of this questionnaire maturity model (QMM) are dedicated to identifying status quo and mapping gaps which need to be filled in order to reach the planned state and are further used as inputs for the creation of the I4.0 readiness MM. Moreover, our approach includes the methodological recommendations in form, e.g., how to measure progress in product modularity and process modularity using quantitative indicators. In addition, we propose a generic organizational model of mass customized manufacturing as a condition for reaching an advanced stage (Level 4) of the proposed MM.

8.4 Methodology

The aim of this section is to guide you through the process of developing the managerial and organizational maturity models. This process starts with a structural analysis of the existing literature related to I4.0 MMs (Sect. 8.2). It helped us to identify what methods already exist, to understand the relationships between them, to find out which domains are essential for I4.0 introduction, and so on. Based on the obtained findings, it was easier to specify categories and levels of QMM for mapping of requirements of SMEs to meet higher maturity levels in the context of the strategy for Industry 4.0. The method used for this purpose is described in detail in Sect. 8.5.1. Subsequently, application of the QMM is presented in Sect. 8.5.2. Respondents were represented by 10 selected SMEs. The next step in our approach was aimed at specifying differences between current states and required states and identifying the key requirements of SMEs on the bases of the obtained results from the questionnaire. In order to validate the obtained results, the overall internal consistency of the questionnaire was measured by Cronbach's alpha coefficient (see Sect. 8.5.3). Development of the I4.0 readiness MM (I4.0 RMM) that is shown in Sect. 8.6, followed methodological recommendations from the relevant scientific publications dealt with in Sect. 8.2. Finally, based on empirical experiences, the generic organizational model of mass customized manufacturing was proposed (see Sect. 8.6.1). The whole methodological framework in the form of a step-by-step guide is available in Fig. 8.3.

8.5 Proposed Approaches and Solutions

This section will provide readers with substantial research outputs that are outlined in Fig. 8.3.

8.5.1 Development of QMM

The QMM method was selected with the aim of applying the collaborative approach by involvement of selected SMEs in order to identify their current status and define future targets in the context of I4.0 challenges. In order to map the requirements of SMEs to meet planned maturity levels in the context of the strategy I4.0, the categories of the QMM for investigation of the managerial and organizational model, were firstly defined. For this purpose, five categories were empirically selected in our previous work, which are business strategy, business models related to product, innovation culture, organizational production model, and knowledge management (Modrak et al. 2019). For each of the categories, five maturity levels were specified in descriptive form as shown in Appendix 8.1.

After the SMEs' self-assessment, obtained questionnaire outputs resulted in specification of their key requirement. The questionnaire form contained options scaled from the lowest level (L1) to the highest level (L5) as shown in Appendix 8.4.

The obtained results from the fulfilled questionnaire forms are presented in the next section and considered further in Sect. 8.5.3.

8.5.2 Application of the QMM

Results from the mapping of requirements using the QMM method to identify the current situation and future targets of the 10 companies are graphically depicted in Fig. 8.4.

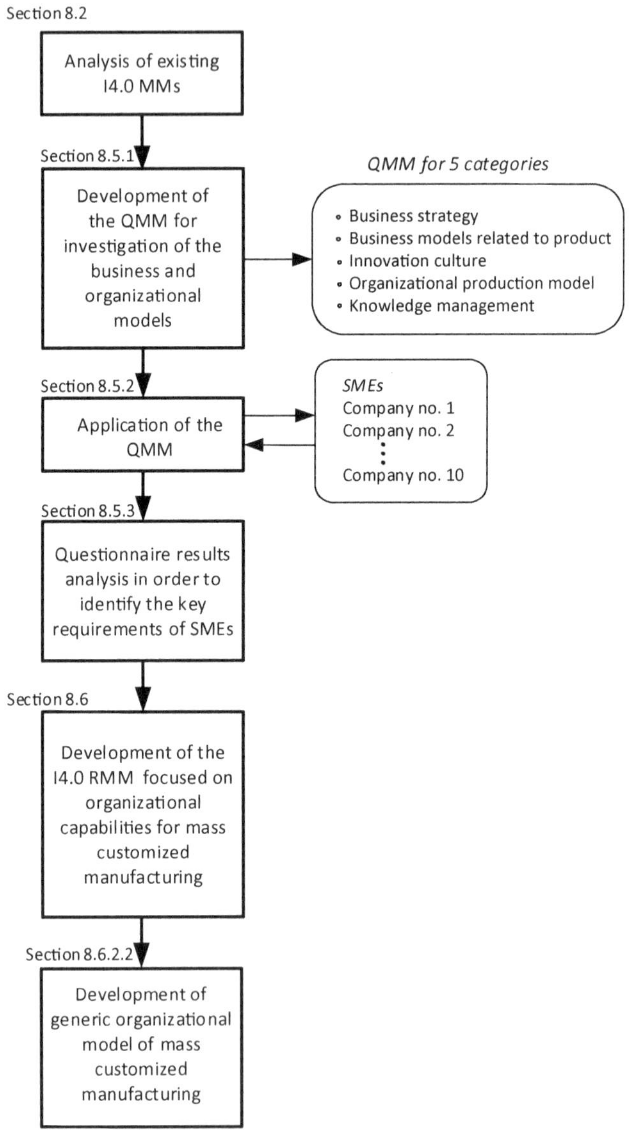

Fig. 8.3 Research methodological framework

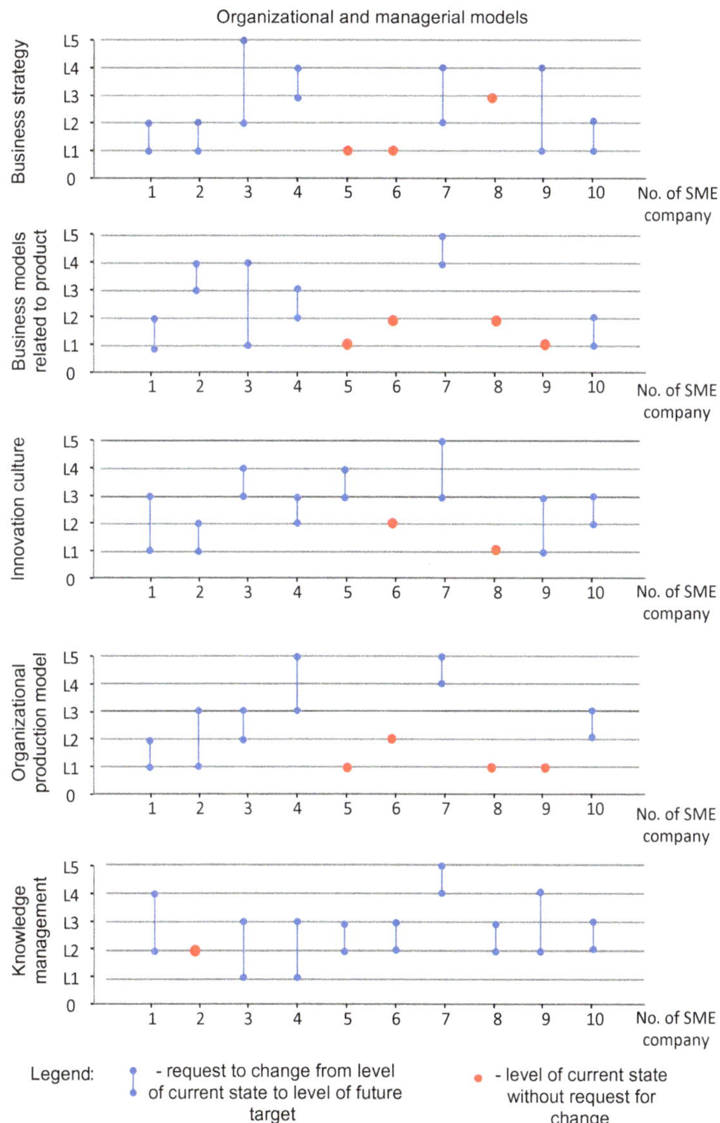

Fig. 8.4 Results from mapping of individual requirements

According to the obtained results shown in Fig. 8.4, it is possible to start with their processing in order to identify the key requirement(s) of SMEs in line with given research objectives.

8.5.3 Identification of the Key Requirement of SMEs

In this step, questionnaire results were processed in the following way.

For determination of the order of significance of assessed categories, the weight coefficient (*V*) was used:

$$V = \sum_{i=1}^{10} R_i \cdot W_i \tag{8.1}$$

where,

R_i—the Rate of the change of *i*-th SME, while if the current state is the same as future target then R_i equals 0, and vice versa, R_i equals 1.

When $R_i = 1$, then for each gap between current state and future target a Weighting value (*W*) is assigned. The weighting value depends on the level of change.

When the extent of the gap equals: 1, then $W_i = 1.2$

2, then $W_i = 1.4$
3, then $W_i = 1.6$
4, then $W_i = 1.8$
5, then $W_i = 2$.

The order of categories of significance based on the result values of the coefficient *V* calculated using Eq. (8.1) are as follows: Category No. 5 ($V = 11.6$); Category No. 3 ($V = 10.2$); Category No. 1 ($V = 9.4$); Category No. 2 ($V = 7.6$) and Category No. 4 ($V = 7.6$).

Then, identification of the key requirement(s) will be determined as follows:

Firstly, the Average level of the current state levels (CL_A) for each category is enumerated using the arithmetical mean from 10 values of the level numbers:

$$CL_A = \frac{\sum_{i=1}^{10} L_i}{10}. \tag{8.2}$$

Secondly, the Average level of the future target levels (RL_A) for each category is determined analogically:

$$RL_A = \frac{\sum_{i=1}^{10} L_i}{10}. \tag{8.3}$$

Finally, the average gap for each category is obtained as the difference between RL_A and CL_A. The obtained average gaps are graphically shown in Fig. 8.5.

However, the overall internal consistency of the questionnaire has to be measured by Cronbach's alpha (Cortina 1993) to validate results from fulfilled questionnaires from the population sample represented by 10 SMEs (S1, S2, ..., S10). This chapter contains results from a questionnaire survey in the domain of business and organizational models,

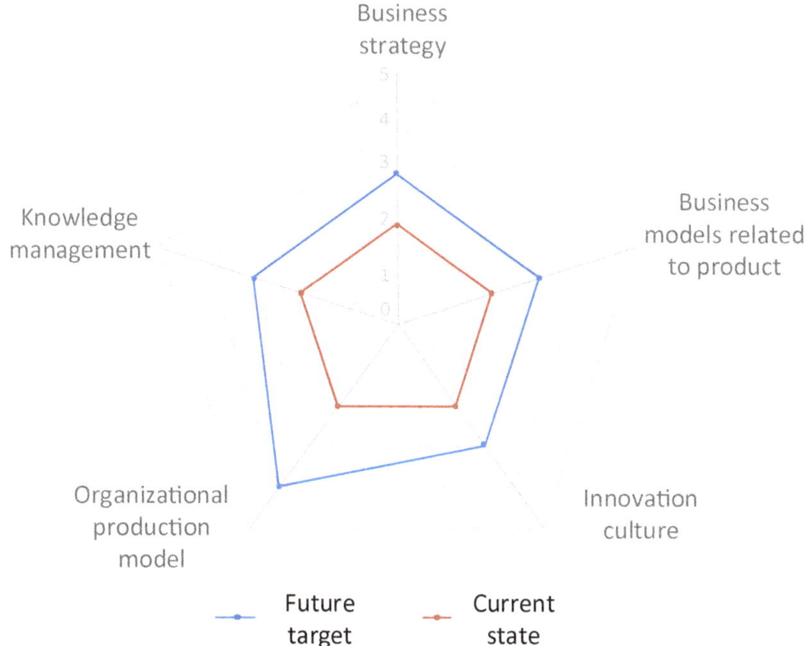

Fig. 8.5 Spider graph of differences between current states and future targets

which was a part of the QMM used for the other two domains, i.e., smart logistics and smart manufacturing with a total number of 15 questions (see Appendices 8.2 and 8.3). Due to this fact, overall internal consistency of the questionnaire used in the QMM will be tested for all three domains. For this purpose, the obtained input data needed to calculate Cronbach's alpha coefficient were arranged into Table 8.2.

Subsequently, Cronbach's alpha coefficients were separately calculated for the current states and the future targets by using the formula (Machin et al. 2007):

$$\alpha = \left(\frac{k}{(k-1)}\right) * \left(1 - \left(\frac{\sum(s_i^2)}{s_t^2}\right)\right), \tag{8.4}$$

where,

k = number of items—questions in questionnaire (Q),
S_i = SD of ith item,
S_t = SD of sum score.

Table 8.2 Input data for calculations of Cronbach's alpha coefficients

	Q1		Q2		Q3		Q4		Q5		Q6		Q7		Q8		Q9		Q10		Q11		Q12		Q13		Q14		Q15	
S1	1	2	2	3	1	1	1	2	2	2	2	4	2	4	1	3	3	5	2	3	1	2	1	2	1	3	1	2	2	4
S2	2	3	2	3	2	2	2	3	2	2	1	2	2	3	1	3	2	4	2	2	1	2	3	4	1	2	1	3	2	2
S3	1	3	2	4	1	2	2	4	1	2	2	3	2	4	2	4	2	4	2	4	2	5	1	4	3	4	2	3	1	3
S4	2	4	3	4	2	3	3	4	2	3	2	4	4	5	2	4	3	5	3	5	3	4	2	3	2	3	3	5	1	3
S5	2	2	2	2	2	2	2	2	2	2	2	2	3	3	3	3	2	2	3	3	1	1	1	1	3	4	1	1	2	3
S6	1	1	2	3	1	1	2	2	1	1	1	2	2	3	2	3	2	2	2	2	1	1	2	2	2	2	2	2	2	3
S7	4	4	3	4	2	2	3	4	2	3	4	5	4	4	3	4	2	5	3	4	2	4	4	5	3	5	4	5	4	5
S8	2	2	2	2	2	2	2	2	2	2	1	2	2	2	1	2	2	3	2	2	3	3	2	2	1	1	1	1	2	3
S9	1	3	2	5	1	5	1	5	1	4	1	5	1	5	1	5	1	5	1	4	1	4	1	1	1	3	1	1	2	4
S10	1	2	2	3	2	3	2	3	2	2	1	2	3	4	3	3	2	3	3	4	1	2	1	2	2	3	2	3	2	3

Current state

Required state

Then, Cronbach's alpha coefficient for the current states is 0.92 and the future targets equal 0.94. Based on a commonly accepted rule for describing internal consistency using Cronbach's alpha, in both cases, the internal consistencies are excellent.

According to the obtained results, the category titled "organizational production model" was specified by 10 SMEs as the key requirement (see Fig. 8.6) and for this key requirement will be further proposed an I4.0 readiness MM with an orientation toward organization capabilities for mass customized manufacturing.

8.6 Maturity Model of Organizational Capabilities for Mass Customized Manufacturing

In this Section, the I4.0 readiness MM is proposed including a recommended specification for SMEs as preconditions for successful implementation of mass customized manufacturing. Presented I4.0 RMM is divided into 5 stages: conventional, starting, moderate, advanced, and optimized, as depicted in Fig. 8.6. The structure of its characteristics can be divided into two groups.

The first one includes the main features of the stages such as: product standardization, product modularity, process modularity, integration of product configurator into process planning, and optimization of intelligent technologies and products. The main features of the first group can be formally modeled by using the arithmetic recursive formula:

$$a_n = a_{n-1} + d \tag{8.5}$$

where n are integers 1–5,

$a_0 = 0$,

and d is common difference (in our case "one step up").

Then, the 1st step up is represented by product standardization, and the 5th step up relates to optimization of intelligent technologies and products.

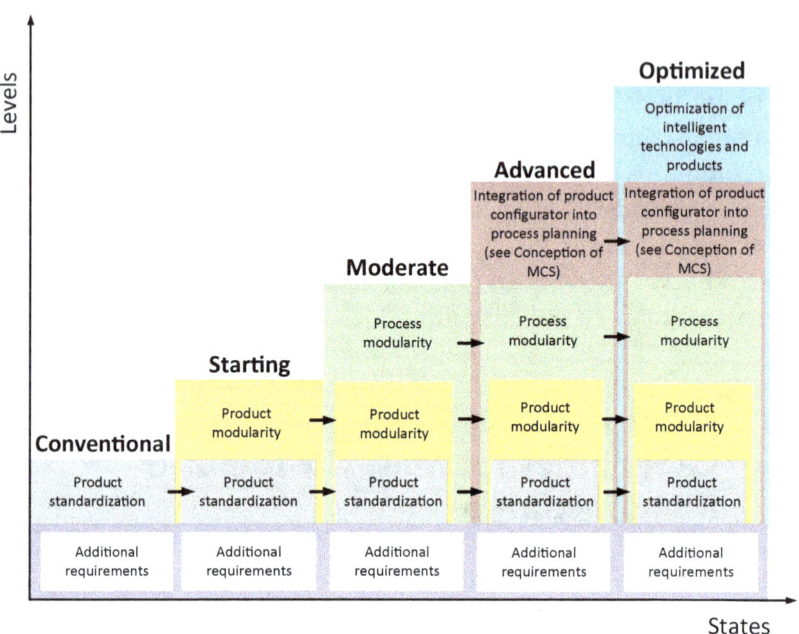

Fig. 8.6 I4.0 RMM of organizational capabilities for mass customized manufacturing

The second group consists of additional requirements that are described in Sect. 8.6.1. The list of additional requirements is only informative, and can be subject to variations.

8.6.1 Additional Requirements of the Maturity Model

The additional requirements of the I4.0 RMM might include at least the following characteristics.

At the Conventional stage:

- Traditional approach based on product standardization of particular products offered for different markets
- Common operational planning methods, communication with suppliers using basic ICT technology, manual processing of orders and logistics, no monitoring of logistics

- Production not connected with ERP system through manufacturing execution systems (MES)
- Physical product without digital functions, sporadic combination of products and digital services.

At the Starting stage:

- Orientation of product modularization
- Assessment of suppliers coordinated with logistics and production, ERP system connected with production through MES
- Internally integrated system-based planning, optimization of logistics operations
- Data analysis for mass customized production needs
- Groups of standardized products are shipped to different markets according to local needs.

At the Moderate stage:

- Orientation on process modularization
- Integrated customer solutions across supply chain boundaries, collaboration with external logistics providers, transfer of product characteristics to the ERP system for marketing purposes
- Individualized customer approach and interaction with supply chain partners by using specific ICT technology
- IT integration with suppliers through ERP system
- Partial focus on the development of intelligent technologies and products.

At the Advanced stage:

- Using product configurators to enhance communication with clients who are ordering customized products
- Adaptation of organizational model of production for mass customized production
- Transition to one-piece flow production in order to increase effectiveness of manufacturing processes

- Application of modern ICT and automatic identification technologies for monitoring and tracking of parts, components, modules, and final products in manufacturing plant and supply chains.

At the Optimized stage:

- Optimization of the organizational model of production for mass customized products
- Optimization of product modularity and process modularity by using quantitative metrics
- Communication with suppliers is completely digitalized, intensive optimization of warehouses, real-time transparency of supply chain
- Application of tools for digital marketing and sales
- Optimization of intelligent technologies and products.

8.6.2 Description of the Main Features of the Maturity Model

The first important features of preconditions of organizational capabilities for mass customized manufacturing that are indicated in our I4.0 readiness MM, are product standardization, product modularity, and process modularity. As is well-known, standardization of the internal components simplifies their assembly into many different products according to a customer's needs (George 2003). It can also be stated that modularity-based approaches in manufacturing practice such as product modularity and process modularity can improve mass customization capability and are strong enablers of this marketing and manufacturing strategy (Kotha 1995; Gilmore 1997). Accepting these statements is one thing, but when companies want to follow the steps shown in our I4.0 readiness MM, they need to manage improvements of these features. However, we also have to accept a common rule: if you can't measure it, you can't manage it. Therefore, it is recommended that effective quantitative measurement for this purpose is applied; this is described below.

8.6.2.1 Product Modularity and Process Modularity

There are several approaches to measure product modularity, which are available in the literature. A comprehensive overview of them has been offered by Ulrich (1994). Hölttä-Otto and De Weck (2007) proposed a product modularity metric called the Singular Value Modularity Index (SMI) to quantify the degree of modularity of a product on its internal structure. SMI is theoretically bounded between 0 and 1, while SMI closer to 1 indicates a higher degree of modularity and vice versa. For its calculation, the following expression is used:

$$\mathrm{SMI} = 1 - \frac{1}{N \cdot \sigma_1} \sum_{i=1}^{N-1} \sigma_i(\sigma_i - \sigma_{i+1}), \tag{8.6}$$

where:

N—is the number of components of the system
σ_i—represents singular values, $i=1, 2, \ldots, N$ ordered in decreasing magnitude.

The authors of this measurement approach provide in the above mention literature sources useful examples of how the SMI measure can be applied.

According to Calcagno (2002), not only is a product modularity measurement important, but it is also necessary to measure a degree of modularity of manufacturing systems. On the other hand, there is a dearth of process modularity measures, which could be simply applied for managerial purposes. For this reason, we proposed to adopt the SMI to measure the degree of modularity of manufacturing processes (Modrak and Soltysova 2018a).

Process modularity issues are important, particularly in the context of optional components entering into an assembly station with a human presence. As a consequence, such a station might be divided into two or more substations to minimize complexity of the operation in order to eliminate the tendency to make mistakes. The important precondition to applying SMI for process modularity is transformation of real process operations into models using graph theory. A methodological procedure of how to apply SMI to measure process modularity can be found in the mentioned literature.

8.6.2.2 Integration of Product Configurator into Process Planning

The proposed generic model of how to organize manufacturing and marketing activities in terms of mass customization is depicted in Fig. 8.7. The model can be divided into four systems, namely the product configuration system, product arrangement and process planning system, manufacturing system, and final product assembly system. The product life cycle in this segment starts in the configuration system, where the product is specified according to customer needs, and which consists of a product definition module. Outputs from this system are, at a minimum, characterized by article codes (ACs), quantity to be assembled (Q), and production due dates (PDD). Moreover, functional requirements are also specified in this phase. When the final product is defined, component separation is performed according to a bill of material (BOM). Firstly, components are divided into stable (S), compulsory optional (CO) components, and voluntary optional (VO) components. This classification is important, especially at higher levels of the maturity model where variety-based complexity is a matter of major concern affecting the product and process design (Modrak and Soltysova 2018b). Requirements for the CO and VO components further face a make or buy decision, where they are divided into their own base production or ordered from a supplier. Subsequently, schedules for product manufacturing are generated. Then, based on information about component consolidation, detailed schedules for multi model assembly line(s) are calculated and provided to the product assembly system department. Finally, the product assembly process is triggered and managed. This phase mostly includes the performance of functional requirements.

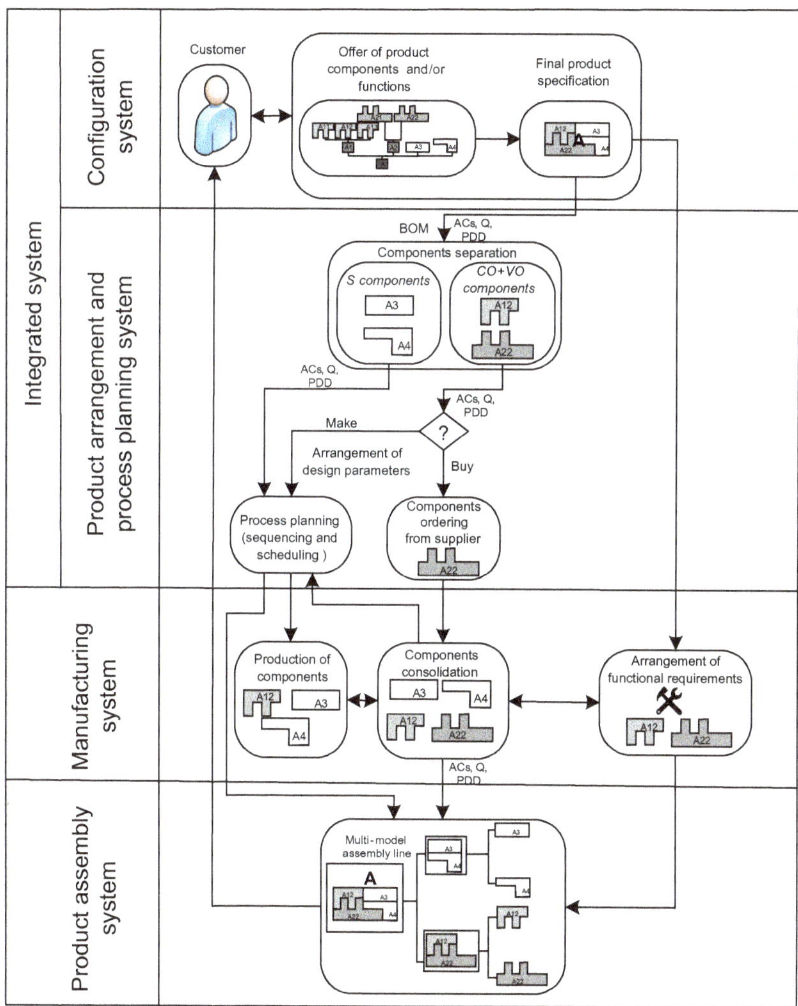

Fig. 8.7 Model of the relations between orders and manufacturing operations

8.7 Future Research Work and Conclusions

Analysis of the results obtained from the application of the QMM confirms and emphasizes that an effort of SMEs in the context of Industry 4.0 challenges has to also be dedicated to organizational and managerial aspects. For this purpose, theoretical concepts such as frameworks or maturity models can be useful to anticipate relevant directions and factors affecting the achievement of this strategic goal. Accordingly, and in line with this statement, this chapter offers a structured approach to carrying out self-assessment for Industry 4.0 implementation, and moreover, proposes the generic I4.0 RMM as the roadmap for helping companies to navigate them and understand their current state in the mass customization environment.

The literature review on existing approaches was mostly based on self-assessment by using a questionnaire with yes or no answers, and oriented towards the identification of a company's current state related to the maturity requirements. The proposed maturity models are focused on organizational and managerial aspects in terms of mass customization. While I4.0 QMM can be used for mapping the three domains: smart manufacturing, smart logistics, and organizational and managerial facets, the I4.0 RMM is dedicated to the readiness of organizational capabilities for mass customized manufacturing.

The given results will be used in our future work for the development of technical solutions and managerial methods that will enable managers to make better decisions for the digital transformation of SMEs from the current state to the targeted state. In this context, the generic conception of mass customized manufacturing based on the integration of a product configurator in process planning will be further developed. For example, it will be necessary to follow the development of marketing tools based on social networks and new communication channels. As is known, the connection of a product configurator with a Facebook site can facilitate not only social connections between existing and potential customers but also help in the codesign of activities (Gownder et al. 2011).

In spite of early skepticism about mass customization, which was seen as not very useful and as a contradictory concept, its penetration in different industries has become a reality. As was almost predicted, MC has become an imperative rather than a choice leading to sustainable success across business sectors (Piller 2010). However, wider acceptance of this strategy in individual industries will strongly depend on the availability of attainable digital manufacturing devices belonging to the smart manufacturing concept.

Appendix 8.1

Maturity levels of QMM in the Organizational and managerial domain

Category title	Level	Description of maturity level
Business strategy	L1	The organization does not have a formal strategy I4.0 as a part of the corporate strategy
	L2	Managers are convinced of the need to develop a strategy for I4.0
	L3	Managers work on a strategy for I4.0 focused on technological aspects
	L4	Business activities for technology change are aligned with company strategy
	L5	The strategy for I4.0 is more focused on people than on production technology
Business models related to product	L1	Earning income from the sale of standardized products
	L2	Groups of standardized products are shipped to different markets according to local needs
	L3	Possibility to customize the product based on group(s) of variant modules
	L4	Possibility to customize the product from a wide range of components
	L5	Mass personalization
Innovative culture	L1	Openness for digital technologies
	L2	Identification with the building of digital enterprise
	L3	Orientation in the development of intelligent technologies and products
	L4	Intelligent technologies and/or products are introduced
	L5	Optimization of intelligent technologies and products
Organizational production model	L1	Traditional approach by type of production type
	L2	Orientation on product modularization
	L3	Orientation on process modularization
	L4	Application of the organizational model of production for mass customized products
	L5	Optimization of the organizational production model for mass customization

Category title	Level	Description of maturity level
Knowledge management	L1	The organization does not have any formal knowledge management strategy (KM)
	L2	Managers are aware of the need to develop their own strategy KM
	L3	Managers develop and implement the KM strategy
	L4	Activities for creation and sharing of knowledge are in line with the KM strategy focused on technology and people
	L5	Activities for creating and sharing knowledge are more people-oriented than on technology. The sustainability of the established KM is permanently monitored

Appendix 8.2

Maturity levels of QMM in the Smart logistics domain

Category title	Level	Description of maturity level
Transport logistics	L1	Decentralized managed transport
	L2	Centralized managed transport
	L3	Predictive centralized transport. Ad hoc managed distribution
	L4	Predictive centralized transport. Optimized management of distribution
	L5	Use of autonomous vehicles
Outbound logistics	L1	Push management of the delivery process (in warehouses)
	L2	Order-based delivery process control
	L3	Order-based delivery process control with sales monitoring
	L4	Automatic control of the delivery process
	L5	Automatic delivery process management with prediction of future order
In-house logistics	L1	Use of manual means in inter-operational traffic
	L2	Use of manually operated trolleys in inter-operational traffic
	L3	Use of automatically guided trolleys in inter-operational traffic on defined routes
	L4	Use of automatically guided trolleys in inter-operational traffic on open production area
	L5	Management of autonomous trolleys through production facilities

Category title	Level	Description of maturity level
Inbound logistics	L1	Push management of the supply process (in warehouses)
	L2	Pull method for managing the supply process (JIT)
	L3	Pull method fot managing the supply process (JIT) provided by the retailer
	L4	Autonomous inventory management
	L5	Predictive inventory management
Warehouse management	L1	Use of manual devices for storage operations
	L2	Use of manually guided forklifts
	L3	Use of automated guided vehicle systems and automated storage systems
	L4	Use of automatic systems with links to superior enterprise management systems
	L5	Use of automatic and/or collaborative transport and storage trolleys

Appendix 8.3

Maturity levels of QMM in the Smart production domain

Category title	Level	Description of maturity level
Data processing in the production	L1	Conventional data processing methods (waybills, etc.)
	L2	Use of optical technologies for data processing (bar codes, etc.)
	L3	Use of radio frequency technologies for data processing (RFID)
	L4	Evaluating and using data for process management and planning
	L5	Use data (monitored in real-time) to automate planning and process management
Man to machine communication	L1	No exchange of information between machine and man
	L2	Using local user connections on the machine.
	L3	Centralized or decentralized monitoring and production control
	L4	Using mobile user interfaces
	L5	Enhanced virtual reality and assisted reality

Category title	Level	Description of maturity level
Machine to machine communication	L1	No exchange of information between machines
	L2	Connect devices using a bus
	L3	Machines have an industrial Ethernet interface (local computer network)
	L4	Machines have internet access.
	L5	Web interfaces and information exchange applications (M2M software)
ICT infrastructures in the production	L1	Exchange information via email/ phone
	L2	Central data servers in production
	L3	Internet portals for data sharing
	L4	Use of ICT to identify statuses in production (e.g., status of order)
	L5	Suppliers and/ or customers have access to a web-supported IS (MES)
Digitalization	L1	Basic level of digitization
	L2	Uniform digitization (horizontal)
	L3	Horizontal and vertical digitization
	L4	Full digitalization
	L5	Optimized full digitalization

Appendix 8.4

The questionnaire structure

No. of category	Category title	Current state	Future target
1	Business strategy	L1 ☐	L1 ☐
		L2 ☐	L2 ☐
		L3 ☐	L3 ☐
		L4 ☐	L4 ☐
		L5 ☐	L5 ☐
2	Business models related to product	L1 ☐	L1 ☐
		L2 ☐	L2 ☐
		L3 ☐	L3 ☐
		L4 ☐	L4 ☐
		L5 ☐	L5 ☐

No. of category	Category title	Current state	Future target
3	Innovative culture	L1 ☐ L2 ☐ L3 ☐ L4 ☐ L5 ☐	L1 ☐ L2 ☐ L3 ☐ L4 ☐ L5 ☐
4	Organizational production model	L1 ☐ L2 ☐ L3 ☐ L4 ☐ L5 ☐	L1 ☐ L2 ☐ L3 ☐ L4 ☐ L5 ☐
5	Knowledge management	L1 ☐ L2 ☐ L3 ☐ L4 ☐ L5 ☐	L1 ☐ L2 ☐ L3 ☐ L4 ☐ L5 ☐

References

Ahuett-Garza, H., and T. Kurfess. 2018. A Brief Discussion on the Trends of Habilitating Technologies for Industry 4.0 and Smart Manufacturing. *Manufacturing Letters* 15: 60–63. https://doi.org/10.1016/j.mfglet.2018.02.011.

Avram, M., C. Bucsan, and T.C. Apostolescu. 2017. Specialised Sensorial Block. In *MATEC Web of Conferences,* vol. 121, 8002. EDP Sciences. https://doi.org/10.1051/matecconf/201712108002.

Barata, J., and P.R. Cunha. 2017. Climbing the Maturity Ladder in Industry 4.0: A Framework for Diagnosis and Action That Combines National and Sectorial Strategies. In *Twenty-Third Americas Conference on Information Systems*, Boston.

Becker, J., R. Knackstedt, and J. Pöppelbuß. 2009. Developing Maturity Models for IT Management. *Business & Information Systems Engineering* 1 (3): 213–222. https://doi.org/10.1007/s12599-009-0044-5.

Belforte, G., and G. Eula. 2012. Smart Pneutronic Equipments and Systems for Mechatronic Applications. *Journal of Control Engineering and Applied Informatics* 14 (4): 70–79.

Bitkom, V., V. Vdma, and V. Zvei. 2016. Implementation Strategy Industrie 4.0. Berlin, Germany.

Brunoe, T.D., and K. Nielsen. 2016. Complexity Management in Mass Customization SMEs. *Procedia CIRP* 51: 38–43. https://doi.org/10.1016/j.procir.2016.05.099.

Calcagno, M. 2002, May. Dynamics of Modularity: A Critical Approach. In *Euram Conference*, 9 (11).

Chromjakova, F. 2016. Flexible Man-Man Motivation Performance Management System for Industry 4.0. *International Journal of Management Excellence* 7 (2): 829–840. https://doi.org/10.17722/ijme.v7i2.269.

Cortina, J.M. 1993. What Is Coefficient Alpha? An Examination of Theory and Applications. *Journal of Applied Psychology* 78 (1): 98. https://doi.org/10.1037/0021-9010.78.1.98.

Cotteleer, M., and B. Sniderman. 2017. *Forces of Change: Industry 4.0.* London: Deloitte Insights. https://doi.org/10.1016/j.compind.2017.04.002.

Curtis, B., W.E. Hefley, and S.A. Miller. 2001. People Capability Maturity Model Version 2.0. *Pittsburgh: Carnegie Mellon Software Engineering Institute* 1 (104): 240–266.

De Carolis, A., M. Macchi, E. Negri, and S. Terzi. 2017, September. A Maturity Model for Assessing the Digital Readiness of Manufacturing Companies. In *IFIP International Conference on Advances in Production Management Systems*, 13–20. Cham: Springer. https://doi.org/10.1007/978-3-319-66923-6_2s.

Fang, C., X. Liu, P.M. Pardalos, and J. Pei. 2016. Optimization for a Three-Stage Production System in the Internet of Things: Procurement, Production and Product Recovery, and Acquisition. *The International Journal of Advanced Manufacturing Technology* 83 (5–8): 689–710. https://doi.org/10.1007/s00170-015-7593-1.

Fraser, P., J. Moultrie, and M. Gregory. 2002. The Use of Maturity Models/Grids as a Tool in Assessing Product Development Capability. *IEEE International Engineering Management Conference.* https://doi.org/10.1109/IEMC.2002.1038431.

Ganzarain, J., and N. Errasti. 2016. Three Stage Maturity Model in SME's Toward Industry 4.0. *Journal of Industrial Engineering and Management (JIEM)* 9 (5): 1119–1128. https://doi.org/10.3926/jiem.2073.

Geissbauer, R., J. Vedso, and S. Schrauf. 2016. Industry 4.0: Building the Digital Enterprise. PwC. https://www.pwc.com/gx/en/industries/industries-4.0/landing-page/industry-4.0-building-your-digital-enterprise-april-2016.Pdf.

George, M. 2003. *Lean Six Sigma for Service, Chapter 1-The ROI of Lean Six Sigma for Services.* New York: McGraw Hill Professional.

Gilmore, J.H. 1997. The Four Faces of Mass Customization. *Harvard Business Review* 75 (1): 91–101.

Gökalp, E., U. Şener, and P.E. Eren. 2017. Development of an Assessment Model for Industry 4.0: Industry 4.0-MM. In *International Conference on Software Process Improvement and Capability Determination*, 128–142. Cham: Springer. https://doi.org/10.1007/978-3-319-67383-7_10.

Gownder, J.P., S. Rotman Epps, A.E. Corbett, C.A. Doty, T. Schadler, and J.L. McQuivey. 2011. Mass Customization Is (Finally) the Future of Products. Forrester Report. https://doi.org/10.1108/09544789910246615.

Hofmann, E., and M. Rüsch. 2017. Industry 4.0 and the Current Status as Well as Future Prospects on Logistics. *Computers in Industry* 89: 23–34. https://doi.org/10.1016/j.compind.2017.04.002.

Hölttä-Otto, K., and O. De Weck. 2007. Degree of Modularity in Engineering Systems and Products with Technical and Business Constraints. *Concurrent Engineering* 15 (2): 113–126. https://doi.org/10.1177/1063293X07078931.

Jæger, B., and L.L. Halse. 2017. The IoT Technological Maturity Assessment Scorecard: A Case Study of Norwegian Manufacturing Companies. In *IFIP International Conference on Advances in Production Management Systems*, 143–150. Cham: Springer. https://doi.org/10.1007/978-3-319-66923-6_17.

Kagermann, H., W. Wahlster, and J. Helbig. 2013. Umsetzungsempfehlungen für das Zukunftsprojekt Industrie 4.0. In *Abschlussbericht des Arbeitskreises Industrie* 4 (5). Frankfurt: Forschungsunion.

Katsma, C., H. Moonen, and J. van Hillegersberg. 2011. Supply Chain Systems Maturing Towards the Internet-of-Things: A Framework. In *24th Bled eConference eFuture: Creating Solutions for the Individual, Organisations and Society Proceedings*, 478–494.

Kese, D., and S. Terstegen. 2017. Industrie 4.0-Reifegradmodelle. Institut für angewandte Arbeitswissenschaft. https://www.arbeitswissenschaft.net/uploads/tx_news/Tool_I40_Reifegradmodelle.pdf, Accessed on May 25, 2018.

Klötzer, C., and A. Pflaum. 2017. Toward the Development of a Maturity Model for Digitalization Within the Manufacturing Industry's Supply Chain. In *2017 50th Hawaii International Conference on System Sciences*, 4210–4219. https://doi.org/10.24251/hicss.2017.509.

Kotha, S. 1995. Mass Customization: Implementing the Emerging Paradigm for Competitive Advantage. *Strategic Management Journal* 16 (S1): 21–42. https://doi.org/10.1002/smj.4250160916.

Kumar, Ravi, and Enose Nampuraja. 2018. Making Industry 4.0 Real—Using the Acatech I 4.0 Maturity Index. A Systematic Methodology for Manufacturing Enterprises to Assess Current Readiness and Strategize Their Industry 4.0 Journey. Infosys.

Leyh, C., K. Bley, T. Schäffer, and S. Forstenhäusler. 2016. SIMMI 4.0—A Maturity Model for Classifying the Enterprise-Wide It and Software Landscape Focusing on INDUSTRY 4.0. In *2016 Federated Conference on Computer Science and Information Systems (FedCSIS)*, 1297–1302. https://doi.org/10.15439/2016F478.

Leyh, C., T. Sch, K. Bley, and S. Forstenh. 2017. Information Technology for Management. *New Ideas and Real Solutions* 277: 103–119. https://doi.org/10.1007/978-3-319-53076-5.

Li, Q., M. Brundage, D. Brandl, and S. Do Noh. 2017. Improvement Strategies for Manufacturers Using the MESA MOM Capability Maturity Model. *In IFIP International Conference on Advances in Production Management Systems*, 21–29. Cham: Springer. https://doi.org/10.1007/978-3-319-66923-6_3.

Lichtblau, K., V. Stich, R. Bertenrath, M. Blum, M. Bleider, A. Millack, K. Schmitt, E. Schmitz, and M. Schröter. 2015. IMPULS-Industrie 4.0-Readiness. Impuls-Stiftung des VDMA, Aachen-Köln.

Machin, D., M.J. Campbell, and S.J. Walters. 2007. Reliability and Method Comparison Studies. In *Medical Statistics*, 203–212.

Mettler, T. 2011. Maturity Assessment Models: A Design Science Research Approach. *International Journal of Society Systems Science* 3 (1/2): 213–222. https://doi.org/10.1504/IJSSS.2011.038934.

Modrak, V., eds. 2017. *Mass Customized Manufacturing: Theoretical Concepts and Practical Approaches*. CRC Press. https://doi.org/10.1201/9781315398983.

Modrak, V., and Z. Soltysova. 2018a. Axiomatic Design Based Complexity Measures to Assess Product and Process Structures. In *MATEC Web of Conferences*, 223. EDP Sciences. https://doi.org/10.1051/matecconf/201822301019.

Modrak, V., and Z. Soltysova. 2018b. Process modularity of Mass Customized Manufacturing Systems: Principles, Measures and Assessment. *Procedia CIRP* 67: 36–40. https://doi.org/10.1016/j.procir.2017.12.172.

Modrak, V., Z. Soltysova, and R. Poklemba. 2019. Mapping Requirements and Roadmap Definition for Introducing I 4.0 in SME Environment. In *Advances in Manufacturing Engineering and Materials*, 183–194. Cham: Springer. https://doi.org/10.1007/978-3-319-99353-9_20.

Paulk, M.C., B. Curtis, M.B. Chrissis, and C.V. Weber. 1993. Capability Maturity Model for Software. Version 1.1. Software Engineering Institute. https://doi.org/10.1109/52.219617.

Pessl, E., S.R. Sorko, and B. Mayer. 2017. Roadmap Industry 4.0–Implementation Guideline for Enterprises. *International Journal of Science, Technology and Society* 5: 193–202. https://doi.org/10.11648/j.ijsts.20170506.14.

Piller, F.T. 2010. *Handbook of Research in Mass Customization and Personalization*, 1. World Scientific. https://doi.org/10.1142/p7378.

Rauch, E., D.T. Matt, C.A. Brown, W. Towner, A. Vickery, and S. Santiteerakul. 2018. Transfer of Industry 4.0 to Small and Medium Sized Enterprises. *Advances in Transdisciplinary Engineering* 7: 63–71. https://doi.org/10.3233/978-1-61499-898-3-63.

Rockwell Automation. 2014. The Connected Enterprise Maturity Model. Rockwell Automation, 12, USA.

Schumacher, A., S. Erol, and W. Sihn. 2016. A Maturity Model for Assessing Industry 4.0 Readiness and Maturity of Manufacturing Enterprises. *Procedia CIRP* 52: 161–166. https://doi.org/10.1016/j.procir.2016.07.040.

Singh, P.M., M. van Sinderen, and R. Wieringa. 2017. Smart Logistics: An Enterprise Architecture Perspective. CAiSE-Forum-DC, 9–16.

Sommer, L. 2015. Industrial Revolution-Industry 4.0: Are German Manufacturing SMEs the First Victims of This Revolution? *Journal of Industrial Engineering and Management* 8 (5): 1512–1532. https://doi.org/10.3926/jiem.1470.

Sternad, M., T. Lerher, and B. Gajsek. 2018. Maturity Levels for Logistics 4.0 Based on Nrw'S Industry 4.0 Maturity Model. *Business Logistics in Modern Management* 18: 695–708.

Tavana, M., eds. 2012. *Decision Making Theories and Practices from Analysis to Strategy*: IGI Global. https://doi.org/10.4018/978-1-4666-1589-2.

Ulrich, K. 1994. Fundamentals of Product Modularity. In *Management of Design*, 219–231: Dordrecht: Springer. https://doi.org/10.1007/978-94-011-1390-8_12.

Weber, C., J. Königsberger, L. Kassner, and B. Mitschang. 2017. M2DDM—A Maturity Model for Data-Driven Manufacturing. *Procedia CIRP* 63: 173–178. https://doi.org/10.1016/j.procir.2017.03.309.

Werner-Lewandowska, K., and M. Kosacka-Olejnik. 2018. Logistics Maturity Model for Service Company—Theoretical Background. *Procedia Manufacturing* 17: 791–802. https://doi.org/10.1016/j.promfg.2018.10.130.

Wong, B.K., and J.A. Monaco. 1995. Expert System Applications in Business: A Review and Analysis of the Literature: (1977–1993). *Information & Management* 29 (3): 141–152. https://doi.org/10.1016/0378-7206(95)00023-P.

9

Implementing Industry 4.0 in SMEs: A Focus Group Study on Organizational Requirements

Guido Orzes, Robert Poklemba and Walter T. Towner

9.1 Introduction

After the first use of the Industry 4.0 label at the Hannover Fair in 2011, the interest for the topic among managers and policy-makers has grown exponentially. Besides Germany, many countries have launched their own plans to foster the transition toward this new manufacturing

---

G. Orzes (✉)
Faculty of Science and Technology, Free University of Bozen-Bolzano, Bolzano, Italy
e-mail: guido.orzes@unibz.it

R. Poklemba
Faculty of Manufacturing Technologies, Technical University of Košice, Prešov, Slovakia
e-mail: robert.poklemba@tuke.sk

W. T. Towner
Center for Innovative Manufacturing Solutions, Worcester Polytechnic Institute, Worcester, USA
e-mail: fabman@wpi.edu

D. T. Matt et al. (eds.), *Industry 4.0 for SMEs*,
https://doi.org/10.1007/978-3-030-25425-4_9

paradigm: Plattform Industrie 4.0 (Austria), China 2025, Impresa 4.0 (Italy), Thailand 4.0 just to cite a few examples. Similarly, consultancy companies have exploited the trend, publishing a wide set of reports on Industry 4.0 (see Rüßmann et al. 2015; McKinsey Digital 2015; Geissbauer et al. 2016; among others). While a clear-cut definition of the phenomenon is still missing (Culot et al. 2018), authors agree that Industry 4.0 is based on the application of cyber-physical systems (CPS) and internet technologies in the manufacturing processes, leading to a convergence between the physical and the virtual world (Kagermann et al. 2013).

Over the last few years, the number of scientific papers on Industry 4.0 has significantly grown (Liao et al. 2017). The literature has shown that Industry 4.0 also offers significant opportunities to small- and medium-sized enterprises (SMEs) which can use these technologies to increase their flexibility, productivity, and competitiveness (Kagermann et al. 2013; Wenking et al. 2016). At the same time, this industrial revolution brings some challenges regarding data security, finding the needed capital, developing a strategy for implementing it and finding qualified employees (Schröder 2016). Extant research has, however, mainly focused on technical aspects of Industry 4.0 (Liao et al. 2017). Consequently, a detailed analysis of the implementation strategies, barriers faced, as well as on the organizational requirements, is missing (Holmström et al. 2016).

This chapter aims, therefore, to address the aforementioned research gap by empirically investigating the main organizational issues faced by SMEs in Industry 4.0 implementation. We focus on SMEs for various reasons: (a) they are the backbone of economies of many European countries; (b) they are expected to face more difficulties in adopting Industry 4.0 than large firms due to the lack of resources and knowledge (Müller et al. 2017; Sandberg and Aarikka-Stenroos 2014); but (c) they can more easily change toward the Industry 4.0 paradigm if a suitable roadmap is available (due to their higher flexibility; Mohnen and Rosa 2002).

We first analyze the existing Industry 4.0 literature. This allows us to highlight a set of potential organizational issues for Industry 4.0 implementation, such as the lack of skilled employees, the lack of monetary resources, and the lack of a systematic approach for implementation. In order to verify whether additional issues should be considered, we also analyzed the broader literature on the barriers to innovation. We concluded that while the extant literature provides some interesting results, it is still characterized by a significant set of gaps and limitations.

In order to refine and empirically validate the set of organizational issues in Industry 4.0 implementation, we then organized some focus groups in four different countries within the research project SME 4.0, funded by the European Commission (H2020 program). These focus groups lasted one full working day each and involved 13–25 CEOs or technical managers of 7–10 SMEs each, who were asked, after a small introduction about the topic, to write on post-its and discuss several issues they faced during Industry 4.0 adoption and implementation.

Our empirical analyses (focus groups) confirmed most of the organizational requirements identified by previous literature. They also allowed us to highlight a set of additional requirements not considered by previous studies. Our study has, therefore, significant implications for researchers, managers, and policy-makers working in the Industry 4.0 field.

The chapter is organized as follows. In Sect. 9.2, we summarize the two relevant streams of study for our work: (a) organizational issues in Industry 4.0 implementation and (b) barriers and problem for innovation. In Sect. 9.3, we formulate the problem and in Sect. 9.4, we explain the adopted methodological approach. Results are then presented in Sect. 9.5 and discussed in Sect. 9.6. Finally, we summarize the contributions to management theory and practice as well as the main limitations in Sect. 9.7.

9.2 Background

In this section, we summarize two main streams of studies that are of interest for our research: (1) organizational obstacles and barriers for Industry 4.0 implementation and (2) barriers for innovation. Despite the second stream of studies not being focused on Industry 4.0, we considered it to analyze whether general barriers to innovation apply also to Industry 4.0 (which is based on a set of innovations).

9.2.1 Organizational Barriers to Industry 4.0 Implementation

In order to identify all the relevant papers dealing with **organizational obstacles and barriers for Industry 4.0 implementation**, we performed a keyword search in the most important electronic database (Elsevier's Scopus). We used a combination of two sets of keywords:

a. Industry 4.0-related terms (e.g., industry 4.0, industrial internet, fourth industrial revolution, 4° industrial revolution, Internet of Things, Smart manufacturing, cyber-physical production systems); and
b. Barrier-related terms (barrier*, obstacle*, challenge*, problem*, SME*, small and medium enterprise*).

This keyword search led us to identify 6029 contributions. After this search, we applied a set of inclusion–exclusion criteria to screen the papers based first on the title and abstract and then on the full text. In greater detail, we excluded papers that do not provide insights on the obstacles and barriers in Industry 4.0 implementation and which were written in other languages than English and German (we included papers in German as the Industry 4.0 concept was initially conceptualized in this country). The final sample consisted of 17 papers. We added to this sample two additional works by consultancy companies and international organizations since they provided relevant inputs for our study (World Economic Forum 2014; IBM 2015). We finally coded the papers based on the obstacles/barriers highlighted.

The results of the literature review are summarized in Table 9.1, in which we also highlight the type of finding (i.e., conceptual vs. supported by empirical data) and the language of the paper (English vs. German). We identified a total of 19 obstacles/barriers that were classified into 6 categories:

- *Economic/financial* (high investments required, lack of monetary resources, lack of clearly defined economic benefits)
- *Cultural* (lack of support by top management; preferred autonomy)

Table 9.1 Organizational obstacles and barriers to Industry 4.0 implementation (Adapted from Orzes et al. 2018)

		Qiao and Wang (2012)	Hatler (2012)	Kagermann et al. (2013)	Koch et al. (2014)	World Economic Forum (2014)	Geissbauer et al. (2014)	Schlaepfer and Koch (2014)	Heng (2014)	IBM (2015)	Dixit et al. (2015)	Otuka et al. (2014)	Meißner et al. (2017)	Wenking et al. (2016)	Jäger et al. (2016)	Müller et al. (2017)	Nylander et al. (2017)	Zawra et al. (2017)	Schröder (2016)	Müller et al. (2018)	Total
Economic/financial	High investments required	C	E		E		E			E				C	E					E	8
	Lack of monetary resources									E				C		E			C		4
	Lack of clearly defined economic benefit				E	E	E														3
Cultural	Lack of support by top management				E	E	E			E				C							5
	Preferred autonomy															E					1
Competencies/ resources	Lack of skilled employees				E	E		E					E	C	E						6
	Lack of technical knowledge		E	E			E			E			E	C		E	C		C	E	10
	Complexity		E							E											2
	Need to find suitable research partner												E			E					2
Legal	Data security concerns		E	C	E	E		E		E	C	E		C	E				C	E	12
Technical	Lack of standards	C	E	C	E	E	E		C		C	E	E	C				C	C		13
	Uncertainty about the reliability of the systems		E									E									2
	Weak IT infrastructure					E		E		E		E	E	C	E						7
	Storage data				E						C							C			3
	Difficult interoperability/compatibility								C			E						C			3
	Technology immaturity				E	E	E														3
Implementation Process	Need for new business models					E									E	E	C				4
	Lack of methodical approach for implementation						E						E						C		3
	High coordination effort															E					1

Note C: Conceptual; E: Empirical

- *Competencies/resources* (lack of skilled employees, lack of technical knowledge; complexity of the Industry 4.0 application both technical and practical, need to find suitable research partner)
- *Legal* (data security concerns)
- *Technical* (lack of standards, uncertainty about the reliability of the systems, weak IT infrastructure, difficult interoperability/compatibility, technology immaturity)
- *Implementation process* (need for new business models, lack of methodical approach for implementation, high coordination effort).

Authors point out that Industry 4.0 has created some opportunities for SMEs which can use these technologies to increase their flexibility, productivity, and competitiveness (e.g., Kagermann et al. 2013; Wenking et al. 2016). They also emphasize, however, that in order to obtain such benefits, *high investments* are often required (Hatler 2012; IBM 2015). Sometimes it is therefore not easy, in particular for SMEs, to see the potential *economic benefits* of Industry 4.0 adoption (Koch et al. 2014; World Economic Forum 2014). A need therefore exists to assess results (such as increase in flexibility, productivity, and market competitiveness) in order to then measure the return on the investment (ROI).

Schröder (2016) argues that Industry 4.0 brings many opportunities but also some significant requirements: data security, finding the needed *monetary resources*, developing an *implementation approach*, and finding *skilled employees*. This sentence is confirmed by our literature review, in which *data security* appeared as a significant issue with which companies must deal in implementing Industry 4.0. To overcome this issue, standards in cryptography and security models should be developed (Kagermann et al. 2013), since with the extension of the boundaries of the company, the traditional security systems are no longer sufficient (Chen and Zhao 2012).

The development of standards and legal regulations is also essential (Wenking et al. 2016). They should be developed not only to address security concerns, but also for the rapid implementation and diffusion of Industry 4.0. Companies tend, in fact, to work on their own solutions (Wenking et al. 2016) also because there is often the fear—due to a lack of trust existing—that sharing knowledge with other companies can reduce profitability (Müller et al. 2017). Such a lack of standards leads, however, to very complex interoperability and compatibility between machines, companies, and infrastructures.

Finally, despite the huge number of articles published on Industry 4.0, the attention that has been given so far to the development of implementation models is not sufficient (Liao et al. 2017). Three articles in Table 9.1 mention indeed that a *methodical approach for implementation* is missing (Geissbauer et al. 2014; Meißner et al. 2017; Schröder 2016). To overcome this obstacle, companies need to cooperate and work together to develop compatible automation solutions, which will result in modular factory structures (Weyer et al. 2015).

9.2.2 Barriers to Innovation

The success of SMEs is strictly related to their capacity to deal with innovation. Companies that successfully incorporate innovation in their business strategy actually increase productivity and competitiveness (Cefis and Marsili 2006). The other side of the medal is that in the implementation process of (radical) innovation, companies must face several organizational obstacles and challenges, the so-called innovation barriers (IB) (e.g., D'Este et al. 2012; Madrid-Guijarro et al. 2009).

Considering that the adoption of Industry 4.0 can, to some extent, be considered a radical innovation (since it might imply a significant modification of processes, relationships with the customers and the suppliers, value proposition, or even of the business model), we believed it useful to consider in our literature review not only the papers focusing on organizational obstacles and barriers for Industry 4.0 implementation but also the broader stream of studies on **organizational barriers to innovation**. Considering the wide number of studies on this topic and its lower centrality to our analysis, we started from two recent reviews (Sandberg and Aarikka-Stenroos 2014; Madrid-Guijarro et al. 2009) rather than conducting a new keyword search.

Table 9.2 Organizational barriers to innovation

Category	Barrier	Exemplary references
Economic/financial	Lack of monetary resources	Kelley (2009)
	High investments required	Martinez and Briz (2000) and Frenkel (2003)
	Innovation cost difficult to control	Hadjimanolis (1999) and Martinez and Briz (2000)
Cultural	Lack of support from customer/supplier	Hewitt-Dundas (2006) and Mohen and Roller (2005)
	Unsupportive government	Hadjimanolis (1999) and Freel (2000)
	Paucity of external finance	Minetti (2010)
	Excessive risk	Hewitt-Dundas (2006) and Galia and Legros (2004)
	Preferred autonomy	Lynn et al. (1996)
	Unsupportive organizational structure	Baldwin and Lin (2002) and Martinez and Briz (2000)
	Restrictive mindset	Wolfe et al. (2006)
	Restrictive local culture	Riffai et al. (2012)
Technical	Technological immaturity	Chiesa and Frattini (2011)
Lack of competencies	Lack of discovery competencies	O'Connor and DeMartino (2006)
	Lack of incubation competencies	O'Connor and DeMartino (2006)
	Lack of acceleration and commercialization competencies	O'Connor and DeMartino (2006) and Story et al. (2009)
	Lack of qualified employees	Mohen and Roller (2005) and Galia and Legros (2004)
	Lack of information about technologies	Galia and Legros (2004) and Frenkel (2003)
	Inappropriate infrastructure	Iyer et al. (2006)

We identified 18 barriers that we brought back to the categories already introduced for the organizational barriers and obstacles in Industry 4.0 implementation (see Table 9.2).

One of the common problems in technological changes are **economic/financial** issues, especially for SMEs (Sandberg and Aarikka-Stenroos 2014; Mohnen and Rosa 2002). This has been confirmed by the review both on innovation and on Industry 4.0 (see Tables 9.1 and 9.2). **Cultural** issues (e.g., *unsupportive organizational structure,* *restrictive mindset*, and *preferred autonomy*) also play an important role in the introduction of new practices. Wolfe et al. (2006) emphasize that the resistance to change is due to the fact that innovation brings changes, which generate in the employees the fear of losing their job.

One barrier highlighted for innovation, but not for Industry 4.0 implementation, is the *unsupportive government*. This might be explained by the fact that many governments have launched significant investment plans to support the transition toward Industry 4.0. We have already mentioned in the introduction section the Italian plan Impresa 4.0, the Austrian Plattform Industrie 4.0, China 2025, and Thailand 4.0.

9.3 Problem Formulation

Extant Industry 4.0 literature has shed light on a wide set of organizational barriers and problems in Industry 4.0 implementation (see Sect. 9.2.1). The literature is, however, characterized by at least two significant limitations. First, most papers (58%) are published in conference proceedings or reports (not subject to a rigorous peer-review process). Second, if we compare the list of barriers highlighted in Industry 4.0 literature (see Table 9.1) with the broader set of barriers in innovation adoption highlighted by the innovation management literature (see Table 9.2), we notice that various barriers are missing in Industry 4.0 literature (such as *the unsupportive government* and *excessive risks*). The comprehensiveness of the list of barriers to Industry 4.0 identified by extant literature is therefore called into question.

The aim of this chapter is therefore to identify through a rigorous empirical analysis the main organizational barriers and issues faced by SMEs in Industry 4.0 implementation, in order to find possible solutions to the identified barriers and to propose some directions for future research. This represents a fundamental step toward the diffusion of Industry 4.0 among SMEs.

9.4 Methodology

9.4.1 Focus Group Method

Considering the novelty of the topic and the need for an in-depth exploration (Stewart and Shamdasani 1990), we adopted the focus group methodology. This research method, which was developed in medical and marketing research, is now frequently used as well in social sciences research (Parker and Tritter 2006). It has been argued to be particularly suitable for providing trustworthy insights about human behavior based on naturalistic data (Grudens-Schuck et al. 2004) and therefore, fits very well with the goals of our paper (i.e., to shed light on the organizational issues faced by SMEs in Industry 4.0 implementation).

Focus groups are typically composed of small groups of 5–12 people, in order to give everyone a chance to express his/her opinion about the topic (Krueger and Casey 2000). The participants have similar characteristics, like the knowledge of the topic or the field, so that they can provide quality data in a focused discussion. In order to be defined as a focus group, the discussion needs to have the following five characteristics: (1) participants should have similar characteristics (e.g., job role, experience, and/or culture); (2) the group should be small; (3) there should be the presence of a moderator (often a researcher; Morgan and Spanish 1984) to keep the group "focused" and generate a productive discussion; (4) the interaction among participants should be allowed; and (5) the topic should be presented before asking the questions (Krueger and Casey 2000). One of the advantages of this methodology is that it can encourage contributions from people who initially feel they have nothing to say but then participate in the discussion generated by other members of the group (Kitzinger 1995).

Four focus groups (lasting one full working day each) were organized in Italy, Austria, USA, and Thailand under the EU research project 'SME 4.0 – Industry 4.0 for SMEs'. These focus groups were scheduled on different days but in the same period and the attendees took part in them physically (not through video conferences). A standardized protocol for the focus groups was defined in order to guarantee comparability of the findings (see Sect. 9.4.2).

9.4.2 Sample Selection and Data Collection

Each focus group was attended by 13–25 CEOs or managers of 7–10 SMEs belonging to different manufacturing sectors, including electronics, industrial and agricultural equipment, furniture, and metal carpentry. Having an overview over different manufacturing sectors allowed us to identify the general issues in the implementation of Industry 4.0 in SMEs, independently from the specific sector of the company.

The reason why CEOs and technical managers were invited is that they have an overall knowledge about the topic and about the problems their company face when introducing changes in its organizational structure.

After a brief introduction by the researchers about Industry 4.0 and related concepts, the participants took part in some brainstorming sessions in which they were asked to reflect on various topics related to Industry 4.0 implementation: (1) adaptable manufacturing systems design; (2) intelligent manufacturing through information and communication technology (ICT) and cyber-physical systems (CPS); (3) automation and human–machine interaction; and (4) main barriers and difficulties for SMEs. During these sessions, the participants also wrote, on some post-its, the most important issues. After these brainstorming sessions, the issues which emerged were then discussed in detail among the participants.

9.4.3 Data Analysis

The data which emerged from the four focus groups were then coded by two researchers among the authorial team. We identified 108 elementary barriers and problems in Industry 4.0 implementation, which were then manually screened to check their validity. Five barriers were eliminated at this stage since they were not clear or too general (i.e., SMEs' risk of losing the lead, missing automated measuring systems; solving, problems when problems are over; culture → people base; technology based; lack of systems to prevent bottlenecks in single point of failure production line).

We then classified the barriers according to the six categories already introduced in the literature review section (*economic/financial, cultural, competencies/resources, legal, technical,* and *implementation process*) and reported all the results of the four countries in a single table (see Table 9.3). In such a table, we also specified if the barrier was already highlighted by previous studies both on Industry 4.0 (I4.0) and on IB, in order to have a clear idea of what is new and what is already present in the existing literature. Some barriers were assigned to more than one category since they included two or more concepts. For instance, the barrier high investments with uncertain ROI refers both to *high investments required (high investments)* and to *lack of clearly defined economic benefits (uncertain ROI)*. Similarly, the barrier product characteristics was included both in the economic/financial and in the implementation process category since in one case, the workshop participants emphasized that for low value-added products the investment in Industry 4.0 is not worthwhile, while in the other case, they highlighted that during the implementation, it is sometimes not easy to combine the need for high flexibility with higher automated processes.

The final list consisted of 103 organizational barriers and problems in Industry 4.0 adoption. These barriers will be analyzed in detail in Sect. 9.5.

9.5 Results

The focus groups highlighted several barriers and problems for Industry 4.0 implementation in SMEs (see Table 9.3). As mentioned above, we classified them according to the six categories used in the literature review (*economic/financial, cultural, competencies/resources, legal, technical,* and *implementation process*).

Most of the participants in the four countries pointed out that the investments required for the implementation of Industry 4.0 are very high, both in terms of money and time required. Italian and Thai managers and CEOs emphasized that not only are the required investments high, but also that the ROI is often not very clear. This can be due to the unclear potential of the different technologies or to the difficulties

Table 9.3 Organizational barriers and problems for Industry 4.0 implementation

Barriers		Liter.	Italy	Austria	Thailand	USA
Economic/ financial	High investments required	I4.0/IB	High investments (with uncertain ROI) High cost and high effort For what size of company does an investment makes sense?	Investment (machineries, construction, …)	Investment in production systems and training	High tool costs and time investments
	Lack of monetary resources	I4.0/IB			Capital/need a capital support	Capital to invest
	Lack of clearly defined economic benefit	I4.0	Measurement of results is difficult (High investments with) uncertain ROI	Investments/ amortization	Do we recognize the impact on our company?	
	Product characteristics				Value of the product (low value is not worth)	
Cultural	Lack of trust between partners	I4.0	Lack of intersectoral cooperation/exchange			
	Lack of support by top management	I4.0/IB	Clear direction in the company is necessary Commitment of the management is important Mentality of the enterprise Self-discipline Communication or transparency is missing Courage for new things Lack of total vision of Industry 4.0 for logistics	Top management support Mostly an unclearly defined part of the corporate strategy Indecisive top management	Top management has no awareness in Industry 4.0 Mindset overview > process, plan, customer …	

(continued)

Table 9.3 (continued)

Barriers		Liter.	Italy	Austria	Thailand	USA
	Preferred autonomy	I4.0				Lack of easy "best practice sharing with other companies"
	Restrictive mindset	IB	Lack of willingness to take risks People must "want" to introduce I4.0 instead of being forced			"the way we've always done it" Changing way of thinking to modern methods
	Unsupportive organiz. structure	IB			HQ's decision-making management system Complexity of the organization.	
	Acceptance of employees	IB	Acceptance of employees Communication to employees Employees are not yet aware of the changes Employees are afraid of losing jobs due to new tech.	Acceptance of new technologies Integration of employees No acceptance (open rejection)	Attitude of workers to accept and change	Company culture due to fear of automation
	Lack of support from customer/ supplier				Customers who are ready to support new systems Don't know the true needs of customers or the market	Squeezed in the middle of supply chain, not seen as strategic partner
	Focus on day-to-day operations					No time to sit back and strategize

(continued)

Table 9.3 (continued)

Barriers		Liter.	Italy	Austria	Thailand	USA
	Awareness about the potential of robots					Need to believe that if I buy a robot for a job, I will still be able to use it when the job is done
	Lack of support from the IT department				We are just a production base. We follow decisions of the parent company	
Competencies/ resources	Lack of skilled employees/lack of technical knowledge	I4.0/IB	Lack of qualifications and training of employees	Know-how	Lack of expertise of personnel for supporting Industry 4.0	Resources (people)
				ICT barriers for employees		Training requirements
	Complexity	I4.0	Complexity of Industry 4.0			
	High coordination effort	I4.0	Implementation requires major changes from suppliers			
			Not all suppliers are prepared			
	Lack of knowledge of Industry 4.0 technologies and technical providers		A complete overview of the market is not yet available		Solution provider	
			Missing overview of what makes sense for the SME		Lack of access to the source of information and technology	
			Analysis of the needs for Industry 4.0		How SMEs access the source of available technology	
			SMEs do not have an own "department" for Industry 4.0		Knowledge of the technologies	

Table 9.3 (continued)

Barriers		Liter.	Italy	Austria	Thailand	USA
	Factory layout constraints		Small spaces and sometimes confined conditions No space for automation of the logistics or internal transports Current buildings are not designed for automating the internal transports Future factory planning needs to be adapted in the future			
Legal	Data security concerns	I4.0		Data security	Data security	
	Lack of support from government	IB			SMEs problems are human resources, capital and policy	
	Bureaucracy		Bureaucracy as a hurdle for a dynamic implementation of I4.0 (certifications, customs, taxes)			
	Restrictive laws and regulations			Legal restrictions	Restricted laws and regulations	
Technical	Uncertainty about the reliability of the systems	I4.0	Uncertainty about the reliability of automated transport or storage systems			
	Weak IT infrastructure	I4.0	Consistency of the IT still does not exist	Integration of existing infrastructure		

(continued)

Table 9.3 (continued)

Barriers		Liter.	Italy	Austria	Thailand	USA
	Difficult interoperability/ compatibility	I4.0	Data silos without communication between each other	Old data -> need for adaption	Collaboration between SMEs and logistics companies. Each company's system is differently interlinked	Integration of new technology with old equipment
			Missing interfaces with suppliers	Interfaces/ communication		
	State of machine park					Current state of the machine park
Implementation process	Lack of methodical approach for implementation	I4.0	Missing toolset for the introduction of Industry 4.0	Few best-practice-examples	What are the initial steps to improve or implement?	Lack of formalized information on Industry 4.0 implementation
			There are no methods and approaches for the correct introduction of Industry 4.0	Clear formulation of objectives vs. solutions	When an automatic machine is introduced. How do humans work?	
	Required time for implementation	I4.0 IB	Required time for implementation The market and competition will not stop Be aware that time is needed to deal with it and implement industry 4.0	Time capacities for project implementation		High tool costs and time investments
	Changes required for implementing Industry 4.0	IB	Adjustment of the company			
	Difficulties in demand forecasting			Fluctuation of order volume	Difficult demand forecasting	
	Product characteristics		Flexibility is the highest priority			Low quantity/frequency of orders from customers
			Modification of the process according to changing customer needs (flexibility, individual products) Difficult use with lot size 1 or individual products			

faced by SMEs in measuring the results. Another interesting **economic/financial** issue reported during the workshops is the *value of the product*. According to some participants, it is not worth adopting Industry 4.0 if the products produced have a low value. This is particularly true in countries characterized by low labor costs, like Thailand.

Cultural issues (such as *lack of support by top management, lack of trust between partners, unsupportive organizational structure, acceptance of employees,* and *focus on day-to-day operations*) appear to be particularly important for Industry 4.0 implementation. Around one-third of the barriers highlighted in the focus groups belong, in fact, to this category.[1] Among these barriers, we noticed the corporate culture/mentality in Austria, Italy, and Thailand (e.g., the *lack of cooperation among functions/departments*), employee resistance, and missing top management vision on Industry 4.0 in all four focus groups (due to their poor knowledge of Industry 4.0 and their fear of losing work), and risk aversion in Austria and Thailand. Our focus groups highlighted that the *lack of support by top management (clear direction of the company is necessary, lack of communication and transparency, lack of total vision of Industry 4.0)* is even more important than *the resistance (acceptance) of employees.*

As far as **competencies/resources** barriers are concerned, in all the four analyzed countries, SMEs struggle to find qualified employees with the required technical competences. This means that the *lack of technical knowledge* is a common factor for SMEs independently of the economic and cultural environment. Another significant barrier highlighted by the focus group organized in Italy is the *high coordination effort.* This barrier emphasizes the fact that Italian SMEs perceive it to be important to cooperate and develop common solutions for Industry 4.0. US companies reported that they prefer to work autonomously at their own solutions. This could be due to cultural differences between Italy and USA as well as to the different resource endowment of SMEs in the

[1]The share has been calculated by dividing the number of barriers included in the category "cultural" by the total number of barriers reported in Table 9.3. Barriers which were repeated in more than one country have been counted only once.

two countries. Another important issue concerns the *capital/investments* required.

Moving to the **legal** barriers, another difference among the analyzed countries can be observed. In Austria, Italy, and Thailand *bureaucracy* and *restrictive laws and regulations* are seen as a hurdle for the implementation of Industry 4.0 while in the USA, no managers/CEOs reported this issue.

SMEs in Thailand, Italy, and USA highlight that they have different ICT systems and their data are stored in different silos that often do not communicate with each other. Furthermore, they argue that their buildings are not designed for automating internal transports. These **technical** issues make the implementation of Industry 4.0 more difficult.

Finally, a very important barrier highlighted in the four countries is that a methodical approach for implementing industry 4.0 is missing (**implementation process**). This is due in part to the novelty of the topic, but also to the fact that each company has its own needs, and these kinds of systems need to be adapted to them.

9.6 Discussion

Our empirical analyses (focus groups) confirmed most of the barriers identified by previous literature (e.g., Qiao and Wang 2012; Hatler 2012; Koch et al. 2014; Zawra et al. 2017; Müller et al. 2018) (see Table 9.4).

Previous studies (e.g., Müller et al. 2017) highlighted that SMEs struggle to obtain the resources and tools needed in order to implement Industry 4.0. This has been confirmed by our focus groups, in which participants cited the difficulty of finding skilled employees and the struggle to find the required capital as crucial issues for the implementation of Industry 4.0. These barriers emerged in all four countries (Austria, Italy, Thailand, and USA), meaning that they are independent of the cultural and economic environment.

Another barrier which has been confirmed by the focus group is the *lack of a methodical approach for implementation* (Liao et al. 2017). CEOs and managers of SMEs located in all four countries reported that

Table 9.4 Confirmed organizational barriers and problems for Industry 4.0 implementation

Category	Barrier
Economic/financial	High investments required Lack of clearly defined economic benefit
Cultural	Lack of support by top management Preferred autonomy
Competencies/resources	Lack of skilled employees Lack of technical knowledge Complexity
Legal	Data security concerns
Technical	Weak IT infrastructure Difficult interoperability/compatibility
Implementation process	Lack of methodical approach for implementation High coordination effort

a model for implementing Industry 4.0 is missing. The words used during the workshops were: *"There are no methods and approaches for the correct introduction of Industry 4.0," "There are limited support resources and a lack of formalized, distilled information on how to implement industry 4.0," "Few Best-Practice-Examples."*

A wide set of new barriers have also been identified through our empirical analysis. After significant work to compare the barriers emerging from the focus groups to the ones highlighted by previous literature (even if the terminology used was different), we concluded that 11 new organizational barriers should be considered (see Table 9.5).

Companies report the desire and the need to *cooperate with customers and suppliers* in order to develop common solutions based on Industry 4.0 (Müller et al. 2017). They also reported that it is very difficult to coordinate themselves with other companies and do joint investments. This can be due to a lack of innovation mentality, or a very rigid organizational structure. Some focus group participants also mentioned that the "real" needs of their customers are sometimes not clear/known and this makes cooperation more difficult.

The second result which emerged from the workshop is that SMEs have some problems in implementing Industry 4.0, because they have to *focus on day-to-day operations*. This can be also related to the lack of

Table 9.5 Proposed organizational barriers and problems (not highlighted by previous Industry 4.0 literature)

Category	Barrier
Cultural	Lack of support from customer/supplier
	Focus on day-to-day operations
	Awareness about the potential of robots
	Lack of support from the IT department
Competencies/resources	Lack of knowledge of Industry 4.0 technologies and technical providers
	Factory layout constraints
Technical	State of machine park
Implementation process	Required time for implementation
	Changes required for implementing Industry 4.0
	Difficulties in demand forecasting
	Product characteristics

monetary resources and to the fact that they do not have a specialized department dedicated to the topic.

Another significant issue is related to the *factory layout*. This barrier is present only in Italy, especially in South Tyrol, perhaps partly due to the low availability of building land and its high costs. As a result, SMEs cannot easily enlarge their factories. Most of the SMEs are also located in old buildings in which some space constraints are present: *small spaces and confined space, no space for automation of logistics and internal transport*.

Furthermore, US participants reported that the current state of the machine park is sometimes an obstacle in the introduction of IoT and CPS. There are companies which have already seen the opportunity in this challenge and established a new successful business model, i.e., to modify old machines by equipping them with sensors and connecting them to the network (Wenking et al. 2016).

Finally, a set of new barriers was related to the implementation process (*time required for implementation, changes required, difficulties in demand forecasting*, and *product characteristics*).

9.7 Conclusions

The interest devoted by managers, policy-makers, and researchers to the Industry 4.0 topic has grown exponentially during the last few years (Liao et al. 2017). Despite this increasing interest, a methodical approach for implementation is still missing.

The main objective of this study was to shed empirical light on the main organizational requirements for Industry 4.0 implementation in SMEs. We first reviewed the relevant literature. Considering the novelty of the topic, we considered not only the studies on organizational obstacles and barriers for Industry 4.0 implementation but also the broader literature on barriers to innovation. We then conducted some focus groups in four countries (Italy, Austria, Thailand, and USA) in order to empirically validate the list of barriers and issues emerging from the literature review. The focus groups confirmed most of the barriers identified by extant literature (see Table 9.4). They also allowed us to highlight a set of additional barriers not considered by previous studies (see Table 9.5).

We contributed to the scientific debate in at least three significant ways. First, to the best of our knowledge, our study is among the first to empirically highlight a comprehensive set of barriers and problems for Industry 4.0 implementation. This way we might open a debate on a topic that is expected to rise significantly in the next few years. Second, we identified 11 new barriers not highlighted by previous literature. Third, we showed that SMEs perceive a strong need for methodical approaches for Industry 4.0 implementation, thus calling for future research in this area.

Our findings also have strong implications for managers and policy-makers. The identified list of barriers and problems in Industry 4.0 implementation can, for instance, be used by managers to define a set of organizational requirements that should be fulfilled for an efficient and effective implementation of Industry 4.0. Similarly, policy-makers can identify a set of measures—such as incentives, roadmaps, consultancy services—to facilitate SMEs in Industry 4.0 adoption.

The results of our study are characterized by two limitations. First, we adopted a focus group research methodology. Despite several actions being performed to enhance validity and reliability, our findings cannot be generalized to a broader population. Second, our sample consisted of 37 SMEs from four countries (Italy, Austria, Thailand, and USA). Caution is therefore needed in extending our results to other contexts. Future research could empirically test our findings on a wider and more heterogenous sample.

References

Baldwin, J., and Z. Lin. 2002. Impediments to Advanced Technology Adoption for Canadian Manufacturers. *Research Policy* 31 (1): 1–18.

Cefis, E., and O. Marsili. 2006. Innovation Premium and the Survival of Entrepreneurial Firms in the Netherlands. In *Entrepreneurship, Growth, and Innovation*, 183–198. Boston, MA: Springer.

Chiesa, V., and F. Frattini. 2011. Commercializing Technological Innovation: Learning From Failures in High-Tech Markets. *Journal of Product Innovation Management* 28 (4): 437–454.

Culot, G., G. Nassimbeni, G. Orzes, and M. Sartor. 2018. Industry 4.0: Why a Definition Is Not Needed (Just Yet). In *Proceedings of the 25th International EurOMA Conference*, 51.

D'Este, P., S. Iammarino, M. Savona, and N. von Tunzelmann. 2012. What Hampers Innovation? Revealed Barriers Versus Deterring Barriers. *Research Policy* 41 (2): 482–488.

Dixit, M., J. Kumar, and R. Kumar. 2015. Internet of Things and Its Challenges. In *2015 International Conference on Green Computing and Internet of Things (ICGCIoT)*, 810–814. IEEE.

Freel, M.S. 2000. Barriers to Product Innovation in Small Manufacturing Firms. *International Small Business Journal* 18 (2): 60–80.

Frenkel, A. 2003. Barriers and Limitations in the Development of Industrial Innovation in the Region. *European Planning Studies* 11 (2): 115–137.

Galia, F., and D. Legros. 2004. Complementarities Between Obstacles to Innovation: Evidence From France. *Research Policy* 3 (8): 1185–1199.

Geissbauer, R., S. Schrauf, V. Koch, and S. Kuge. 2014. Industrie 4.0–Chancen und Herausforderungen der vierten industriellen Revolution. *PricewaterhouseCoopers (PWC)* 227: 13.

Geissbauer, R., J. Vedso, and S. Schrauf. 2016. *Industry 4.0: Building the Digital Enterprise*. PwC, UK.

Grudens-Schuck, N., B.L. Allen, and K. Larson. 2004. *Methodology Brief: Focus Group Fundamentals*, 12. Ames: Extension Community and Economic Development Publications.

Hadjimanolis, A. 1999. Barriers to Innovation for SMEs in a Small Less Developed Country (Cyprus). *Technovation* 19 (9): 561–570.

Hatler, M. 2012. Industrial Wireless Sensor Networks: Trends and Developments. *InTech* 59: 9–10.

Heng, S. 2014. *Industry 4.0: Upgrading of Germany's Industrial Capabilities on the Horizon*. Available at SSRN. https://ssrn.com/abstract=2656608.

Hewitt-Dundas, N. 2006. Resource and Capability Constraints to Innovation in Small and Large Plants. *Small Business Economics* 26 (3): 257–277.

Holmström, J., M. Holweg, S.H. Khajavi, and J. Partanen. 2016. The Direct Digital Manufacturing Evolution: Definition of a Research Agenda. *Operations Management Research* 9 (1): 1–10. https://doi.org/10.1007/s12063-016-0106-z.

Hölzl, W., and J. Janger. 2012. *Innovation Barriers Across Firms and Countries*. WIFO Working Papers (No. 426).

IBM. 2015. *Was kann Industrie 4.0? Und können Sie das auch?* Available at https://www-935.ibm.com/services/multimedia/Whitepaper_Industrie_4.0_screen.pdf.

Iyer, G.R., P.J. LaPlaca, and A. Sharma. 2006. Innovation and New Product Introductions in Emerging Markets: Strategic Recommendations for the Indian Market. *Industrial Marketing Management* 35 (3): 373–382.

Jäger, J., O. Schöllhammer, M. Lickefett, and T. Bauernhansl. 2016. Advanced Complexity Management Strategic Recommendations of Handling the 'Industrie 4.0' Complexity for Small and Medium Enterprises. *Procedia CIRP* 57: 116–121. https://doi.org/10.1016/j.procir.2016.11.021.

Kagermann, H., J. Helbig, A. Hellinger, and W. Wahlster. 2013. *Recommendations for Implementing the Strategic Initiative INDUSTRIE 4.0, Securing the Future of German Manufacturing Industry, Final Report of the Industrie 4.0 Working Group*. Frankfrut: Forschungsunion.

Kelley, D. 2009. Adaptation and Organizational Connectedness in Corporate Radical Innovation Programs. *Journal of Product Innovation Management* 26 (5): 487–501.

Koch, V., S. Kuge, R. Geissbauer, and S. Schrauf. 2014. *Industry 4.0: Opportunities and Challenges of the Industrial Internet*. Strategy & PwC, UK.

Kitzinger, J. 1995. Qualitative Research: Introducing Focus Groups. *BMJ* 311: 299–302. https://doi.org/10.1136/bmj.311.7000.299.

Krueger, R.A., and M.A. Casey. 2000. *Focus Groups: A Practical Guide for Applied Research*. Thousand Oaks, CA: Sage.

Liao, Y., F. Deschamps, E.D.F.R. Loures, and L.F.P. Ramos. 2017. Past, Present and Future of Industry 4.0—A Systematic Literature Review and Research Agenda Proposal. *International Journal of Production Research* 55 (12): 3609–3629. https://doi.org/10.1080/00207543.2017.1308576.

Lynn, G.S., J.G. Morone, and A.S. Paulson. 1996. Marketing and Discontinuous Innovation: The Probe and Learn Process. *California Management Review* 38 (3): 8–37.

Madrid-Guijarro, A., D. Garcia, and H. Van Auken. 2009. Barriers to Innovation Among Spanish Manufacturing SMEs. *Journal of Small Business Management* 47 (4): 465–488. https://doi.org/10.1111/j.1540-627X.2009.00279.x.

Martinez, M.G., and J. Briz. 2000. Innovation in the Spanish Food & Drink Industry. *The International Food and Agribusiness Management Review* 3 (2): 155–176.

McKinsey Digital. 2015. Industry 4.0: How to Navigate the Digitalization of the Manufacturing Sector. https://www.mckinsey.com/~/media/mckinsey/business%20functions/operations/our%20insights/industry%2040%20how%20to%20navigate%20digitization%20of%20the%20manufacturing%20sector/industry-40-how-to-navigatedigitization-of-the-manufacturing-sector.ashx. Accessed on 20 Dec 2018.

Meißner, A., R. Glass, C. Gebauer, S. Stürmer, and J. Metternich. 2017. Hindernisse der Industrie 4.0–Umdenken notwendig? *ZWF Zeitschrift Für Wirtschaftlichen Fabrikbetrieb* 112 (9): 607–611. https://doi.org/10.3139/104.111787.

Minetti, R. 2010. Informed Finance and Technological Conservatism. *Review of Finance* 15 (3): 633–692.

Mohen, P., and L. Roller. 2005. Complementarities in Innovation Policy. *European Economic Review* 49: 1431–1450.

Mohnen, P., and J.M. Rosa. 2002. Barriers to Innovation in Service Industries in Canada. In *Institutions and Systems in the Geography of Innovation*, 231–250. Boston, MA: Springer and Kluwer Academic Publishers. https://doi.org/10.1007/978-1-4615-0845-8_11.

Müller, J.M., O. Buliga, and K.I. Voigt. 2018. Fortune Favors the Prepared: How SMEs Approach Business Model Innovations in Industry 4.0. *Technological Forecasting and Social Change* 132: 2–17. https://doi.org/10.1016/j.techfore.2017.12.019.

Müller, J.M., L. Maier, J. Veile, and K.I. Voigt. 2017. Cooperation Strategies Among SMEs for Implementing Industry 4.0. In *Proceedings of the Hamburg International Conference of Logistics (HICL)*, 301–318. https://doi.org/10.15480/882.1462.

Nylander, S., A. Wallberg, and P. Hansson, P. 2017. Challenges for SMEs Entering the IoT World: Success Is About so Much More Than Technology. In *Proceedings of the Seventh International Conference on the Internet of Things*, 16. https://doi.org/10.1145/3131542.3131547.

O'Connor, G.C., and R. DeMartino. 2006. Organizing for Radical Innovation: An Exploratory Study of the Structural Aspects of RI Management Systems in Large Established Firms. *Journal of Product Innovation Management* 23 (6): 475–497.

Orzes, G., E. Rauch, S. Bednar, and R. Poklemba. 2018. Industry 4.0 Implementation Barriers in Small and Medium Sized Enterprises: A Focus Group Study. *2018 IEEE International Conference on Industrial Engineering and Engineering Management (IEEM),* 1348–1352. https://doi.org/10.1109/IEEM.2018.8607477.

Otuka, R., D. Preston, and E. Pimenidis. 2014. The Use and Challenges of Cloud Computing Services in SMEs in Nigeria. *Proceedings of the European Conference on Information Management and Evaluation* 43 (10): 47–55. https://doi.org/10.5281/zenodo.1127458.

Parker, A., and J. Tritter. 2006. Focus Group Method and Methodology: Current Practice and Recent Debate. *International Journal of Research & Method in Education* 29 (1): 23–37. https://doi.org/10.1080/01406720500537304.

Qiao, H.S., and G.L. Wang. 2012. An Analysis of the Evolution in Internet of Things Industry Based on Industry Life Cycle Theory. *Advanced Materials Research* 430: 785–789. https://doi.org/10.4028/www.scientific.net/AMR.430-432.785.

Riffai, M.M.M.A., K. Grant, and D. Edgar. 2012. Big TAM in Oman: Exploring the Promise of On-line Banking, Its Adoption by Customers and the Challenges of Banking in Oman. *International Journal of Information Management* 32 (3): 239–250.

Rüßmann, M., M. Lorenz, P. Gerbert, M. Waldner, J. Justus, P. Engel, and M. Harnisch. 2015. Industry 4.0: The Future of Productivity and Growth in Manufacturing Industries. *Boston Consulting Group* 9 (1): 54–89.

Sandberg, B., and L. Aarikka-Stenroos. 2014. What Makes It so Difficult? A Systematic Review on Barriers to Radical Innovation. *Industrial Marketing Management* 43 (8): 1293–1305. https://doi.org/10.1016/j.indmarman.2014.08.003.

Schlaepfer, R.C., and M. Koch. 2014. Industry 4.0 Challenges and Solutions for the Digital Transformation and Use of Exponential Technologies. Finance, Audit Tax Consulting Corporate, Zurich, Swiss.

Schröder, C. 2016. *The Challenges of Industry 4.0 for Small and Medium-Sized Enterprises.* Bonn, Germany: Friedrich-Ebert-Stiftung.

Stewart, D.W., and P.N. Shamdasani. 1990. *Focus Groups: Theory and Practice*, 20. Sage. https://doi.org/10.2307/3172875.

Story, V., S. Hart, and L. O'Malley. 2009. Relational Resources and Competences for Radical Product Innovation. *Journal of Marketing Management* 25 (5–6): 461–481.

Wenking, M., C. Benninghaus, and T. Friedli. 2016. Umsetzungsbarrieren und-lösungen von Industrie 4.0. *ZWF Zeitschrift Für Wirtschaftlichen Fabrikbetrieb* 111 (12): 847–850.

Weyer, S., M. Schmitt, M. Ohmer, and D. Gorecky. 2015. Towards Industry 4.0-Standardization as the Crucial Challenge for Highly Modular, Multi-vendor Production Systems. *IFAC-Papersonline* 48 (3): 579–584.

Wolfe, R., P.M. Wright, and D.L. Smart. 2006. Radical HRM Innovation and Competitive Advantage: The Moneyball Story. *Human Resource Management 45* (1): 111–145.

World Economic Forum. 2014. *Industrial Internet of Things, Unleashing the Potential of Connected Products and Services*. Geneva, Switzerland: World Economic Forum Industrial Internet Survey.

Zawra, L.M., H.A. Mansour, A.T. Eldin, and N.W. Messiha. 2017. Utilizing the Internet of Things (IoT) Technologies in the Implementation of Industry 4.0. In *International Conference on Advanced Intelligent Systems and Informatics*, 798–808. https://doi.org/10.1007/978-3-319-64861-3_75.

10

Smart SME 4.0 Implementation Toolkit

Apichat Sopadang, Nilubon Chonsawat and Sakgasem Ramingwong

10.1 Introduction

The term SMEs is normally used to describe businesses that are small or medium in size by which their personnel numbers or investment fall below certain limits (EIP 2005; MOBIE 2014; OSMEP 2017). For example, the European SMEs are those who employ fewer than 250 persons and have an annual turnover not exceeding 50 million EUR, and/or an annual balance sheet total not exceeding 43 million EUR. In Australia, an SME has fewer than 200 employees, while in Thailand, SMEs are those having total asset value of not more than 200 million THB (less than 5 million EUR) and fewer than 200 employees.

A. Sopadang · S. Ramingwong (✉)
Center of Excellence in Logistics and Supply Chain Management, Chiang Mai University, Chiang Mai, Thailand
e-mail: sakgasem.ramingwong@cmu.ac.th

N. Chonsawat
Faculty of Engineering, Chiang Mai University, Chiang Mai, Thailand
e-mail: nilubon_chon@cmu.ac.th

D. T. Matt et al. (eds.), *Industry 4.0 for SMEs*,
https://doi.org/10.1007/978-3-030-25425-4_10

The definition of an SME may vary from each socioeconomic perspective and policy development. However, it is common that SMEs are important to the economy in terms of number, employment, and export. For example, 99.3% of UK private sector businesses are SMEs. SMEs in Poland generate almost 50% of the GDP, while Australian SMEs makeup 97% of all Australian businesses, produce one-third of total GDP and employ 4.7 million people. SMEs represent 90% of all goods exporters and over 60% of services exporters. In the case of Thailand, currently, there are more than 3 million SME operators, accounting for 42.2% of Thailand's GDP, expanding 4.8% annually. Thai SMEs account for 99% of Thailand's enterprises and 78% of the total employment in the country. More than 90% of Thai exports are from SMEs.

SMEs are an important contribution to the creation of new jobs. SMEs are often characterized as reactive, resource limitations, informal strategies, and flexible structures (Hudson et al. 2001; Qian and Li 2003). SMEs are usually characterized by a high level of environmental uncertainty. The OECD report suggests that competitiveness of SMEs is dependent on the role of the owner or manager, intelligence management, technologically suitable equipment, and strategic capability (innovation and flexibility). It is noted that, while technology plays an increasingly important role in all aspects of competitiveness, management methods, the organization of the firm and the training of its staff are also very significant (OECD 1993).

The chapter aims at examining the readiness of SMEs toward the SME 4.0 concept. The research uses Thai SMEs as case study.

10.2 Background and Literature Review

SME 4.0 is a new, modified version of Industry 4.0 (I4) for SMEs. By which the term "Industry 4.0" refers to modern industrial concepts, empowering by technological advancement. Industry 4.0 concepts encourage the industrial systems to be connected and interacted and, thus, make appropriate decisions based on gathered and analyzed data across the manufacturing processes. The production can be faster, more flexible, and more efficient (Lee et al. 2015; Rüßmann et al. 2015). The concepts of Smart Factory, Cyber-physical Systems, and self-organization are among the key drivers (Lasi et al. 2014; Stock and Seliger 2016; Pereira and Romero 2017).

In order to assess SMEs with Industry 4.0 concept, the scope of SMEs 4.0 is now of interest. The basic idea of SMEs 4.0 is captured per Industry 4.0 dimensions and characteristics and modeled as "SMEs 4.0 assessment modules."

At first, the keyword "Industry 4.0" is selected to search the published papers from 2015 to 2018. The literature review is conducted considering by following electronic databases: Google Scholar, Web of Science, and Elsevier.

As there are many perspectives related to Industry 4.0; however, the chapter chooses to focus on key factors related to Organization and Management. Organization Management actions are divided into three levels, i.e., Strategic Management, Management Planning, and Management Control/Operational Control. The actions include objective/goal setting, resource determination/allocation, and task assignment/resources utilization, while management functions include, planning, organizing/staffing/directing, and controlling.

Figure 10.1 illustrates the Organization Management scope used in this chapter.

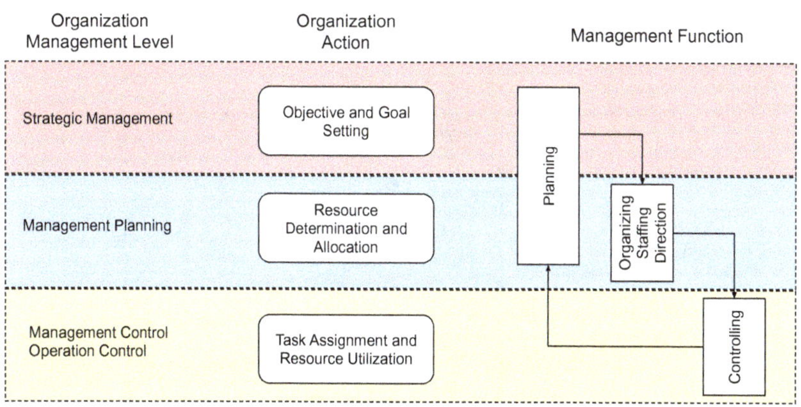

Fig. 10.1 Organization management scope

Hence, the chapter focuses on technological resource and human resource managements in relationship to the Industry 4.0 concept. Here, areas of interest are divided into four categories, i.e., Information Technology, Production and Operations, Automation, and Human Resources.

The significance of different factors is described in the following subsections.

10.2.1 Information Technology

Information Technology (IT) enables an environment that controls the physical operation and allows the collection of data. Advance IT is a new factor in the Industry 4.0 concept. It adopts and requires IT infrastructure for data acquisition, collection, and sharing excellent performance in the manufacturing system (Zhong et al. 2017). IT dimension comprises of four factors, i.e., equipment infrastructure, IT system, information sharing, and cloud based (see Fig. 10.2).

10.2.1.1 Equipment Infrastructure

The primary infrastructure includes IT resource, networking equipment, and hardware considered as a piece of vital equipment for

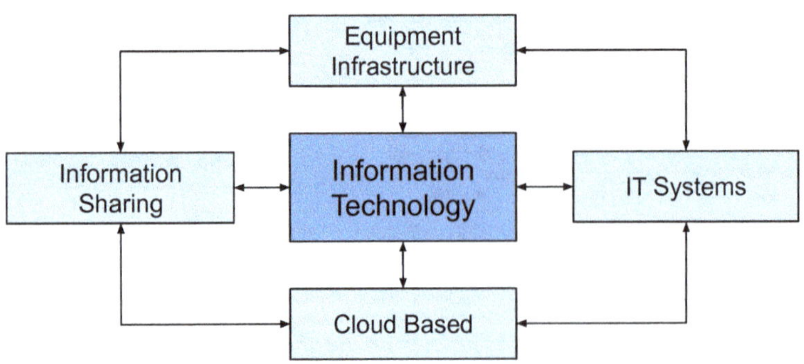

Fig. 10.2 Information technology factors

implementing and adopting new technology (Dombrowski et al. 2017). The equipment is able to be flexible and changes to adapt to better value creation (Stock and Seliger 2016). All machines and devices are already being prepared to support Industry 4.0 and future requirements (Agca et al. 2015; Lichtblau et al. 2015).

10.2.1.2 IT System

IT systems are fundamental to control the potential and effectiveness of Industry 4.0. It supports and integrates all the organization and includes the operation, production, and process (Agca et al. 2015; Lichtblau et al. 2015). The company has the readiness of IT technology to support the business. It will also use the IT security for data protection (Dombrowski et al. 2017). Modern IT, such as Big Data can provide optimized decision-making in the production planning, process, and management (Schumacher et al. 2016; Qian et al. 2017).

10.2.1.3 Information Sharing

Intelligent manufacturing uses the advantage of information to achieve flexible manufacturing processes. This process requires real-time data and collection with a collaboration between the production department, workers, and information systems (Lichtblau et al. 2015; Leyh et al. 2016). It can allow the information flows in all processes and, as such, delivers in the manufacturing and across the supply chains (Zhong et al. 2017). Therefore, the organization shares information and data resources more efficiently.

10.2.1.4 Cloud Based

Cloud-based manufacturing is a requirement in the concept of Industry 4.0 with intelligent management (Schumacher et al. 2016). It enables data to be generated in multiple locations and can transfer to data center stores for analysis (Agca et al. 2015; Leyh et al. 2016); the system covers all the production and resources (Dombrowski et al. 2017; Zhong et al. 2017). Cloud system allows for remote use of all devices, machines, and production communication. This system provides a different service. The company can operate cloud based in the field of Industry 4.0 to increase effectiveness and efficiency (Leyh et al. 2016).

10.2.2 Production and Operations

Industry 4.0's vision is the interconnecting of intelligent systems. It has self-control in the processes and manufacturing system. Consequently, the operation and production process, based on technology under Industry 4.0, will provide an innovative value-added process. As such, it provides more flexibility, data reliability, and increases operational efficiency (Dombrowski et al. 2017). The Production and Operations dimension comprises of four factors, i.e., innovation management, data analytics, horizontal/vertical data integration, and expert system (see Fig. 10.3).

10.2.2.1 Innovation Management

Industry 4.0 offers the opportunity to develop business models which use advanced technology and innovation management (Lichtblau et al. 2015).

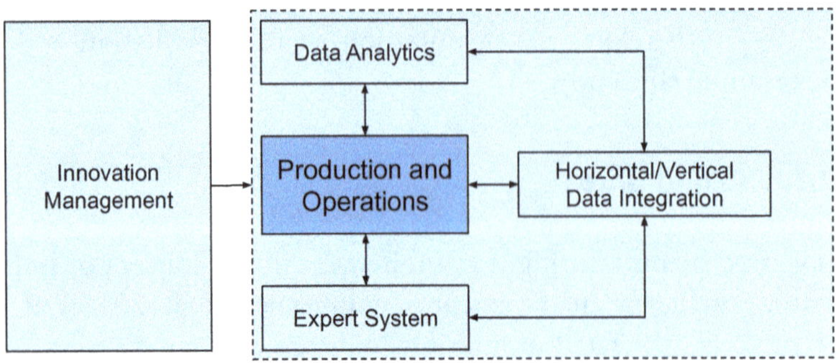

Fig. 10.3 Production and operations factors

10.2.2.2 Data Analytics

Data analytics is a gathering and assessment of data from many different sources. The data are enabled to collect and be comprehensive to make decisions in the business and operations (Lichtblau et al. 2015).

It includes manufacturing systems and infrastructure, such as information systems (Zhong et al. 2017). The systems must be standard for supporting real-time decision-making and management (Dombrowski et al. 2017).

10.2.2.3 Horizontal/Vertical Data Integration

Companies plan to integrate the system to link processes and traceability solutions. Horizontal/vertical data integration can be, for example, Enterprise Resource Planning (ERP), a system that has efficiency in Industry 4.0. It is a concept of an interconnected and intelligent factory in the production system that communicates directly with the overlying IT systems and value-adding process (Lichtblau et al. 2015; Leyh et al. 2016)

10.2.2.4 Expert Systems

Expert systems (ESs) are a knowledge base that involves the knowledge and experience from experts and expressed in specific structured formats. Applications are developed to solve complex problems in manufacturing. In Industry 4.0, an expert system allows the worker to monitor the organization's activity and control repercussions of the manufacturing (Pan et al. 2015).

10.2.3 Automation

In the factory, operations and automation are the main focus of the Industry 4.0 vision. It is a vision of autonomous production in self-optimization and factory management—an environment in an enterprise

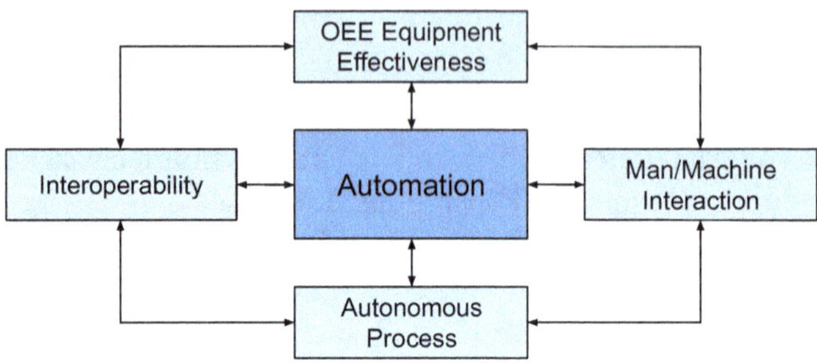

Fig. 10.4 Automation factors

in which the physical and cyber are combined as one (Stock and Seliger 2016). The automation dimension comprises of four factors, i.e., OEE equipment effectiveness, man-machine interaction, autonomous process, and M2M machine connectivity (see Fig. 10.4).

10.2.3.1 OEE Equipment Effectiveness

Overall Equipment Effectiveness (OEE) is a method for assessing the total equipment performance and efficiency. It shows the degree that the equipment is doing and what it is supposed to do. In the current environment for SMEs, the reliability of tools and OEE are the main components for increasing profitability and performance of manufacturing systems. OEE is also a suitable analytical performance evaluation tool for SMEs (Yazdi et al. 2018).

10.2.3.2 Man-Machine Interaction

Industry 4.0 has become more complex and operates as an automatic device. Workers can work together collaboratively with the advanced machinery. For the higher complexity and more control structures, a better quality of cooperation and communication between human and machine is required. The technology and equipment are able to support the change in other work tasks flexibly (Stock and Seliger 2016).

10.2.3.3 Autonomous Process

The production and operation environment in the manufacturing systems are self-organized without human intervention (Lichtblau et al. 2015). The manufacturing equipment and tools can be characterized by the application of advanced automatic machines and robotics (Stock and Seliger 2016). All machines and operating systems can be controlled through smart devices and an automation process (Agca et al. 2015).

10.2.3.4 M2M Machine Connectivity

Machine-to-machine communication or interoperability is where the systems consist of the interaction between intelligent production systems (Qian et al. 2017). It can link with each other device for easy, secure, and fixed data exchange. In other words, controlling, integrating, and coordinating processes. It provides that data accessing and processing of all machines and systems are fully integrated (Agca et al. 2015; Schumacher et al. 2016).

10.2.4 Human Resource

The requirement in manufacturing jobs will contain more knowledge of work. The workers must have potential in both short-term and hard-to-plan tasks. They can integrate the knowledge with the intelligent system, such as data analytics, decision-making, and engineering activities as end-to-end engineering (Stock and Seliger 2016). The human resource dimension comprises of two groups, i.e., technical and non-technical (see Fig. 10.5).

10.2.4.1 Technical

In intelligent systems which are self-controlled and guide the employees in the job task, human skill, and knowledge are required in Industry

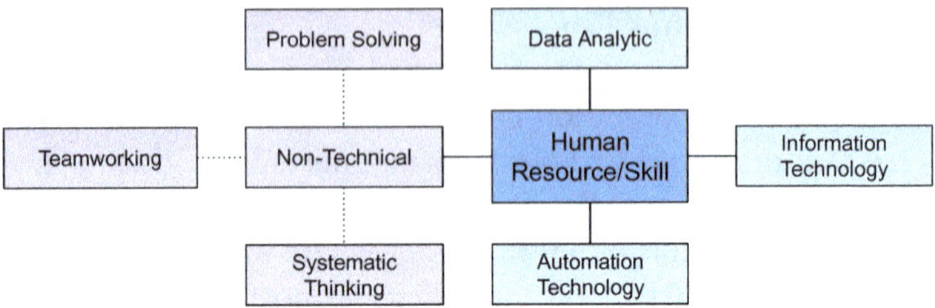

Fig. 10.5 Information technology factors

4.0. These are (1) Data Analytics, (2) Information Technology—in the IT system, the human skill set in the personnel resource is required for transformation of the organizational management, and (3) Automation Technology. The workers will monitor the automated devices and equipment, which requires qualification of highly specialized experts and knowledge based on automation technology.

10.2.4.2 Non-Technical

The employee is the core element of Industry 4.0. They consider the different tasks in the current professional and scientific discussion. These are (1) Problem Solving, (2) Teamworking, and (3) Systematic Thinking. These skills will create the value in Industry 4.0.

The main factors of Industry 4.0, as reviewed, can be found in Table 10.1.

10.3 Problem Formulation

It is obviously difficult to apply all Industry 4.0 concepts to SMEs due to the limitation of human resources, technology, and financial potential. Thus, SMEs should start their implementation of SMEs 4.0 concept with prioritized and appropriate measures. Therefore, the Smart SMEs 4.0 Implementation Toolkit is developed.

Table 10.1 Main factors of industry 4.0

Factors	Agca et al. (2015)	Pan (2015)	Lichtblau et al. (2015)	Schumache et al. (2016)	Leyh et al. (2016)	Stock and Seliger (2016)	Qian et al. (2017)	Zhong et al. (2017)	Dombrowski et al. (2017)	Farahani et al. (2017)	Fatorachian and Kazemi (2018)
Autonomous process	X		X			X					
Cloud based	X			X	X			X	X		X
Data analytics			X					X	X		
Horizontal/vertical data integration			X		X						
Expert system		X									
Equipment infrastructure	X		X			X					
Human resource				X		X	X		X		
IT system	X		X	X			X		X		
Information sharing			X					X			
Innovation management			X								X
Overall Equipment Effectiveness (OEE)										X	
Man-machine interaction						X					
M2M machine connectivity	X			X							X

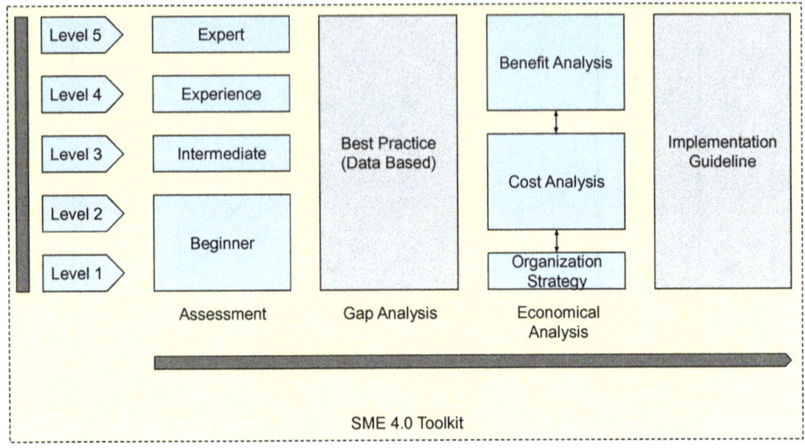

Fig. 10.6 SMEs 4.0 toolkit framework

The idea is to suggest appropriate implementation guidelines for SMEs, in responding to the SMEs 4.0 concept. The guideline can be strategies, projects idea or investment, depending on the level of implementation readiness. However, the guideline must align benefit and cost of the idea with the organizational strategies. In advanced firms, further analysis can be applicable, for example, sustainability analysis.

The guideline can be developed using consultancy or expert opinion or, at best, learning from best practice. In the case of future work where there are sufficient number of assessed companies in the toolkit database, in the primary framework, the assessment can be divided into beginner, intermediate, experience, and expert levels (see Fig. 10.6).

10.4 Methodology

The methodology of the research is how to design the Smart SMEs 4.0 Implementation Toolkit. Firstly, the related literature is reviewed to address the scope of Smart SMEs. As discussed, the scope of the toolkit is divided into four dimensions, i.e., (1) Information Technology, (2) Production and Operations, (3) Automation, and (4) Human Resources. This will then be used as the assessment module.

In addition to the assessment module, the analysis, and implementation phases are added to the toolkit to assist in the assessment and reflect the requirement of the company. The ideal methodology is starting with the company profile study and site visit with audit checklist. Then, the gap analysis is made using the assessment module of four SMEs 4.0 dimensions. Once complete, the module will be evaluated and the implementation guideline as the appropriate decision can be made accordingly (see Fig. 10.7).

Thus, the Smart SMEs 4.0 Implementation Toolkit is structured and divided into four phases, i.e., organizational analysis, gap analysis, economic analysis, and implementation guideline (see Fig. 10.8).

Phase 1 refers to the Organizational Analysis. The aim is to investigate the assessee on the organizational level. The analysis can be subjective, descriptive or structured into any business assessment. Of interest are type, size, product, process, business position, supply chain relationship, as well as the business strategy.

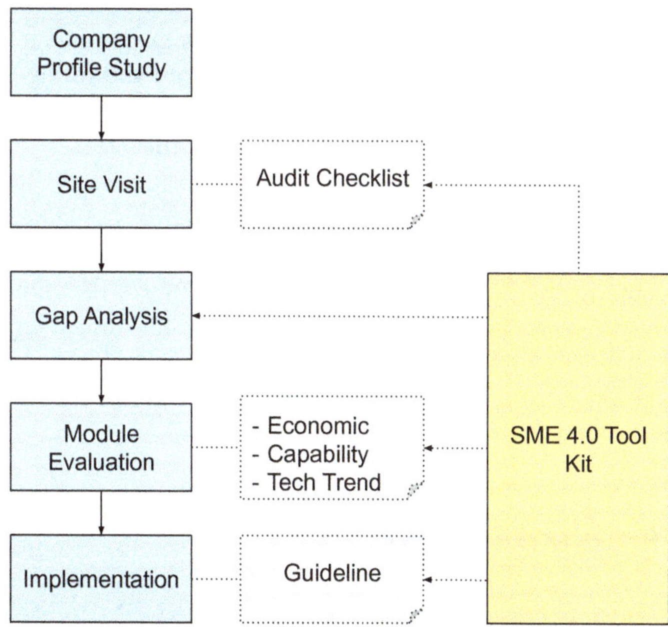

Fig. 10.7 Smart SMEs 4.0 implementation toolkit methodology

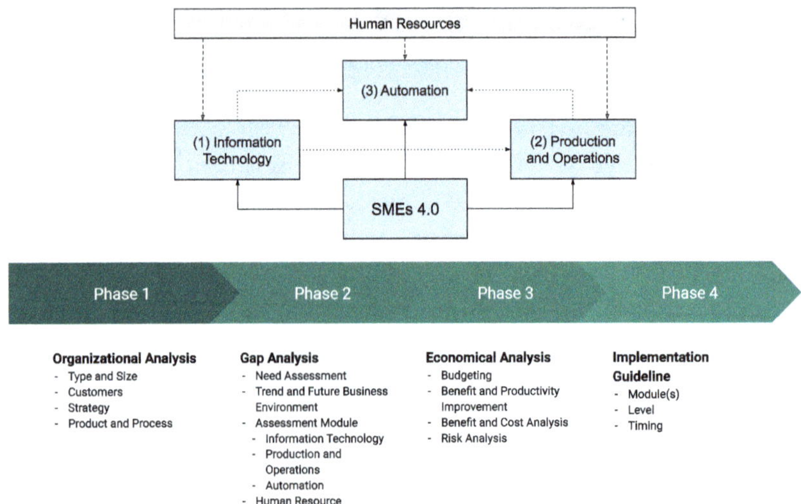

Fig. 10.8 Phases of smart SMEs 4.0 implementation toolkit

Phase 2 refers to the Gap Analysis. This phase starts with need assessment. This can be interview, addressing strategy, target, and limitation of the organization. Then, the investigation of trend and future business environment will be made. Finally, the assessment of four Industry 4.0 modules will be made to identify the "gap" for further steps.

Phase 3 refers to Economic Analysis in which the gap is identified and appropriate measure should be suggested. Here, budgeting, benefit and productivity improvement, benefit and cost analysis, and risk analysis are among the factors of interest. Then, business decision can be made if any measure is suitable and, thus, selected for implementation.

Phase 4 refers to the Implementation guideline. This is the phase to respond to the selected measures from Phase 3. It will address the module, level, and timing for each measurement.

10.5 Problem Solution

The case study of this developed toolkit is four Thai SMEs in Northern Thailand region. The country is part of pilot areas supported by the project "Industry 4.0 for SMEs" from the European Union's Horizon

2020 research and innovation program under the Marie Skłodowska-Curie grant agreement. Several researches were conducted to identify the potential of the country in terms of Organization and Management, at a national level (Ramingwong and Manopiniwes 2019; Ramingwong et al. 2019).

The first SME is a medium-sized make-to-order snack factory. The second SME belongs to a service industry, a small coffee shop. The third SME is a small plastics manufacturing company. And the fourth SME is a medium-sized multinational company. The company is a supplier of the automotive industry, producing wire mesh and conveyors.

10.5.1 SMEs 4.0 for Make-to-Order Snack Factory

The first case study SME is a make-to-order snack factory. The factory was founded as a joint venture with a Japanese investor in 1991 in a Northern Thailand Industrial Estate. The products were initially exported to Japan as rice crackers using Thai rice. Today, 20% of the products are consumed within Thailand through modern trade channels.

Although the products come with variety, the production processes are quite common due to the raw material preparation and cooking method. However, the shape and size can be varied. For all product, the first phase of the production process is in a closed automated system, including preparing, cutting, and baking. Then, the flavoring is added manually depending on flavoring type and can be coating, powdering or filling. For example, the filling of the flavoring core is done manually by hand. The process is expensive and time-consuming. Finally, the packaging is done by machine.

The company has participated in the study, starting with organizational analysis. It was found that the customers are segmented, the company has positioned itself to different customers and the market is continuously studied. New product development and R&D are the main focus of the company to expand the market and to better respond to customer satisfaction. The strategies are directed by top management and communicated to all personnel. Supplier relationship and customer relationship are the most important key success factors of the company.

Table 10.2 illustrates the assessment of four Smart SMEs 4.0 dimensions of the case study, make-to-order snack factory.

The case study company has assessed four Smart SMEs 4.0 dimensions in terms of Significance and Readiness, as shown in Table 10.2.

The company shows interest to many Smart SMEs 4.0 factors, especially automation. IT, production and operations, and human resource dimensions are comparatively considered low to medium significance. The readiness of the company is also assessed and found to be in the low and medium levels in all factors.

Thus, the company should focus on the automation dimension as a result of the low readiness but high significance. Further investigation should be conducted and the results aligned where the company has struggled with labor cost and labor availability. The autonomous process and machine connectivity can improve the productivity.

Therefore, the suggestion and priority are on automation of the process. The company is surveying on the feasibility of machine investment. Thus, the budgeting, expected benefit and productivity

Table 10.2 SMEs 4.0 assessment of the case study: make-to-order snack factory

Factors	Significance	Readiness
1. Information technology		
Equipment infrastructure	Medium	Medium
IT system	Medium	Medium
Information sharing	Low	Low
Cloud based	Low	Low
2. Production and operations		
Innovation management	Medium	Low
Data analytics	Medium	Low
Horizontal/vertical data integration	Medium	Medium
Expert system	Low	Low
3. Automation		
Overall Equipment Effectiveness (OEE)	High	Medium
Man-machine interaction	Medium	Medium
Autonomous process	High	Low
M2M Machine connectivity	High	Medium
4. Human resource		
Technical	Medium	Low
Non-technical	Low	Low

improvement, expected benefit and cost analysis, and risk analysis can then be done. Moreover, the level of implementation and timing can be strategized.

10.5.2 SMEs 4.0 for Service Industry—A Coffee Shop

The second case study SMEs is a coffee shop, a representative of the Thai service industry. While promoted as Chiang Mai Coffee City, there are more than 4000 coffee shops in Chiang Mai, a capital city of Northern Thailand region. The key value chain activities of the case study coffee shop are inbound logistics, operations, and service. Enjoying good coffee bean as raw material, the procurement and purchasing are critical, yet opportune. There are more than 20,000 rais (3200 hectares) of coffee cultivated area in Chiang Mai. The gross production of Chiang Mai coffee is more than 20,000 tons per year. Many are organic, GAP, and GMP certified.

Moreover, with sophisticated coffee machines and good raw material, the coffees are distinctive. The case study coffee shop is rated high in social media. There are many loyal favorite customers and onetime tourists. The service is also a key success factor of this shop. The owner is a trained barista and serves customers himself. Customer behavior is inspected directly by the owner and coffee formula is adjusted accordingly.

The case study company has assessed four Smart SMEs 4.0 dimensions in terms of Significance and Readiness, as shown in Table 10.3. The company shows interest to many Smart SMEs 4.0 factors, especially the IT and non-technical skills of human resource. The readiness of the company is mostly low and medium. This is not surprising for SMEs. Thus, the interests are in IT as the biggest gap. The investigation was conducted in the IT dimension and it was found that the data are collected yet not properly processed to the information level. For example, sales are collected but not analyzed, best seller items cannot be identified, transaction times are not collected and cost of each items are not known. Therefore, the case study shop is suggested to have an appropriate IT system such as a Point of Sale (POS) system. Then, the transaction can be analyzed and proper strategies can then be made.

Table 10.3 SMEs 4.0 assessment of the case study: service industry

Factors	Significance	Readiness
1. Information technology		
Equipment infrastructure	Medium	Low
IT system	High	Low
Information sharing	Medium	Low
Cloud based	Low	Low
2. Production and operations		
Innovation management	Medium	Low
Data analytics	Low	Low
Horizontal/vertical data integration	Low	Low
Expert System	Low	Low
3. Automation		
Overall Equipment Effectiveness (OEE)	Low	Low
Man-machine interaction	Medium	Medium
Autonomous process	Low	Low
M2M machine connectivity	Low	Low
4. Human resource		
Technical	Low	Low
Non-technical	High	Medium

Figure 10.9 illustrates the developed mobile application for the case. The mobile application performs POS and budgeting functions.

10.5.3 SMEs 4.0 for Small Fabrication Company

The third case study company is a plastic shoemaking company. The factory is considered small size with 20 employees. The production is both make-to-stock and make-to-order. The products are both sold domestically and exported to neighboring countries. The processes of shoemaking are discontinuous and costly. The production is low-technology and labor intensive.

Currently, the company is facing a price war and the competition is higher due to the ASEAN Economic Community's single production area (ASEAN and ASEAN Secretariat 2008).

The case study company lacks capability in the dimension of Production and Operations, and Information Technology, by which

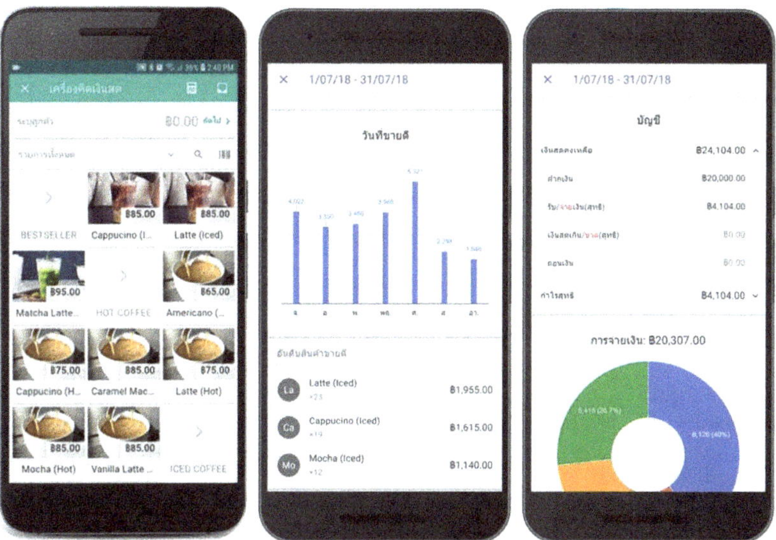

Fig. 10.9 Developed mobile application for coffee shop (Reproduced with permission from Chiang Mai University, Department of Industrial Engineering)

they are important to the company (see Table 10.4). The key concerns of the company are the information sharing and communication within the company and toward its supply chain members. The company agility is low as the information are disconnected and offline. Thus, the decision-making is ineffective.

The company was suggested to pay attention to an Information Technology system. The platform of simple electronic data interchange, e.g., Google Docs, where the data can be updated and accessed openly and freely, can help sharing necessary information at a required time. The template sheets, including inventory and production tracking, and the standard procedures for input and analysis of the data are also designed.

The company was also suggested to develop an information sharing platform within its supply chain, both customer and supplier sides. Thus, the production and other resources can be planned responsively.

Table 10.4 SMEs 4.0 assessment of the case study: small fabrication company

Factors	Significance	Readiness
1. Information technology		
Equipment infrastructure	Low	Low
IT system	Medium	Low
Information sharing	Medium	Low
Cloud based	Low	Low
2. Production and operations		
Innovation management	Low	Low
Data analytics	Low	Low
Horizontal/vertical data integration	High	Low
Expert system	Low	Low
3. Automation		
Overall Equipment Effectiveness (OEE)	Medium	Low
Man-machine interaction	Medium	Medium
Autonomous process	Low	Low
M2M machine connectivity	Low	Low
4. Human resource		
Technical	Low	Low
Non-technical	Medium	Low

10.5.4 SMEs 4.0 for Multinational SMEs

The last case study company is a medium-sized supplier of the automotive industry. The company is a wire mesh and conveyor production site of a Japanese mother company. There are also similar production sites in Japan, China, Singapore, the United States, and Australia. The factory in Thailand is considered a medium-sized and production only. The production plan and the Research and Development are done only by overseas.

After having assessed the company by the Smart SMEs 4.0 Implementation Toolkit (see Table 10.5), the company is suggested mainly to improve their Expert System and Human Resource areas. While the production can be done effectively and efficiently, the knowledge management is limited. The expert system can help improve the knowledge sharing and collect the tacit knowledge, present yet limitedly transferred. Defect and 7-waste management are suggested to be a pilot theme of the project idea. Then, the human resource can be managed accordingly.

Table 10.5 SMEs 4.0 assessment of the case study: multinational SMEs

Factors	Significance	Readiness
1. Information technology		
Equipment infrastructure	High	High
IT system	High	High
Information sharing	Medium	Medium
Cloud based	Medium	Medium
2. Production and operations		
Innovation management	Low	Low
Data analytics	Low	Low
Horizontal/vertical data integration	High	Medium
Expert system	High	Low
3. Automation		
Overall Equipment Effectiveness (OEE)	Medium	High
Man-machine interaction	Medium	Medium
Autonomous process	Medium	Medium
M2M Machine Connectivity	Medium	Medium
4. Human resource		
Technical	Medium	Low
Non-technical	Medium	Low

10.6 Discussion

The chapter presents the initial use of the Smart SMEs 4.0 Implementation Toolkit to four case study SMEs in Thailand. Where the requirement differs from company to company, the toolkit can reflect the needs by assessing the significance and readiness and identify the gap of improvement. The implementation guidelines are initially created, yet, at this stage, only suggestive. Further development is needed to concreate the methodology and validate the toolkit.

10.7 Conclusions

Smart SMEs 4.0 Implementation Toolkit is a modified and implementable version of Industry 4.0 for SMEs. Divided into four phases, i.e., organizational analysis, gap analysis, economic analysis, and implementation guideline, the company can use the toolkit to reflect the gap and the implementation suggestions can be made. With the scope of assessment identified into four dimensions of interest, i.e., (1) Information Technology, (2) Production and Operations, (3) Automation, and (4) Human Resource, the significance and readiness of each SME can be aligned and the gap can be analyzed. In this manuscript, four case study companies, i.e., make-to-order snack factory, coffee shop, shoe-making factory, and multinational mesh/conveyor factory in Northern Thailand, are used as examples of the toolkit usage. Implementation guidelines are preliminarily suggested. Further study is needed to validate the toolkit.

References

Agca, O., J. Gibson, J. Godsell, J. Ignatius, C.W. Davies, and O. Xu. 2015. An Industry 4 Readiness Assessment Tool. In *International Institute for Product and Service Innovation.* Coventry: University of Warwick.

ASEAN, and ASEAN Secretariat. 2008. *ASEAN Economic Community Blueprint.* Jakarta: Association of Southeast Asian Nations.

Dombrowski, U., T. Richter, and P. Krenkel. 2017. Interdependencies of Industrie 4.0 & Lean Production Systems: A Use Cases Analysis. *Procedia Manufacturing* 11: 1061–1068. https://doi.org/10.1016/j.promfg.2017.07.217.

EIP. 2005. *The New SME Definition, User Guide and Model Declaration.* Luxembourg: Enterprise and Industry Publications.

Farahani, P., C. Meier, and J. Wilke. 2017. Digital Supply Chain Management Agenda for the Automotive Supplier Industry. In *Shaping the digital enterprise,* 157–172. Cham: Springer.

Fatorachian, H., and H. Kazemi. 2018. A Critical Investigation of Industry 4.0 in Manufacturing: Theoretical Operationalisation Framework. *Production Planning & Control* 29 (8): 633–644.

Hudson, M., A. Smart, and M. Bourne. 2001. Theory and Practice in SME Performance Measurement Systems. *International Journal of Operations & Production Management* 21 (8): 1096–1115. https://doi.org/10.1108/EUM0000000005587.

Lasi, H., P. Fettke, H.G. Kemper, T. Feld, and M. Hoffmann. 2014. Industry 4.0. *Business & Information Systems Engineering* 6 (4): 239–242. https://doi.org/10.1007/s11576-014-0424-4.

Lee, J., B. Bagheri, and H.A. Kao. 2015. A Cyber-Physical Systems Architecture for Industry 4.0-Based Manufacturing Systems. *Manufacturing Letters* 3: 18–23. https://doi.org/10.1016/j.mfglet.2014.12.001.

Leyh, C., K. Bley, T. Schäffer, and S. Forstenhäusler. 2016. SIMMI 4.0-A Maturity Model for Classifying the Enterprise-Wide It and Software Landscape Focusing on Industry 4.0. In *2016 Federated Conference on Computer Science and Information Systems (FedCSIS)*, 1297–1302. http://dx.doi.org/10.15439/2016F478.

Lichtblau, K., V. Stich, R. Bertenrath, M. Blum, M. Bleider, A. Millack, K. Schmitt, E. Schmitz, and M. Schröter. 2015. *IMPULS—Industrie 4.0-Readiness*. Aachen-Köln: Impuls-Stiftung des VDMA.

MOBIE. 2014. *The Small Business Sector Report 2014. Ministry of Business.* Ministry of Innovation and Employment.

OECD. 1993. *Small and Medium-Sized Enterprises: Technology and Competitiveness*. Paris: Organization for Economic Corporation and Development.

OSMEP. 2017. *SME4.0 The Next Economic Revolution, White Paper on Small and Medium Enterprises 2017*. The Office of Small and Medium Enterprises Promotion of Thailand.

Pan, M., J. Sikorski, C.A. Kastner, J. Akroyd, S. Mosbach, R. Lau, and M. Kraft. 2015. Applying Industry 4.0 to the Jurong Island Eco-Industrial Park. *Energy Procedia* 75: 1536–1541. https://doi.org/10.1016/j.egypro.2015.07.313.

Pereira, A.C., and F. Romero. 2017. A Review of the Meanings and the Implications of the Industry 4.0 Concept. *Procedia Manufacturing* 13: 1206–1214. https://doi.org/10.1016/j.promfg.2017.09.032.

Qian, G., and L. Li. 2003. Profitability of Small-and Medium-Sized Enterprises in High-Tech Industries: The Case of the Biotechnology Industry. *Strategic Management Journal* 24 (9): 881–887. https://doi.org/10.1002/smj.344.

Qian, F., W. Zhong, and W. Du. 2017. Fundamental Theories and Key Technologies for Smart and Optimal Manufacturing in the Process Industry. *Engineering* 3 (2): 154–160. https://doi.org/10.1016/J.ENG.2017.02.011.

Ramingwong, S., and W. Manopiniwes. 2019. Supportment for Organization and Management Competences of ASEAN Community and European Union Toward Industry 4.0. *International Journal of Advanced and Applied Sciences* 6 (3): 96–101. https://doi.org/10.21833/ijaas.2019.03.014.

Ramingwong, S., W. Manopiniwes, and V. Jangkrajarng. 2019. Human Factors of Thailand Toward Industry 4.0. *Management Research and Practice* 11 (1): 15–25.

Rüßmann, M., M. Lorenz, P. Gerbert, M. Waldner, J. Justus, P. Engel, and M. Harnisch. 2015. Industry 4.0: The Future of Productivity and Growth in Manufacturing Industries. *Boston Consulting Group* 9 (1): 54–89.

Schumacher, A., S. Erol, and W. Sihn. 2016. A Maturity Model for Assessing Industry 4.0 Readiness and Maturity of Manufacturing Enterprises. *Procedia CIRP* 52: 161–166. https://doi.org/10.1016/j.procir.2016.07.040.

Stock, T., and G. Seliger. 2016. Opportunities of Sustainable Manufacturing in Industry 4.0. *Procedia CIRP* 40: 536–541. https://doi.org/10.1016/j.procir.2016.01.129.

Yazdi, P.G., A. Azizi, and M. Hashemipour. 2018. An Empirical Investigation of the Relationship Between Overall Equipment Efficiency (OEE) and Manufacturing Sustainability in Industry 4.0 with Time Study Approach. *Sustainability* 10: 1–28. https://doi.org/10.3390/su10093031.

Zhong, R.Y., X. Xu, E. Klotz, and S.T. Newman. 2017. Intelligent Manufacturing in the Context of Industry 4.0: A Review. *Engineering* 3 (5): 616–630. https://doi.org/10.1016/J.ENG.2017.05.015.

Part V

Case Studies and Methodical Tools for Implementing Industry 4.0 in SMEs

11

The Digitization of Quality Control Operations with Cloud Platform Computing Technologies

Kamil Židek, Vladimír Modrák, Ján Pitel and Zuzana Šoltysová

11.1 Introduction

The main scope of this chapter is to present experimental results from the implementation of selected vision technologies and a UHF RFID system for automatic identification and inspection of product parts before and after the assembly operations. Digitization of quality control processes requires a new approach to how data is captured, stored, analyzed, and used. The presented approach is based on maximizing

K. Židek (✉) · V. Modrák · J. Pitel · Z. Šoltysová
Faculty of Manufacturing Technologies,
Technical University of Košice, Prešov, Slovakia
e-mail: kamil.zidek@tuke.sk

V. Modrák
e-mail: vladimir.modrak@tuke.sk

J. Pitel
e-mail: jan.pitel@tuke.sk

Z. Šoltysová
e-mail: zuzana.soltysova@tuke.sk

D. T. Matt et al. (eds.), *Industry 4.0 for SMEs*,
https://doi.org/10.1007/978-3-030-25425-4_11

data collection from the mentioned technologies integrated into the assembly process. For the purpose of data storage and analysis it is proposed that Big Data technologies are based on Cloud Platforms. The goal of quality control is to prevent components' faults and to identify any defects after the components have been assembled.

The object of quality control is the rapid prototyping feeder mechanism with stepper motor, plastic, and mounting parts, including 10 individual parts positioned on the assembly fixture. The experimental inspection and identification system consists of a conveyor belt with PLC control system and servomotor to be able to fluently change conveyor speed depending on production process status. The problem-solution starts with a selection of suitable contactless technologies for fast data mining and digitization in line with the Industry 4.0 concept. These technologies are described in Sect. 11.4.2. The vision inspection system consists of 3 separate modules. The first one is aimed at checking shapes of components, the second one serves to measure components' dimensions, and the last one is used for component surface "macro" inspection. The RFID system consists of two power adjustable circular/vertical antennas and different UHF tags (labels and transponders). Cloud Platforms are used for data analysis and visualization (data mining). Some of them are open source and are suitable for Fog computing (Thinger.IO). The others are commercial-based platforms for Cloud computing (IBM Watson IoT, Microsoft Azure IoT, Siemens MindSphere). We chose Cloud Platform Siemens Mindsphere, because of compatibility with the control system used.

In order to create a digital twin of the experimental inspection and identification system for remote quality control optimization, all of the devices on the experimental system including parts of products were converted into a 3D model by using Technomatix software. A digital twin of the experimental inspection and identification system with simulation functions by using virtual reality software is described in Sect. 11.4.4.

The paper is organized as follows. After this introductory section, section two offers a brief theoretical background on technologies used for contactless inspection and identification. Then, problem formulation methodology will be explained in detail. Subsequently, a problem solution will be described in Sect. 11.4. In the following section, existing problems are disguised, and possible solutions are proposed. Finally, some conclusion notes and future research directions are outlined.

11.2 Background and Literature Review

Technologies for quality control inspection can be divided into two basic groups: contact and contactless. Thanks to contactless technologies there is no need to stop object(s) during a quality control operation. Advanced contact technologies are mostly represented by coordinate measuring systems (CMS), which work in semiautomatic mode. But measuring the time achieved by these technologies is too long for checking production batches, and one of the main requirements of the Industry 4.0 concept is to minimize times between the appearance and the handling of errors (Groggert et al. 2017). Another disadvantage of this method is hard implementation into an automated conveyor line, because measuring table is a part of a CMS machine.

Suitable technologies for contactless quality control inspection are vision systems, and Radio-frequency identification (RFID) systems, and not just because these technologies work well together. RFID technologies can also be employed to track items such as pallets or products within a supply chain, and are additionally capable of ensuring full component process history for end users (Velandia et al. 2016). Both of these technologies can be used for certain tasks such as dimension measuring, errors detection, dynamic status identification of the product, and presence detection of the correct part for assembly process. These technologies can provide digital data acquisition from every part of production. The captured data can help to update and extend digital twins created from 3D models of devices or products. These digital data can be accumulated in Programmable Logic Controllers (PLC) or Supervisory Control and Data Acquisition (SCADA) system, but they can't be stored in a long-term horizon, because industrial control and monitoring systems are quite limited in storage space. The possible solution is using Cloud Platform combined with the PLC system used as an IoT device. The subsequent data can be acquired from other sources such as RFID, etc. Data analyzes or knowledge extraction (data mining) is the main task of a cloud system. Cloud Platforms can provide user-friendly data representation by timelines, day/weeks or months automated reports, and alarm systems for critical production status usually as a message by email or SMS.

Industry 4.0 (commonly referred to as the fourth industrial revolution) is the current trend of automation, control, monitoring, and data exchange in manufacturing technologies. It includes cyber-physical systems (CPS), the Internet of things (IoT), cloud computing, cognitive computing, and other related disciplines (Xu and Duan 2019). There are many research papers concerned with to CPS in the Industry 4.0 concept (Xu and Duan 2019), Big Data processing (Li et al. 2019), and combination of CPS with IoT (Sanin et al. 2019). Industry 4.0 specifies many methods and technologies usable in product customization, because customer needs are directed toward unique products. The next requirement is a low-cost product with maximized customization. Theory and practical implementation are solved in the area of mass customization. Mass customization is a process of delivering wide-market goods and services which are modified to satisfy a specific customer's needs in the manufacturing and industry services. It combines the flexibility and personalization of custom-made products with the low unit costs. The problem of mass customization is in connection to the Industry 4.0 concept, and variety based complexity issues were discussed by works Schmidt et al. (2015), Modrak (2017), Rauch et al. (2016); research results in complexity metrics problems for mass customization were published for example by works Mourtzis et al. (2017, 2018), Modrak and Soltysova (2018).

Many technologies rapidly developed, especially those which can significantly increase the effectivity of massive customization production:

- rapid prototyping technologies for part printing,
- virtual reality devices for training workers in virtual factory environments,
- collaborative robots for fast cooperation with workers,
- augmented reality devices for training and optimization of assembly process.

Rapid prototype part printing technologies now offer very high product customization with minimal of cost of production preparation (CNC program preparation, semi-product specification, cutting, and handling). Currently, there are no limitations in using this technology only for the plastic parts, because rapid prototyping devices can create parts also from other materials (aluminum, bronze parts, steel, or titan) using laser technologies. Material costs are significantly decreasing with technology expansion. The future proposal is printing a fully assembled product, not only parts for the next assembly process. The usability of rapid prototyping technologies in the concept of Industry 4.0 is described for example in Krowicki et al. (2019), Żabiński et al. (2017).

Customized production cannot use standard automation, because preparation work costs are higher than the profit. This problem is significant in the assembly process, which must usually be manual. A completely manual assembly process is very ineffective, but there is space for implementation of partial automation based on collaborative robots and assisted assembly by virtual devices for training workers and augmented devices for quality control and inoperative assistance.

Collaborative robots can work in the same workspace as human workers and prepare basic manipulations or perform simple monotonous assembly tasks. The main advantage is minimal transport delay of assembly parts between manual and automated operations which is necessary by using standard industrial robots. The next preference is an integrated vision system for additional inspection of manual operation. Collaborative robots usually provide an interface for digital data collection (force sensors data, vision data, and end effector positions data) and communication with external Cloud Platforms. Some research results of the mixed assembly process between humans and collaborative robots are described in works by Malik and Bilberg (2017), Akkaladevi et al. (2018), Realyvásquez-Vargas et al. (2019).

Virtual reality devices are currently available on the market (for example Oculus Rift or HTC Vive) combined with power PC provide enough performance to virtualize complete production process which can be used for employee training. It is possible to explain material flows in movement or to simulate critical situation without real production stop. Some research in the usability of virtual devices for factory simulation is described in Żywicki et al. (2018), Dong and Wang (2018), Gong et al. (2019). Augmented reality devices are usually named "smart glasses" and can significantly increase the effectivity of training new workers during their first assembly task (assisted assembly). Smart glasses can also be used after basic training for quality control of the assembly process. An AR-based worker support system for human-robot collaboration is described in H. Liu and Wang (2017).

A dynamic virtual representation of a physical object or system is a very important part of the Industry 4.0 concept. This digital replica of physical assets is called a digital twin, which continuously learns and updates itself from multiple sources. Case studies on digital twins are described in articles about digital twin ergonomic optimization (Caputo et al. 2019), digital twin commentary (Tomko and Winter 2019), learning experiences by digital twins (David et al. 2018), automatic generation of simulation-based digital twins (Martinez et al. 2018), digital twins for legacy systems (Khan et al. 2018), possibilities of digital twins technology (Shubenkova et al. 2018), and rapid qualification of product by digital twins (Mukherjee and DebRoy 2019).

Industrial vision systems are computer-based systems where software performs tasks for acquiring, processing, analyzing, and understanding digital images usually aimed at industrial quality assurance, defect detection, part recognition, etc. Research articles concerning industrial vision systems describe, for example, an automatic surface detection (Zhou et al. 2019), pre-inspection of steel frames (Martinez et al. 2019), embedded vision systems (Zidek et al. 2016), and stereo vision sensing (OrRiordan et al. 2018). Additional related research aspects of RFID system were presented in works focused on the security of tags (Han et al. 2019), detection of missing tags (Lee et al. 2019), and new searching protocol (Liu et al. 2019).

The availability of high-performance computers, high-capacity storage devices, and high-speed networks as well as the widespread adoption of hardware virtualization and service-oriented architecture has led in recent years to growth in cloud computing solutions (Raihana 2012). Related research areas on cloud systems and data mining are described in the papers concerning Clouds and Big Data connection (Lu and Xu 2019), cloud robotics data (Aissam et al. 2019), clouds in industrial automation (Mahmoud 2019), and chaos theory combination with clouds systems (Hu et al. 2019).

11.3 Problem Formulation and Methodology

This study proposes a practical approach to full digitization of data from the control quality process. The main motivation for this objective is based on the fact that commonly in practical applications gathered data from the product quality control process are not stored and subsequently used for continual quality improvement.

The following three major issues concerning experimental identification and inspection of parts of the product will be addressed in this study:

- The first problem is transformation of data from vision systems to the storage system based on Cloud Platform and RFID technology.
- The next task the bidirectional connection of digital twins (based on a 3D model) to an experimental quality control device.
- The last task is data analyses and data knowledge extraction (data mining) for usable user-friendly report (timetables or alarms).

The simplified scheme of data transition from the vision systems and RFID technology to the control system is shown in Fig. 11.1.

As shown in Fig. 11.1, the quality control process consists of two contactless technologies RFID and vision systems. The methodology concept of inspection by vision system and identification through RFID technology is universally designed for any assembly part or end item which can be placed on the fixture or pallet in the conveyor system. A scheme of components used in an automated quality control system with digital twin and Cloud Platform communication is shown in Fig. 11.2.

The quality control process is divided into two phases:

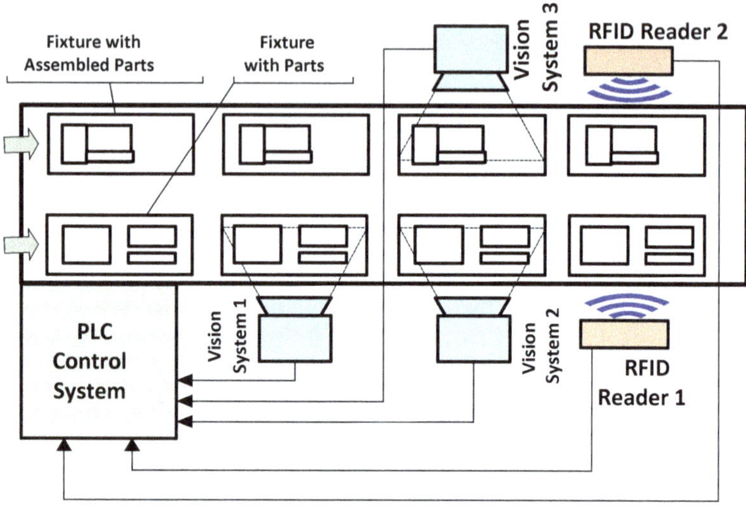

Fig. 11.1 The scheme of quality control digitization

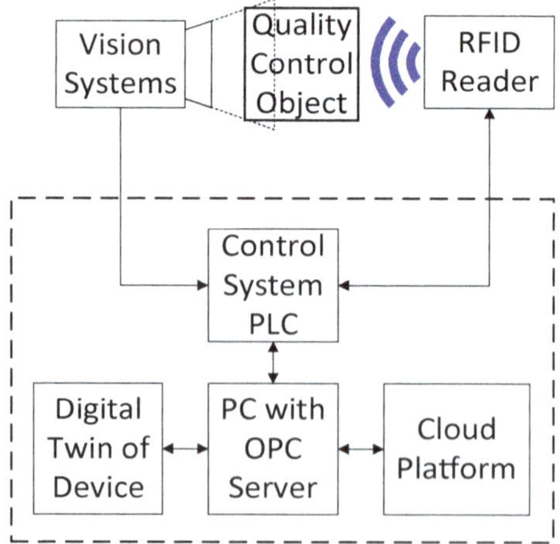

Fig. 11.2 The block diagram data of digitization with Cloud Platform and digital twin

1. The first phase includes part inspection by vision system (dimensions, surface errors, parts localization verification).
2. The second phase is aimed at assembly identification and personalization by using RFID system. The small assembly parts (nuts, screws, washers) can be inspected only by the vision systems.

The digital twin of the system has been developed through the simulation software Tecnomatix Plant simulator. A digital twin of the product is stored in the Cloud Platform due to big data volume for data processing. The following two bidirectional data synchronizations between real system and events in simulation software are expected:

- simulation software can be modified in the real process,
- modifications made in real plant are transformed into simulation model.

For this purpose, the digital twin was connected to OPC Server for data exchange with control system. A description of the digital twin for simulation of experimental inspection and identification system by using virtual reality software is provided in Sect. 11.4.

11.4 Problem Solution

In order to achieve full digitization of the quality control process we propose creating digital twin for inspection and identification system and personalized a digital twins for every assembled product and its parts. Preconditions for creation of digital twins are 3D models of real devices and products (parts, and end item). Under the term "end item" can be understand final output of assembly process comprising of a number of parts or subassemblies put together to perform a specific function. Two main approaches will be used to extend the 3D models with additional data mining and/or processing:

- based on obtaining data from product parts and end items during inspection using the vision system,
- based on data recording to the product during identification using RFID tags.

11.4.1 Quality Control Objects

For experimental purposes a simple subassembly product—extruder filament guide of rapid prototyping printer was selected and it consists of variable parts (see Fig. 11.3), i.e.:

- two plastic parts made from a different material PLA or ABS, and each of them can differ by color,
- a one stepper motor which can have a different step/force and variable spring for filament pressure modification.

3D model object consists of:

- list of the product components (product structure),
- 3D graphical models of each part, and the assembled product (end item),

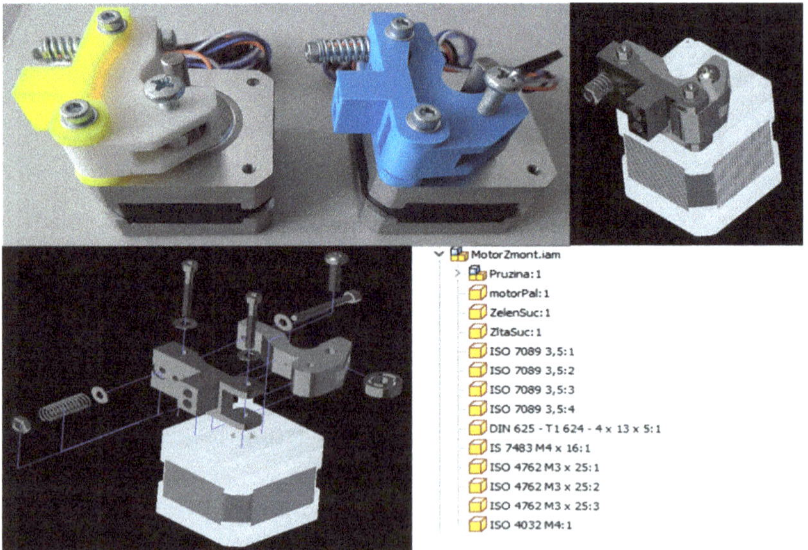

Fig. 11.3 Pictures of subassembly products (top left), 3D digital twin of the product (top right), 3D digital model in exploded view (bottom left), the list of the subassembly product components (bottom right)

Fig. 11.4 The pictures of the fixture and quality control objects (left), along with their 3D model (right)

- 3D model of the fixture,
- RFID tags and fasteners.

The picture and 3D models of the real assembled products are shown in Fig. 11.3.

The empty fixture, fixture with the parts, and fixture with the end item along with their 3D models are shown in Fig. 11.4.

11.4.2 Data Acquisition During Quality Control Process

As mentioned earlier, one of the requirements of Industry 4.0 is full digitization of data and its storage for subsequent analyses. Any data acquired through using vision system or RFID technology has to be stored in suitable type of database. Proposed methods for data mining and processing are described in the next two sections.

11.4.2.1 Data Extraction from Vision Systems

The acquired data about the objects of the quality control on the conveyor line before and after the assembly process was collected by using three vision systems for four tasks of the quality control:

1. Surface inspection of the stepper motor (see Fig. 11.5) by using Omron (vision system 1).
2. Dimension control (see Fig. 11.6) by using Cognex (vision system 2).
3. Parts presence control in the fixture (see Fig. 11.7) by using Sick Inspector I10 (vision system 3).
4. Verification of the end item completeness (see Fig. 11.8) by using Sick Inspector I10 (vision system 3).

An example of macro surface inspection of errors through applying vision system 1 is shown in Fig. 11.5.

As can be seen in Fig. 11.5, four sectors with simple true/false value represented by binary numbers have been defined for surface inspection. In Fig. 11.5 (right), there is detected surface error in one sector depicted by red box. The principle of error detection is based on a pixel color change with a defined range limit.

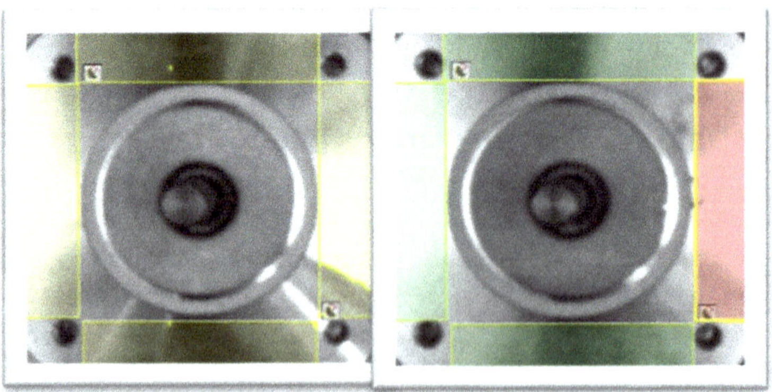

Fig. 11.5 Vision system Omron (left), calibration of the recognition area (middle), detection of surface errors (right)

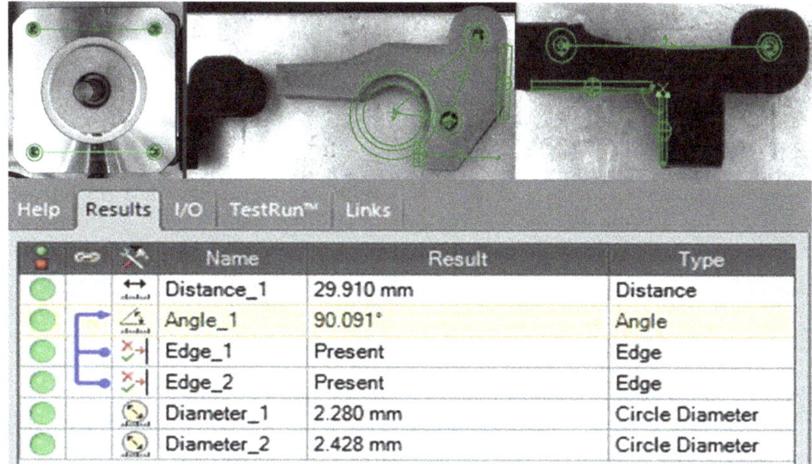

Name	Result	Type
Distance_1	29.910 mm	Distance
Angle_1	90.091°	Angle
Edge_1	Present	Edge
Edge_2	Present	Edge
Diameter_1	2.280 mm	Circle Diameter
Diameter_2	2.428 mm	Circle Diameter

Fig. 11.6 Dimension control of assembly elements (top), measured data interpretation from the Cognex vision system 2 (bottom)

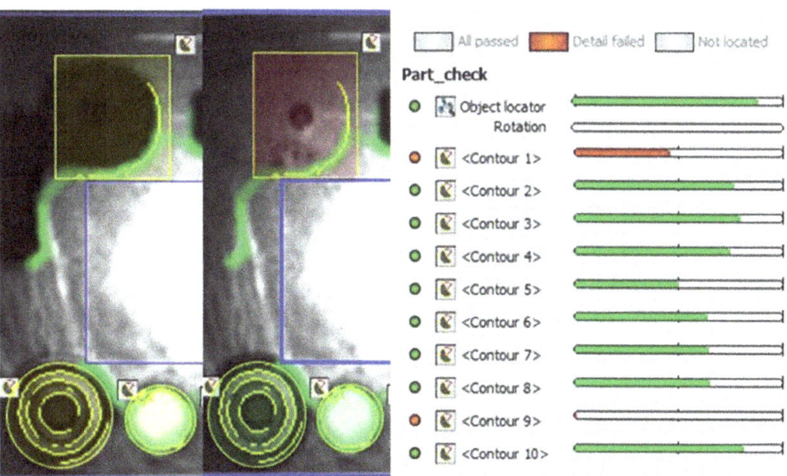

Fig. 11.7 Calibration of the recognition area (left), parts presence control in the fixture along with graphical indication of missing part (right)

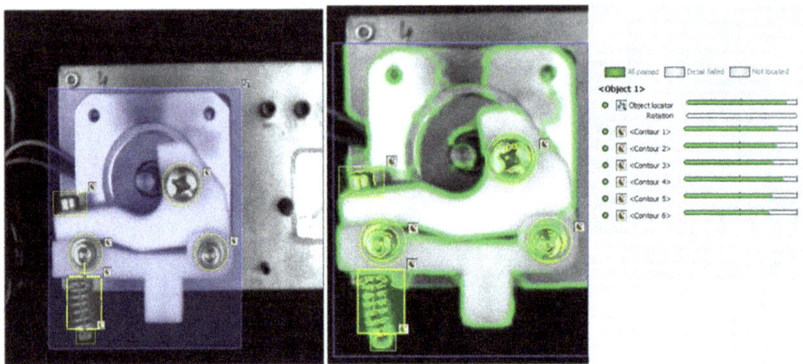

Fig. 11.8 Verification of the end item completeness by Sick Inspector I10

The next vision task is dimension control of every part of the end item. It is intended to measure mainly screw holes and its distances. The dimension inspection of the three assembly parts by using Cognex vision system is shown in Fig. 11.6.

Distances between thread holes of the stepper motor has to be checked on two layers, because this dimension is used for assembly other plastic parts made by rapid prototyping technology. Therefore, two values are measured. The first plastic part has to be checked for three features. There is the need to check two holes, one radius, and three dimensions (together six values are checked).

The second plastic part has to be checked for dimension of features for extruder filament guide. Therefore, four values have to be checked: two holes, one angle, and one dimension.

So, twelve dimensions are required to be checked by three snapshots during conveyor movement. It is impossible to stop the fixture because the conveyor doesn't have fixed stops and a motor converter stops the belt using time ramps. The same position of the snapshot is secured by an optical sensor with a combination of path measured by incremental sensors in three places.

The next operation is check of parts presence control in the fixture before an assembly operation (see Fig. 11.7). The verification is based on the edges detections by using a reference image. There is a need to check the eleven product parts.

After the assembly is completed, it is necessary to verify assembled end item status. Control of assembly completeness is provided by vision system 3 (see Fig. 11.8).

To prove the end item completeness, a reference image (see Fig. 11.8 in left) was used to define regions with mounted parts. The reference image was developed for screws, nuts, and spring parts. Totally six values were checked in order to prove the end item completeness (see Fig. 11.8 in right). As it can be seen from Fig. 11.8, no nonconformity was found.

In order to encode the all available data from the vision systems to one packet, they can be divided into the four groups following the four tasks of the quality control described previously:

- Data about the surface distortion of the stepper motor (four true/false values represented by 4-bit binary number).
- Data from the dimension control (twelve float values of dimensions).
- Data about parts position presence in the fixture (eleven true/false values represented by 11-bit binary number).
- Data from the control of the end item completeness (six true/false values represented by 6-bit binary number).

The form of data encoding from vision systems to one packet is in Table 11.1.

Then, the data from vision systems can be encoded to one binary number as follows:

$$4 + 12*11 + 14 + 11 + 6 = 167 \text{ bits}$$

Table 11.1 Data encoding from vision systems to one packet

Surface inspection	Dimension control 1	...	Dimension control 12	Part presence control	Control of end item completeness
XXXX	XXXX.XXXXXXX		XXXXXXX.XXXXXX0	XXXXXXXXXX00000000000	XXXXXX

This data can be saved to the end item through RFID tag and can serve as full component process history for end users.

11.4.2.2 Data Extraction from RFID System

In order to employ UHF RFID technology to identify assembled parts in the experimental inspection and identification system, one reader with two antennas will be used. The first antenna checks the presence of the part by RFID tags. The second antenna checks the assembly completeness and serves for data write to integrated RFID tag labels. RFID implementation with two antennas creates identification gate (input/output) and it writes the personalized data to main assembly RFID UHF tag. The used RFID system with two antennas, and RFID tag implementation for some assembly of parts and the fixture are shown in Fig. 11.9.

RFID tag transponder assigned in Fig. 11.9 as 0 is used for the fixture. Tags 1–3 are RFID labels placed on the three parts.

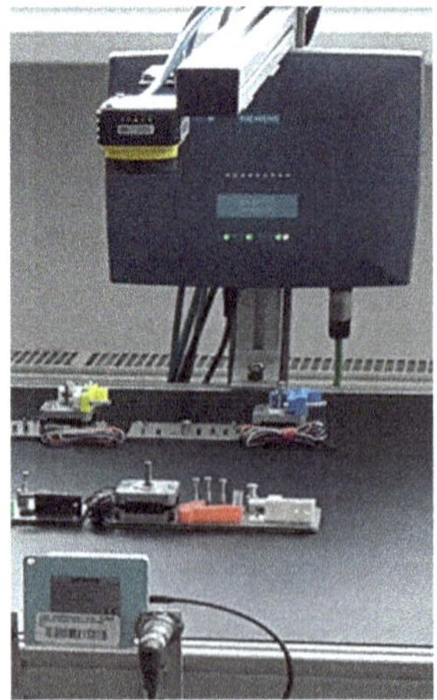

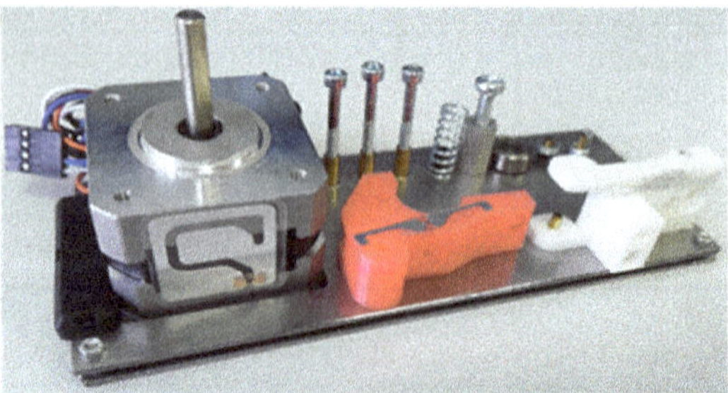

Fig. 11.9 RFID gate (left), RFID tag implementation into the assembling parts and the fixture (right)

For the purpose of position detection of the fixture and the parts before it is captured by SV1, the laser sensor is used. Subsequently, the fixture position prior to it coming to RFID gate is detected by the optical sensor. The RFID gate serves for checking a presence of selected parts of the end item at the fixture. Block schemes of the fixture detections by optical sensor and laser sensor along with RFID detection are shown in Fig. 11.10.

The distance values *x, y, z* are acquired by the incremental sensor providing the exact position for snapshot timing by the vision system. The picture of RFID antennas used is shown in Fig. 11.11.

The RFID storage consists of four banks:

- bank00 reserved memory (Kill/access passwords),
- bank01 EPC (CRC, PC) writable,
- bank10 TID Tag identification (read-only),
- bank11 user memory.

UHF RFID ETSI standard for Europe with frequency 868 MHz has been used for the given purpose. All used tags are RFID-EPC Gen2 standard with 64 KB (524 288 bits) user memory.

The fixture is identified by an RFID tag in a plastic cover, and other three parts are identified by RFID labels. The fixture RFID tags are rewritable according to actual product. The class 2—rewritable passive tags were chosen for this purpose. The data for the end items can be

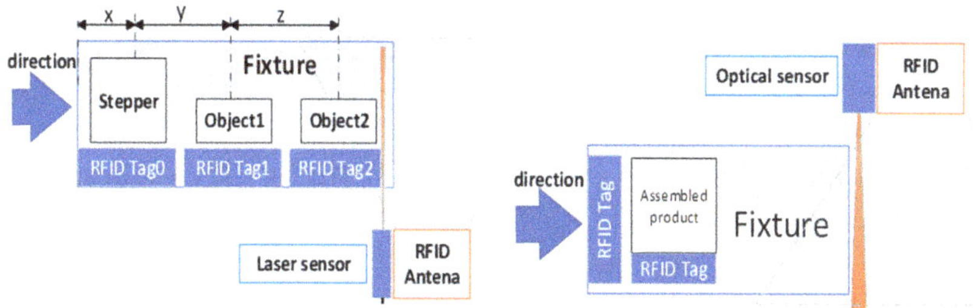

Fig. 11.10 The fixture and parts position detection by laser sensor and incremental encoder for the parts position identification [*x, y, z*] (left), the fixture position detection by RFID and optical sensor (right)

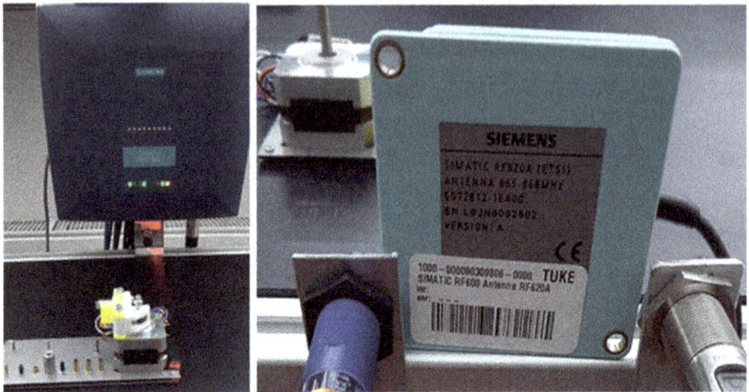

Fig. 11.11 RFID antenna for the fixture with the end item (left) and RFID antenna for fixture with the parts (right)

USER	TID	EPC				RESERVED
Dimensions 167bit	IDENTIFIER 8bit	HEADER 8bit	EPC MANAGER 28bit	OBJECT CLASS 24bit	SERIAL NUMBER 36bit	Access/Kill Password 32bit
		GTIN	Manufacturer Ident	Type of Product	Id of Product	

Fig. 11.12 Modified RFID tag data structure with EPC Gen2 (96 bit)

written only once. The class 1 for the end item with write once read many passive RFID tags were chosen. EPC number contains 96 bits for unique product identification. For our end item can be selected GTIN trade object or CPI component for part identifier as the data structure with 32-bit access password memory lock. All tags after data recording must be locked with a 32-bit password to avoid a customer to rewrite the data. The proposed data frame storing all information acquired from the quality control process is shown in Fig. 11.12.

Data from the control process are transferred to JSON format for simple writing in one step to RFID tag as follows:

"*ps*" "****"	(password)
"*sn*" "0178ff0005575b0f"	(EPC)
"*tid*" "e20034120178ff0005575b0f"	(fixed data)
"*usr*" "00..00"	(167 bit user data)

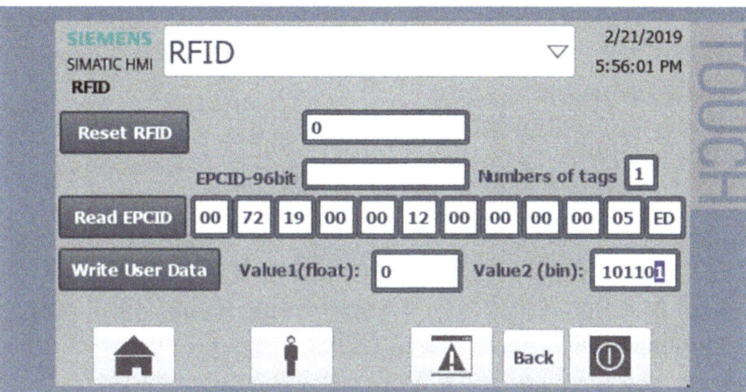

Fig. 11.13 An example of RFID tag read value

SCADA system can read the number of the presented tags along with its EPC number, and write encoded binary/float data to RFID tags.

Graphical panel to read and write data from the vision system to RFID tags is shown in Fig. 11.13.

11.4.3 Data Storing and Analyzing Using Cloud Platform

The next task after data digitization is data storage and representation. The current trend of data storage in industrial applications are Cloud Platforms, since basic data visualization and knowledge extraction can be achieved by timelines. Advanced interactivity in remote monitoring of real manufacturing process can be achieved through its digital twin and virtual reality devices. Cloud Platforms are innovative systems comparing to standard SQL databases or SCADA data storages and they provide extended tools for data knowledge extraction from stored Big Data. Some cloud systems provide only graphical representation with simple alarm system (email, SMS), other ones serve as tools for data transformation and analyses.

In general, the acquired data from quality control operations can provide information including possible degradation of precision in real manufacturing processes. This degradation can predict maintenance time for the production machine(s) and reduce the production of faulty parts. Obtained data from vision systems can be stored in MindSphere Cloud Platform and create extension data in the digital twin model for all produced parts.

With the aim to demonstrate how obtained data from visions systems (see Sect. 11.4.2.1) can be transferred and used for monitoring and quality control, the following example will be shown.

Commonly, quality control activities are supported by recording and analyzing measured dimensions through Shewhart charts. This statistical process control tool was applied for analytical purpose in the following way. Upper control limits and Lower control limits will be used as warning limits for early detection. Acquired data from the vision system 2 are transferred through OPC Server (see an example in Fig. 11.14 top) into MindSphere Cloud Platform, where this data in form of timelines are compared to the warning limit values. Based on that automated reports can be generated, and alarm messages sent by email or SMS. An example of Cognex OPC Server for data transfer to MindSphere Cloud Platform is shown in Fig. 11.14.

If some dimension of the product parts would be out of the setup range, this online information is used to substitute a wrong part by a new one.

11.4.4 Digital Twin with Simulation and Virtual Reality

As was outlined in introductory section of this chapter a digital twin of physical assets can be effectively used for various purposes including for remote quality control. In this intention all the devices of the experimental inspection and identification system including parts of product were converted into 3D models by using Technomatix software.

The 3D model of presented quality control and identification system (see Fig. 11.15) consists of:

- conveyor device,
- Vision systems and RF identification devices,
- PLC control system.

The 3D models were created in Autodesk Inventor.

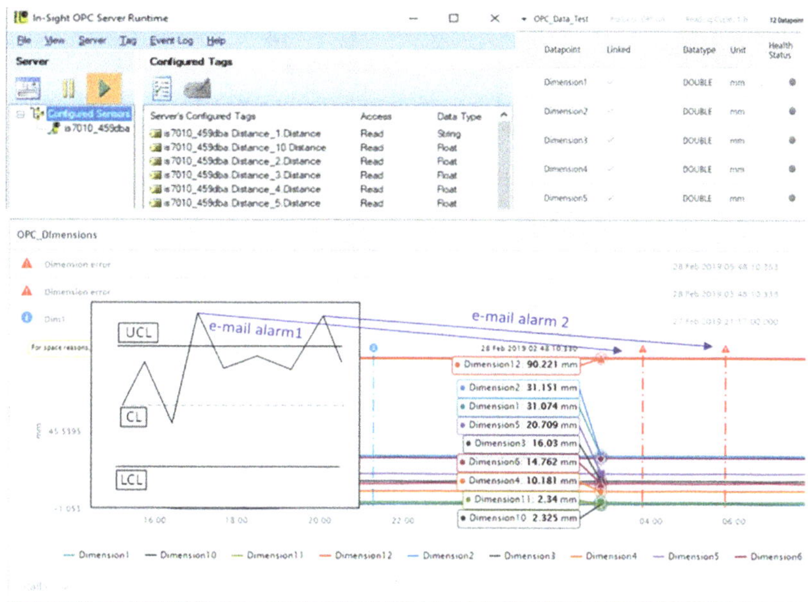

Fig. 11.14 OPC Server data from vision system (top left), MindSphere value list (top right), measurements timeline with alarm message example (bottom)

The simulation of the systems for quality control purposes and is usually executed offline, prior to beginning the production run. In our approach, we connect simulation through digital twin of the system to the PLC of the physical asset and to synchronize any optimization activities in very short time. The principle of bidirectional data synchronization between real device process and simulation is shown in Figs. 11.16 and 11.17.

Python OPC UA server is responsible for data transfer from digital twin simulation to PLC system, and moreover it creates data communication bridge between Cloud Platform and the real system.

The simulation of the inspection/identification process is executed through Plant Simulator software and allows to modify crucial parameters in order to reach improvements in quality of products and productivity of manufacturing. The main advantage of the digital twin is very fast optimization, because we don't need neither to change program in

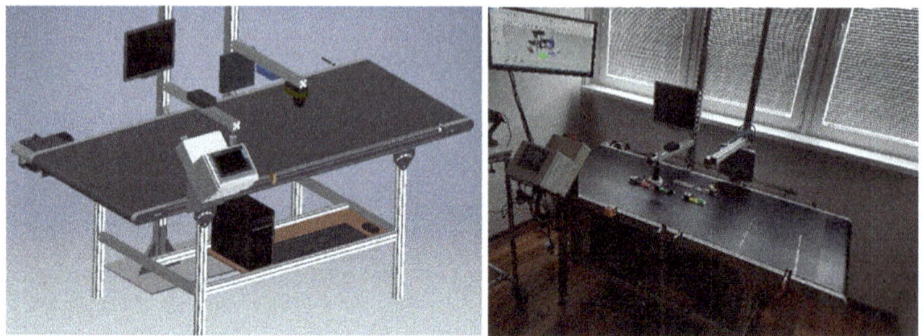

Fig. 11.15 3D digital twin (left) and experimental inspection and identification system (right)

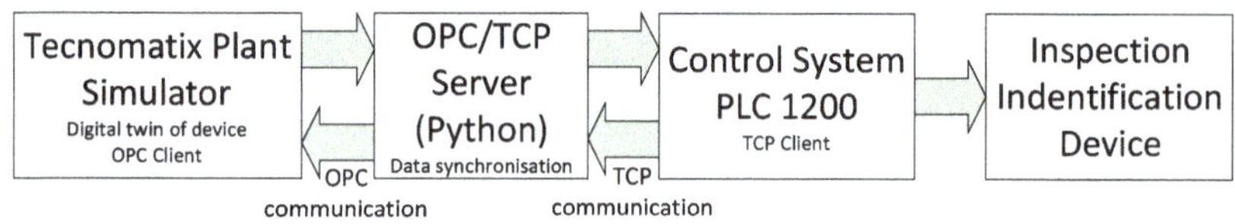

Fig. 11.16 The scheme of bidirectional data synchronization between quality control system and digital twin

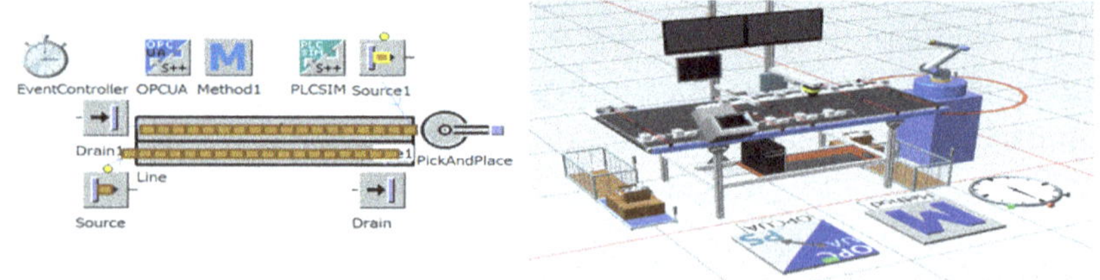

Fig. 11.17 An example of 2D and 3D simulation in Tecnomatix plant simulator with OPC communication

the PLC control system nor by the SCADA system. But, it allows to change optimized parameters straight from simulation software to manufacturing. Remote monitoring and quality control activities are possible thanks to smart server integrated in HMI/SCADA touch panel KTP400 Basic. One can access and control inspection/identification process by Android smartphone or from PC. An example of remote monitoring is shown in Fig. 11.18.

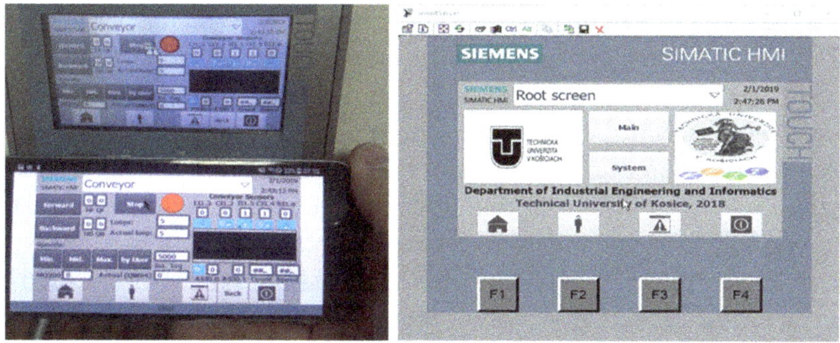

Fig. 11.18 Remote monitoring of inspection device by mobile phone (left) and by PC (right)

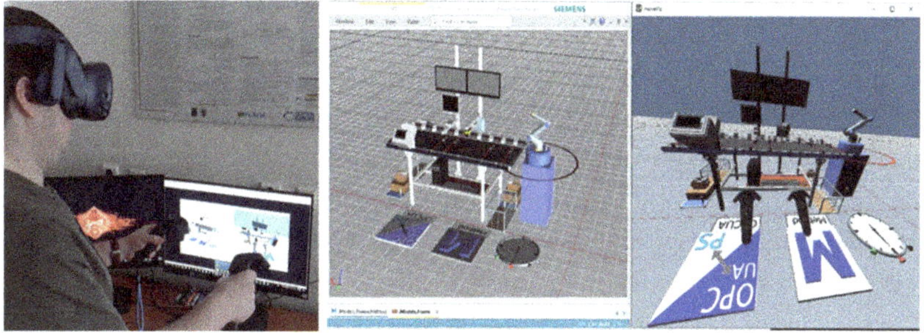

Fig. 11.19 HTC Vive Pro (left), 3D model in plant simulator (middle), digital twin inspection in VR (right)

The simulation of the quality control operations can be transferred into virtual reality devices, e.g., HTC Vive Pro or Oculus Rift. The staff responsible for the quality control can monitor the device with synchronized data from real production in a 3D virtual digital twin. An example of transfer Technomatix simulation into virtual reality by HTC Vive Pro is shown in Fig. 11.19.

11.5 Discussion

The main problem of currently used Clouds Platforms is lack of support for linking 3D personalized models (digital twins) represented by Binary Large Objects (BLOB) with the clouds. Available commercial clouds offer only basic variables such as integer, float, bool, data time, and string. The clouds are primarily focused on data collection, basic graphical representation to timelines, and data knowledge extraction for basic alarm systems with periodical reports (days, week, months, and years). Data are structured in simple frames and defined only by numbers and strings. It is not possible to define advanced structures with hybrid data, for example images, and binary data combined with numerical/string variables. In the current status of clouds technology it is very hard to store a personalized digital twin of product in one storage system (cloud).

The one possible solution is to create a separate database for a personalized digital twin 3D model with a hyperlink to measure data stored in the cloud. This solution can only link data one way from 3D model to cloud, not back. But the main advantage of this approach is increased security for critical data about products and technology, because 3D model is stored in local database.

The next solution can be modification of an open-source cloud system to store BLOB, image data, and measured values in one place. The main advantage can be low operating costs because a whole solution can operate on a local server.

It can be also mentioned that all used identification and inspection devices have their limitations (IoT, Vision systems, RFID tags). A UHF RFID system cannot be used for very small parts, because the tag size is limited by antenna length. LF or HF RFID tags are the solutions for tagging smaller parts but they are not primarily developed for industrial parts (mainly for food industry). An industrial vision system is a closed source system to algorithm modification. The main problems arise in surface errors recognizing very geometrically complicated parts. One of the possible ways to solve this problem can be exploitation of convolutional neural networks with deep learning techniques. This approach is currently tested on experimental devices, but there is a perspective of its fast transfer to commercial devices.

11.6 Conclusions

In this study a computer vision system along with RFID system applications for contactless quality control activities before and after the assembly process has been presented. It has been demonstrated through the experimental inspection and identification system by using a subassembly module for a rapid prototyping printer. Presented experiments aimed to verify the proposed approach to capture and transport data from production process to the extended digital 3D twin model. The bridge for transfer data in both directions was performed by OPC technology, mainly OPC UA Server (OPC DCOM). OPC Server was written in the Python programming language customized for data collection from many sources. The digital twin of inspection and identification system was designed for online connection to synchronize data from a quality control process. Extended data from the digital twin was also synchronized online with Cloud Platform. The main purpose of using the RFID gate was to personalize all products with acquired dimension data stored in the main assembly RFID tag label. RFID system has been used to localize parts on the conveyor line by RSSI signal from tags.

The ambition of this study was to provide ideas for maximizing utilization of gathered data from the product quality control process, because many companies don't store and use this data, e.g., for an analysis purpose. It was proved that the data used so far are only for product classification into two groups, i.e., a good product or noncompliant product, can be utilized much more through the digitization platforms used in our study.

Further experimental works will focus on long-term data collection, reliability verification with implementation to some existing quality check systems in a real production environment.

References

Aissam, Manal, Mohammed Benbrahim, and Mohammed Nabil Kabbaj. 2019. Cloud Robotic: Opening a New Road to the Industry 4.0. In *New Developments and Advances in Robot Control*. Singapore: Springer. https://doi.org/10.1007/978-981-13-2212-9_1.

Akkaladevi, S.L., M. Plasch, S. Maddukur, C. Eitzinge, A. Pichlee, and R. Rinner. 2018. Toward an Interactive Reinforcement Based Learning Framework for Human Robot Collaborative Assembly Processes. *Frontiers in Robotics and AI* 5. https://doi.org/10.3389/frobt.2018.00126.

Caputo, Francesco, Alessandro Greco, Marcello Fera, and Roberto Macchiaroli. 2019. Digital Twins to Enhance the Integration of Ergonomics in the Workplace Design. *International Journal of Industrial Ergonomics* 71: 20–31. https://doi.org/10.1016/j.ergon.2019.02.001.

David, Joe, Andrei Lobov, and Minna Lanz. 2018. Learning Experiences Involving Digital Twins. In *IECON 2018—44th Annual Conference of the IEEE Industrial Electronics Society*, 3681–3686. https://doi.org/10.1109/IECON.2018.8591460.

Da Xu, Li, and Lian Duan. 2019. Big Data for Cyber Physical Systems in Industry 4.0: A Survey. *Enterprise Information Systems* 13 (2): 148–169. https://doi.org/10.1080/17517575.2018.1442934.

Dong, Liang, and Zhifeng Wang. 2018. Plant Layout and Simulation Roaming System Based on Virtual Reality Technology. In *MATEC Web of Conferences*, ed. G. Beydoun, 214. https://doi.org/10.1051/matecconf/201821404001.

Gong, L., J. Berglund, Å. Fast-Berglund, B. Johansson, Z. Wang, and T. Börjesson. 2019. Development of Virtual Reality Support to Factory Layout Planning. *International Journal on Interactive Design and Manufacturing (IJIDeM)*. https://doi.org/10.1007/s12008-019-00538-x.

Groggert, S., M. Wenking, R.H. Schmitt, and T. Friedli. 2017. Status quo and Future Potential of Manufacturing Data Analytics—An Empirical Study. In *2017 IEEE International Conference on Industrial Engineering and Engineering Management (IEEM)*, 779–783. https://doi.org/10.1109/IEEM.2017.8289997.

Han, W., W. Liu, K. Zhang, Z. Li, and Z. Liu. 2019. A Protocol for Detecting Missing Target Tags in RFID Systems. *Journal of Network and Computer Applications* 132: 40–48. https://doi.org/10.1016/j.jnca.2019.01.027.

Hu, Y., F. Zhu, L. Zhang, Y. Lui, and Z. Wang. 2019. Scheduling of Manufacturers Based on Chaos Optimization Algorithm in Cloud Manufacturing. *Robotics and Computer-Integrated Manufacturing* 58: 13–20. https://doi.org/10.1016/j.rcim.2019.01.010.

Khan, Adnan, Martin Dahl, Petter Falkman, and Martin Fabian. 2018. Digital Twin for Legacy Systems: Simulation Model Testing and Validation. In *2018 IEEE 14th International Conference on Automation Science and Engineering (CASE)*, 421–426. https://doi.org/10.1109/COASE.2018.8560338.

Krowicki, Paweł, Grzegorz Iskierka, Bartosz Poskart, Maciej Habiniak, and Tomasz Będza. 2019. Scanπ—Integration and Adaptation of Scanning and Rapid Prototyping Device Prepared for Industry 4.0. In *Intelligent*

Systems in Production Engineering and Maintenance, 574–586. https://doi.org/10.1007/978-3-319-97490-3_55.

Lee, C.-C., S.-D. Chen, C.-T. Li, C.-L. Cheng, and Y.-M. Lai. 2019. Security Enhancement on an RFID Ownership Transfer Protocol Based on Cloud. *Future Generation Computer Systems* 93: 266–277. https://doi.org/10.1016/j.future.2018.10.040.

Li, Gang, Jianlong Tan, and Sohail S. Chaudhry. 2019. Industry 4.0 and Big Data Innovations. *Enterprise Information Systems* 13 (2): 145–147. https://doi.org/10.1080/17517575.2018.1554190.

Liu, Chuan-Gang, I-Hsien Liu, Chang-De Lin, and Jung-Shian Li. 2019. A Novel Tag Searching Protocol with Time Efficiency and Searching Accuracy in RFID Systems. *Computer Networks* 150: 201–216. https://doi.org/10.1016/j.comnet.2019.01.011.

Liu, Hongyi, and Lihui Wang. 2017. An AR-Based Worker Support System for Human-Robot Collaboration. *Procedia Manufacturing* 11: 22–30. https://doi.org/10.1016/j.promfg.2017.07.124.

Lu, Yuqian, and Xu Xun. 2019. Cloud-Based Manufacturing Equipment and Big Data Analytics to Enable On-Demand Manufacturing Services. *Robotics and Computer-Integrated Manufacturing* 57: 92–102. https://doi.org/10.1016/j.rcim.2018.11.006.

Mahmoud, Magdi S. 2019. Architecture for Cloud-Based Industrial Automation. *Robotics and Computer-Integrated Manufacturing*: 51–62. https://doi.org/10.1007/978-981-13-1165-9_6.

Malik, Ali Ahmad, and Arne Bilberg. 2017. Framework to Implement Collaborative Robots in Manual Assembly: A Lean Automation Approach. In *28th DAAAM International Symposium on Intelligent Manufacturing and Automation*, 1151–1160. https://doi.org/10.2507/28th.daaam.proceedings.160.

Martinez, Gerardo Santillan, Seppo Sierla, Tommi Karhela, and Valeriy Vyatkin. 2018. Automatic Generation of a Simulation-Based Digital Twin of an Industrial Process Plant. In *IECON 2018—44th Annual Conference of the IEEE Industrial Electronics Society*, 3084–3089. https://doi.org/10.1109/IECON.2018.8591464.

Martinez, Pablo, Rafiq Ahmad, and Mohamed Al-Hussein. 2019. A Vision-Based System for Pre-Inspection of Steel Frame Manufacturing. *Automation in Construction* 97: 151–163. https://doi.org/10.1016/j.autcon.2018.10.021.

Modrak, V. (ed.). 2017. *Mass Customized Manufacturing: Theoretical Concepts and Practical Approaches*. CRC Press. https://doi.org/10.1201/9781315398983.

Modrak, Vladimir, and Zuzana Soltysova. 2018. Development of Operational Complexity Measure for Selection of Optimal Layout Design Alternative. *International Journal of Production Research* 56 (24): 7280–7295. https://doi.org/10.1080/00207543.2018.1456696.

Mourtzis, Dimitris, Sophia Fotia, and Nikoletta Boli. 2017. Metrics Definition for the Product-Service System Complexity Within Mass Customization and Industry 4.0 Environment. In *2017 International Conference on Engineering, Technology and Innovation (ICE/ITMC)*, 1166–1172. https://doi.org/10.1109/ICE.2017.8280013.

Mourtzis, Dimitris, Sophia Fotia, Nikoletta Boli, and Pietro Pittaro. 2018. Product-Service System (PSS) Complexity Metrics Within Mass Customization and Industry 4.0 Environment. *The International Journal of Advanced Manufacturing Technology* 97 (1–4): 91–103. https://doi.org/10.1007/s00170-018-1903-3.

Mukherjee, Tridibesh, and Tarasankar DebRoy. 2019. A Digital Twin for Rapid Qualification of 3D Printed Metallic Components. *Applied Materials Today* 14: 59–65. https://doi.org/10.1016/j.apmt.2018.11.003.

OrRiordan, Andrew, Thomas Newe, Gerard Dooly, and Daniel Toal. 2018. Stereo Vision Sensing: Review of Existing Systems. In *2018 12th International Conference on Sensing Technology (ICST)*, 178–184. https://doi.org/10.1109/ICSensT.2018.8603605.

Raihana, G.F.H. 2012. Cloud ERP—A Solution Model. *International Journal of Computer Science and Information Technology & Security* 2 (1): 76–79.

Rauch, E., P. Dallasega, and D.T. Matt. 2016. Sustainable Production in Emerging Markets Through Distributed Manufacturing Systems (DMS). *Journal of Cleaner Production* 135: 127–138. https://doi.org/10.1016/j.jclepro.2016.06.106.

Realyvásquez-Vargas, A., K. Cecilia Arredondo-Soto, J. Luis García-Alcaraz, B. Yail Márquez-Lobato, and J. Cruz-García. 2019. Introduction and Configuration of a Collaborative Robot in an Assembly Task as a Means to Decrease Occupational Risks and Increase Efficiency in a Manufacturing Company. *Robotics and Computer-Integrated Manufacturing* 57: 315–328. https://doi.org/10.1016/j.rcim.2018.12.015.

Sanin, C., Z. Haoxi, I. Shafiq, M.M. Waris, C. Silva de Oliveira, and E. Szczerbicki. 2019. Experience Based Knowledge Representation for

Internet of Things and Cyber Physical Systems with Case Studies. *Future Generation Computer Systems* 92: 604–616. https://doi.org/10.1016/j.future.2018.01.062.

Schmidt, R., M. Möhring, R.C. Härting, C. Reichstein, P. Neumaier, and P. Jozinović. 2015. Industry 4.0-Potentials for Creating Smart Products: Empirical Research Results. In *International Conference on Business Information Systems*, 16–27. Cham: Springer. https://doi.org/10.1007/978-3-319-19027-3_2.

Shubenkova, K., A. Valiev, V. Shepelev, S. Tsiulin, and K.H. Reinau. 2018. Possibility of Digital Twins Technology for Improving Efficiency of the Branded Service System. In *2018 Global Smart Industry Conference (GloSIC)*, 1–7. https://doi.org/10.1109/GloSIC.2018.8570075.

Tomko, Martin, and Stephan Winter. 2019. Beyond Digital Twins—A Commentary. *Environment and Planning B: Urban Analytics and City Science* 46 (2): 395–399. https://doi.org/10.1177/2399808318816992.

Velandia, D.M.S., N. Kaur, W.G. Whittow, P.P. Conway, and A.A. West. 2016. Towards Industrial Internet of Things: Crankshaft Monitoring, Traceability and Tracking Using RFID. *Robotics and Computer-Integrated Manufacturing* 41: 66–77. https://doi.org/10.1016/j.rcim.2016.02.004.

Żabiński, Tomasz, Tomasz Mączka, and Jacek Kluska. 2017. Industrial Platform for Rapid Prototyping of Intelligent Diagnostic Systems. In *Polish Control Conference*, 712–721. Cham: Springer. https://doi.org/10.1007/978-3-319-60699-6_69.

Zhou, Q., R. Chen, B. Huang, C. Liu, J. Yu, and X. Yu. 2019. An Automatic Surface Defect Inspection System for Automobiles Using Machine Vision Methods. *Sensors* 19 (3): 644. https://doi.org/10.3390/s19030644.

Zidek, Kamil, Vladimir Maxim, Jan Pitel, and Alexander Hosovsky. 2016. Embedded Vision Equipment of Industrial Robot for Inline Detection of Product Errors by Clustering-Classification Algorithms. *International Journal of Advanced Robotic Systems* 13 (5). https://doi.org/10.1177/1729881416664901.

Żywicki, Krzysztof, Przemysław Zawadzki, and Filip Górski. 2018. Virtual Reality Production Training System in the Scope of Intelligent Factory. In *International Conference on Intelligent Systems in Production Engineering and Maintenance*, 450–458. Cham: Springer. https://doi.org/10.1007/978-3-319-64465-3_43.

12

Implementation of a Laboratory Case Study for Intuitive Collaboration Between Man and Machine in SME Assembly

Luca Gualtieri, Rafael A. Rojas, Manuel A. Ruiz Garcia, Erwin Rauch and Renato Vidoni

12.1 Introduction

"Industry 4.0" is the name given to the ongoing fourth industrial revolution, which is actually transforming worldwide factories. This concept was initially introduced by a German government strategic initiative in 2011 (Kagermann et al. 2013) and represents the current

L. Gualtieri (✉) · R. A. Rojas · M. A. Ruiz Garcia · E. Rauch · R. Vidoni
Faculty of Science and Technology, Free University of Bozen-Bolzano, Bolzano, Italy
e-mail: luca.gualtieri@unibz.it

R. A. Rojas
e-mail: rafael.rojas@unibz.it

M. A. Ruiz Garcia
e-mail: ManuelAlejandro.RuizGarcia@unibz.it

E. Rauch
e-mail: erwin.rauch@unibz.it

R. Vidoni
e-mail: renato.vidoni@unibz.it

D. T. Matt et al. (eds.), *Industry 4.0 for SMEs*,
https://doi.org/10.1007/978-3-030-25425-4_12

evolution of modern industry. Production systems are shifting from mass production to mass customization logic (Pedersen et al. 2016), by adapting their performance to a globalized, interconnected and volatile market (Chryssolouris 2013). Actually, in order to be competitive and profitable, modern manufacturing companies need further production flexibility and efficiency in terms of lot sizes, variants, and time-to-market. For these reasons, the key point of Industry 4.0 is the integration of adaptable and reconfigurable manufacturing systems and technologies introducing innovative and advanced elements such as cyber-physical systems (CPS), Internet of Things (IoT), and cloud computing for manufacturing purposes (Zhong et al. 2017). In particular, the role of cyber-physical production systems (CPPS) is to connect the physical and the virtual manufacturing world in order to satisfy agile and dynamic production requirements. The goal is the union of conventional production technology and information technology (IT) for the mutual communication of machines and products in an IoT environment (Lu 2017; Penas et al. 2017). Industrial collaborative robots (see Fig. 12.1) are particular kinds of enabling CPSs and one essential technology of Industry 4.0, and allow direct and safe physical human–robot interaction (HRI). Collaborative robotics aims to help operators in production activities through different levels of coexistence, cooperation, and collaboration by supporting humans in less ergonomic, repetitive, and alienating tasks, also considering product and process production efficiency. The main potential advantages are:

- Improvement of operators' work conditions
- Better use of production areas (no physical barriers are required)
- Improvement of workspace accessibility
- Enlargement of production capacity
- Improvement of products and process quality
- Better use of skilled labor.

In particular, according to the definition provided by ISO TS 15066, physical HRI entails hybrid operations in a shared workspace, which is defined as the "*space within the operating space where the robot system (including the workpiece) and a human can perform tasks concurrently*

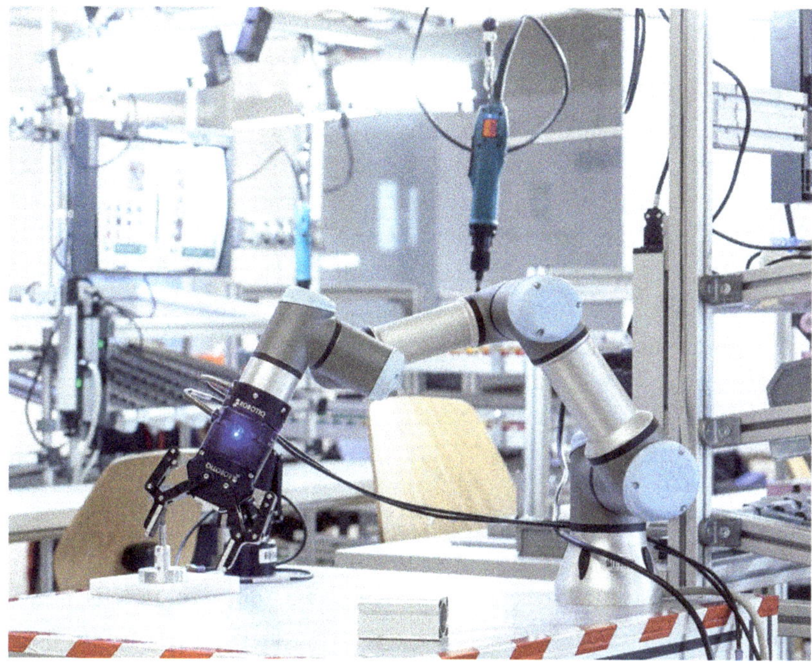

Fig. 12.1 A collaborative robot (UR3 model) in a shared workspace (*Source* Smart Mini Factory, unibz [Reproduced with permission from Smart Mini Factory Lab, unibz])

during production operation" (ISO 2016, p. 1). This involves a fenceless production environment where operators and robots can work together in a safe, ergonomic, and efficient way. According to this definition, conventional protective systems for traditional industrial robotics (such as physical barriers for workspace isolation), no longer apply (Matthias and Reisinger 2016). In fact, modern human–robot collaboration (HRC) requires and allows physical interaction between operators and robots. Considering the nature of mechanical risks related to traditional industrial robotics, possible unexpected and unwanted collisions between a non-collaborative robot arm and an operator could be lethal. Fortunately, if safety systems are properly implemented, collaborative robots exceed this adverse and dangerous condition by allowing safe hand-in-hand HRC. Of course, one of the biggest future challenges in the development of collaborative systems is to ensure operators' psycho-physical well-being in terms of occupational health and safety (OHS) while preserving high robot performance. Due to the novelty of the technology, the complexity of the topic and the limited knowledge of companies about the design and management of collaborative systems, small- and medium-sized enterprises (SMEs) should be supported in the proper integration of safe, ergonomic, and efficient HRI. The proposed methodology aims to improve the adoption of collaborative systems into industrial SMEs by providing an efficient methodology for the conversion of manual assembly workstations into collaborative workcells.

12.2 Theoretical Background

Considering that the industrial collaborative robot market is continuously growing (Djuric et al. 2016), it is reasonable to suppose that collaborative assembly will be a crucial application in the near future. A large part of future collaborative systems will arise from existing manual workstations. For this reason, it is necessary to study a structured methodology, which enables production technicians and managers to simply evaluate if it is possible and reasonable to implement a collaborative assembly workstation starting from an existing one, by considering a set of production criteria. The introduction of collaborative robots aims to support operators' work conditions and production performances by improving physical ergonomics, enlarging production capacity and enhancing product and process quality. Since HRC aims to combine human abilities like flexibility, creativity, and decision-making skills with smart machine strengths like accuracy, repeatability, and payload (Siciliano and Khatib 2016), it is advisable to design new collaborative systems by considering the abilities and constraints of both human and robot resources. As a consequence, the layout and the input/output material flows of the new assembly workstation have to be changed according to the abovementioned considerations. Furthermore, due to the fact that both humans and robots will pick, handle, and assemble different components, the logistics aspects have to be reconsidered by evaluating the new hybrid assembly cycle. Of course, the selection of an adequate and process-oriented collaborative gripping technology will be crucial. In addition, suitable robot sensors for object recognition and situation awareness have to be implemented according to specific production and safety requirements. In addition, it will be fundamental to properly manage the organizational effects of the introduction of collaborative systems by balancing internal (inside the workstation) and external (outside the workstation) production parameters (see Fig. 12.2). In fact, the integration of collaborative systems must not create critical points in a well-structured existing workflow and related production environment.

Safety and ergonomics have necessarily to be incorporated into the preliminary design stage of the collaborative assembly workstation. This will be particularly useful to maximize the design effectiveness and to avoid future useless and time-consuming iterations for the adjustments of the related systems once the development of the workstation is partially completed. In other words, it is necessary to provide all the necessary upgrades to the new assembly workstation in order to facilitate an easy and proper integration of the collaborative robot into the existing production environment. To fill the current gap in terms of design knowledge and skills, guidelines and standards for the implementation

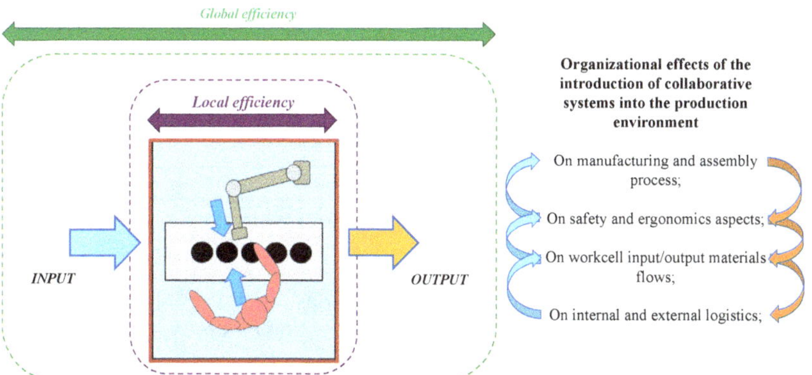

Fig. 12.2 Internal and external effects of the introduction of collaborative robots into existing production systems

of existing and new collaborative systems have to be developed in the near future. This will support an intuitive and barrier-free diffusion of collaborative assembly technologies especially in SMEs.

12.3 Methodology for the Evaluation of Human–Robot Task Allocation

The requirements for the transformation of a manual workstation into a collaborative one should include technical, (physical) ergonomics, qualitative, and finally economic aspects (Gualtieri et al. 2019). The core part of that analysis is to identify which tasks of an existing assembly cycle are more recommended for the robot and which ones for the operator by considering the abovementioned transformation criteria. Preliminary division criteria are provided in Table 12.1. More details about these particular choices and considerations will be discussed in this section.

While designing a transformation process, it is firstly necessary to consider that only certain assembly tasks can be performed efficiently by a robot due to inherent technical limitations (Boothroyd et al. 2010; Boothroyd 2005; Crowson 2006). This is a primary and mandatory constraint which influences all further evaluations. The second constraint will be physical ergonomics. In fact, one of the main purposes of Industry 4.0 is to create anthropocentric factories where the human

Table 12.1 Main guidelines for the preliminary evaluation of human–robot task allocation starting from existing manufacturing activities

Collaborative robot	Operator
Less ergonomic activities which imply physical and/or mental stress for the operator	Activities which imply reasoning ability, interpretation, and responsibility
Activities which imply repetitive tasks and/or which require complex movements for the operators	Activities which imply high handling ability and dexterity
Non value adding (NVA) activities	Value adding (VA) activities
Activities which require standardization and/or quality improvements	Activities which imply flexibility and ability to adapt

factors are the core part of production systems. Finally, it is important to integrate other organizational and economic factors for the development of accurate, flexible, and lean collaborative workstations. The general evaluation workflow and related priorities for the workstation transformation are summarized in Fig. 12.3.

Actually, the main part of the integration of a collaborative into an existing production system will be the division of tasks and activities between the operator and the robot. There are different studies relating to human–robot coordination and the "dynamic" task division in collaborative applications (Chen et al. 2011; Darvish et al. 2018; Liu and Wang 2017), which means a real-time sequencing of activities depending on different operator behaviors and preferences during the assembly cycle. In this situation, the operator can freely choose which task will be the next one indiscriminately. This positive condition of independence could improve cognitive ergonomics conditions, operators' psychological well-being (Gombolay et al. 2015) and production flexibility (Shen et al. 2015). On the other hand, every task is considered efficaciously executable both by human and robot and as a consequence, there are no technical constraints in terms of robot execution feasibility. For these reasons, it could be useful to firstly identify which tasks of a sequence can be efficiently performed by both humans and robots. This preliminary evaluation allows the designer to successively integrate the dynamic task division approach (variable during the process), by considering the real limits of the robot system. That condition permits a real-time scheduling of the identified unconditioned tasks and as a consequence, allows the operator to freely change the assembly sequence according to his needs and preferences. More details will be explained in Sect. 12.5.1. Since the dynamic task division is an early-stage research topic, this part

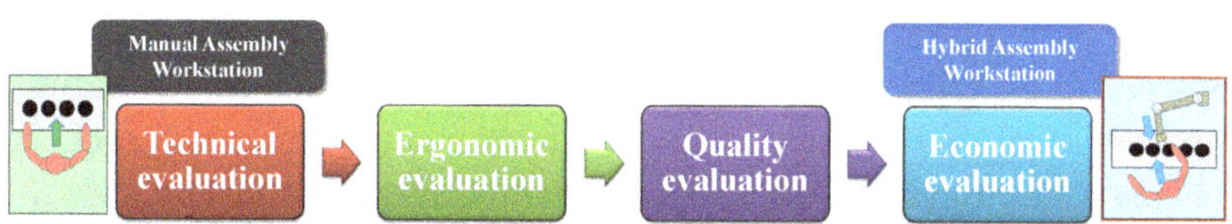

Fig. 12.3 General evaluation workflow and related priorities for the workstation transformation

of the chapter will focus on the preliminary human–robot division of tasks. The proposed discussion will support SME designers to adopt a structured methodology for the preliminary feasibility analysis of the integration of collaborative systems. This involves the evaluation of a manual assembly system in order to decide if a process is suitable or not for collaborative conversion. The main useful data could be: assembly cycle description (including sequences and priorities), task time, task variability, labor and components costs, main geometrical and material features of components, list of value added/not value added activities and physical ergonomics evaluation values. The preliminary task allocation should define if an assembly activity can be performed exclusively by the human (H), exclusively by the robot (R) or equally by the human and robot (H or R) by considering all the proposed considerations. In the following section, a detailed analysis of the evaluation criteria for manual workstation transformation is explained.

12.3.1 Technical Evaluation

The analysis of the technical feasibility of the transformation process aims to investigate if an activity can actually be performed by a robot in an efficient way, by considering its technical limitations of hardware and/or software. In general, it is necessary to verify if a certain type of industrial collaborative robot (equipped with standard commercial devices) is able to perform the feeding, handling and/or assembling of the involved components by using a proper amount of production resources in a suitable time. In this context, the main complexities could arise from product geometry, product dimension, product materials features, assembly location, and assembly sequence organization (Boothroyd et al. 2010; Boothroyd 2005; Crowson 2006). In practice, there are many product or process "technical critical issues" which could complicate or prevent the use of a collaborative robot for assembly or manufacturing activities. Actually, the chance to properly pick and manage a component strictly depends on the type of gripper which it is intended to add to the robot arm (Monkman et al. 2007). A partially completed list of feeding, handling, and assembly critical issues is summarized in Table 12.2.

Table 12.2 Summary of main feeding, handling and assembly critical issues according to the guidelines developed by Boothroyd and Crowson for the design of robotic and automatic assembly (Boothroyd et al. 2010; Boothroyd 2005; Crowson 2006)

Critical issues—Feeding

- The component is magnetic or sticky
- The component is a nest or tangle

Critical issues—Handling

- The component has no symmetry axis
- The component is fragile or delicate
- The component is flexible
- The component is very small or very big (in reference to a human hand)
- The component is light so that air resistance would create conveying problems
- The component is slippery

Critical issues—Assembly

- Components do not have a "datum surface" (reference surface) which simplifies precise positioning during the assembly
- Components cannot be easily orientated
- Components do not include features which allow self-aligning during the assembly
- Components cannot be located before they are released
- Components provide resistance to insertion
- Components do not provide chamfers or tapers that help to guide and position the parts in the correct position
- Components do not have a suitable base part on which to build the assembly
- Components cannot be assembled in layer fashion from directly above (z-axis assembly)
- The assembly is overconstrained
- It is difficult to reach the assembly area/the components access for assembly operations is restricted or not easy to reach
- The component and/or the assembly sequence requires high physical dexterity
- The assembly requires high accuracy and/or demanding insertion tolerances
- The assembly needs to reposition the partially completed subassembly, other components or fixtures
- The assembly requires the partial assembly to be reorientated or previously assembled parts to be manipulated
- Components must be compressed during the assembly
- The component and/or the assembly sequence requires two hands for handling
- The component and/or the assembly sequence requires typical human skills (for example touch perception, hearing, ability to interpret situations…)

In general, it is possible to consider two main categories of activities: feeding and handling tasks; assembly tasks. The main reason for the proposed division is that if an operator or a robot needs to assemble one or more components, it is firstly necessary to pick up and handle them. For this motive, a task which requires the assembly also includes the critical issues related to feeding and handling. On the other hand, a task that requires just the feeding or handling does not have to include the assembly critical issues.

A general workflow for the preliminary evaluation of the technical possibility to use a collaborative robot for assembly activities is represented in Fig. 12.4. This guided procedure will help designers to understand if feeding, handling, and/or assembling activities could actually be performed by a certain collaborative robot system (robot arm, gripper, and sensors) efficaciously. In any case, further detailed analysis is necessary for complete comprehension of the problem.

12.3.2 Physical Ergonomic Evaluation

Ergonomics, or human factors, is the science which aims to study the interactions among humans and other elements of a system in order to optimize human well-being and overall system performance (Salvendy 2012). In this context, one of the main goals of the introduction of collaborative robots into manual production systems is to improve an operator's physical conditions by reducing work-related biomechanical stress. A collaborative workstation should be a practical implementation of the so-called "anthropocentric" or "human centered" design, a method which considers the operators' work conditions the main elements of the production system. According to user needs and requirements, the main goals of this design methodology are the improvement of effectiveness, efficiency, well-being, user satisfaction, accessibility, and sustainability by counteracting possible adverse effects of use on human health, safety and performance at the same time (ISO 2010). In this context, the role of the proposed physical ergonomic evaluation stage is to identify if the integration of a collaborative robot could improve operators' physical work

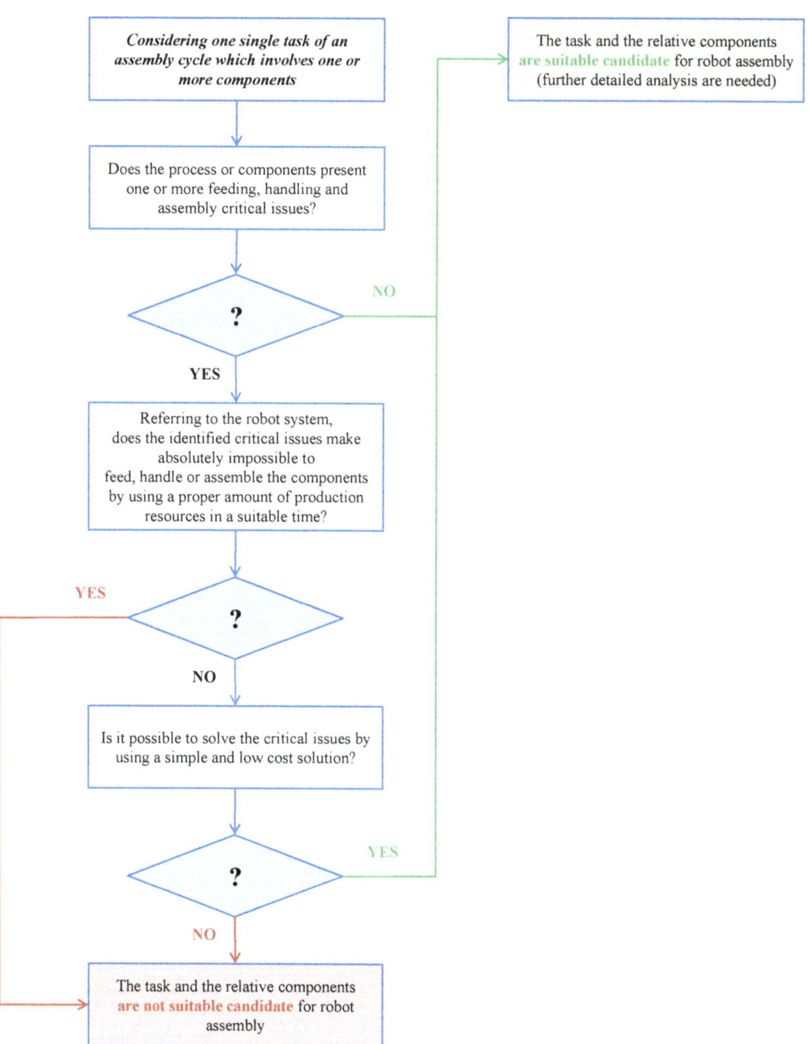

Fig. 12.4 Workflow for the evaluation of the technical possibility to use a collaborative robot for assembly activities

conditions and to quantify the relative benefits. A crucial part of that evaluation is the use of a standard approach for the systematic analysis of the work-related biomechanical stress of the existing workstation. In fact, it is necessary to identify if the future integration of a collaborative robot could really support the operators during manual operations in a physical way. According to the nature of work activities, there are many different recognized methodologies for physical ergonomics evaluation: NIOSH for lifting and carrying (ISO 2003), Snook and Ciriello for whole-body pushing and pulling (ISO 2007a) and Occupational Repetitive Actions (OCRA) for the handling of low loads at high frequency (ISO 2007b). A less detailed, faster and simpler evaluation method is Rapid Upper Limb Assessment (RULA), which can be a useful tool for a preliminary and approximate analysis, particularly focusing on postures. Of course, it is possible to evaluate the physical ergonomics conditions according to other kinds of recognized methodologies or by using the results from different approaches.

12.3.3 Product/Process Quality Evaluation

The product or process quality evaluation aims to investigate if a task or an activity requires improvements in terms of standardization and reduction of process instability or variability. Actually, the concept of quality is often related to the concept of standardization. Standardization improvement means a reduction in related process variability levels. From a manufacturing prospective, variability is defined as an inherent process deviation from a prespecified requirement or nominal value. As a consequence, variability is a negative situation which requires a more controlled condition to achieve the designed process and product quality values (Sanchez-Salas et al. 2017). Obviously, in order to quantify the variability levels, it is necessary to identify one or more process variables to measure. A common possibility could be the task process time of the actual assembly cycle. Once all tasks are mapped and measured, it is useful to identify a list of tasks which present a high level of process variability. Since automation is a useful tool to increase process control, it is advisable to use a collaborative robot for uncontrolled tasks in order to improve the related standardization level.

12.3.4 Economic Evaluation

The economic evaluation aims to recognize the tasks, which can really provide economic value to the final customer according to a cost criteria analysis. Due to the fact that it is necessary to promote easy and fast procedures, a possibility for the implementation of the economic evaluation could be an investigation based on "value added" (VA)/"not value added" (NVA). In industrial management, an NVA task will absorb production resources and/or time by generating unnecessary costs without providing perceived value and satisfaction to the final costumer. In contrast, a VA task will be able to significantly increase the product value and satisfaction to the final customer even if it can generate production costs (Swamidass 2000). In general, in order to reduce and control production costs, it will be advisable to use automation for those activities (and the relative components) which do not provide sufficient economic value to the final customer. In addition, in this case, it will be useful to identify a list of tasks by classifying the related NVA/VA nature through main lean management.

12.3.5 Final Evaluation

Finally, it is necessary to hierarchically relate all the abovementioned concepts in order to achieve a final and all-embracing evaluation of the conversion process. The overall combination of the different evaluation analysis for human–robot task allocation is summarized in Fig. 12.5.

12.4 Application of Intuitive Human–Robot Interaction in the Smart Mini Factory Lab

12.4.1 Introduction to the Smart Mini Factory

The Smart Mini Factory (SMF) is a laboratory of the Free University of Bolzano-Bozen (https://smartminifactory.it/) dedicated to applied research and teaching. Inspired by the concept of a learning factory, it aims to study

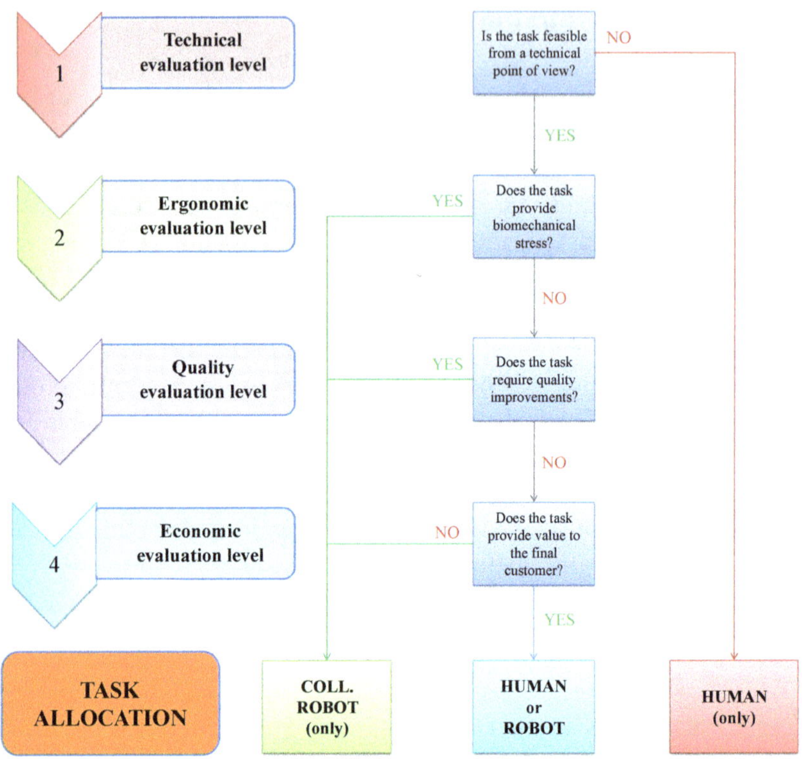

Fig. 12.5 Overall combination of various evaluation analysis for human–robot task allocation

and simulate production systems with a special focus on technologies and methods that enable the fourth industrial revolution. A primary goal is to develop a meeting platform where research, learning, and industry meet to allow common and productive knowledge transfer (Gualtieri et al. 2018). In fact, to achieve knowledge production, diffusion and application through innovation, the laboratory is built to serve three purposes:

- **Research**: company's needs are translated into application-oriented research projects. In addition, research results and know-how are provided for the future.
- **Teaching**: beyond regular lessons and practical sessions, students can develop their study projects as well as final theses and thus gain

valuable experience using state-of-the-art Industry 4.0 systems and automation equipment.
- **Industry**: the SMF is a bridge between industry and research used to supports companies during the implementation of Industry 4.0 concepts by common project collaboration. As a consequence, companies can be involved in the research side as well as in the qualification of their employees via customized industry-oriented seminars for the challenges of Industry 4.0.

Taking these into account, the main requirements of the SMEs in the region and the topics focused on in the SMF lab are the following: Industry 4.0 key enabling technologies, Automation & Robotics, Mechatronics & Electric Drives, Human-Machine Collaboration, Hybrid Assembly Systems, Assistance Systems for Production and Virtual/Augmented Reality. To these, two additional topics that bring the Industry 4.0 concepts and ideas outside the factory are developed or in development: Construction 4.0 and Agro-mechatronics. The topics of human–machine interaction and robotics merge in HRI, which entails physical interaction between operators and collaborative robots. In addition to the Kuka KMR iiwa robotic system, two models of collaborative anthropomorphic robots are available: Universal Robot UR3 and UR10. The main research activities refer to:

- Identification of human-centered robotized solutions for SME
- Development of new methodologies for the evaluation of industrial HRC systems from a safety, economic and technical point of view
- Research on new concepts of collaborative human–robot assembly workstations, taking into account requirements for safety, ergonomics, and production efficiency
- Development of virtual reality solutions for simulation and training for HRI.

12.4.2 Case Study Description

The proposed case study aims to explain the conversion process between an existing manual assembly workstation and a collaborative one.

The manual workstation is located in the assembly simulation line of the SMF laboratory. It is a flexible working area used for training and research in the field of the design of manual and hybrid assembly systems for light industrial products, workplace organization, human-centered design, and ergonomics (see Fig. 12.6). In particular, it is equipped with

Fig. 12.6 Manual assembly workstation for assembly of pneumatic cylinders (Figs. 12.6–12.11 reproduced with permission from Smart Mini Factory Lab, Unibz)

a mobile workbench, a block-and-tackle for lightweight applications, an integrated kanban rack, a working procedures panel, a double lighting system, an industrial automatic screwdriver and a knee lever press where a single operator can completely assemble a pneumatic cylinder (see Fig. 12.7). The main research activities refer to the analysis and optimization of production system performance and operators' work conditions by simulating different assembly circumstances and applications.

Theoretically, it will be advisable to consider the possibility of adopting different types of collaborative robots and grippers in order to identify the more suitable solution according to task sequence and components features. For the proposed simplified case study, a Universal Robot model UR3 (see Fig. 12.9) equipped with a 2-finger Robotiq collaborative gripper (see Fig. 12.8) is used.

The UR3 is the smallest member of the Universal Robots collaborative series. It is a 6-rotating-joint anthropomorphic manipulator suitable

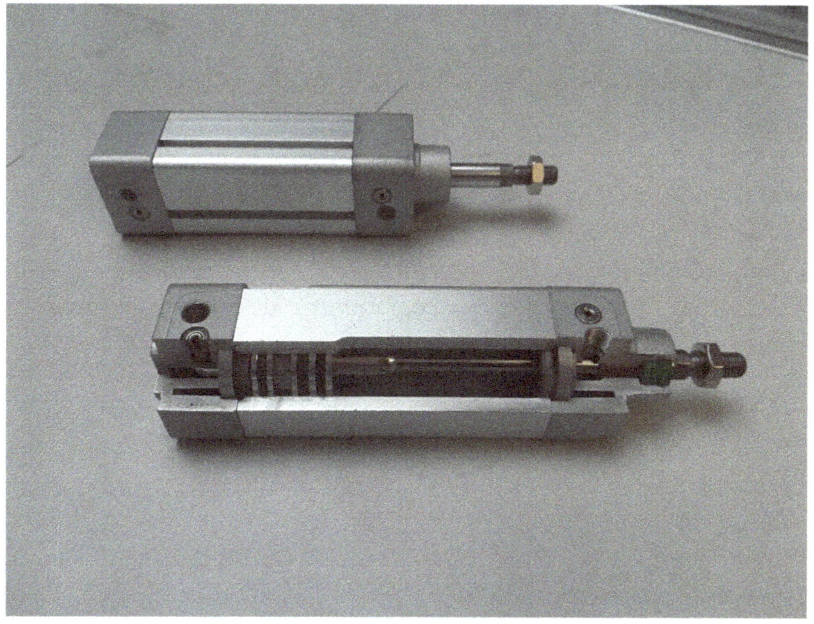

Fig. 12.7 Pneumatic cylinder (As for Figs. 12.6–12.11 the Fig. 12.7 is reproduced with permission from Smart Mini Factory Lab, Unibz)

Fig. 12.8 Robotiq collaborative gripper

for light assembly and high precision tasks. Flexibility and versatility, including an operator-friendly programming device are the main features of this multipurpose robot. Its main technical specifications are (Universal Robot 2019):

- Degrees of freedom: 6 rotating joint
- Payload: 3 kg
- Reach: 500 mm
- Repeatability: ±0.1 mm
- Power consumption: min 90 W; typical 125 W; max 250 W
- Ambient temperature range: 0–50 °C—at high continuous joint speed, ambient temperature is reduced

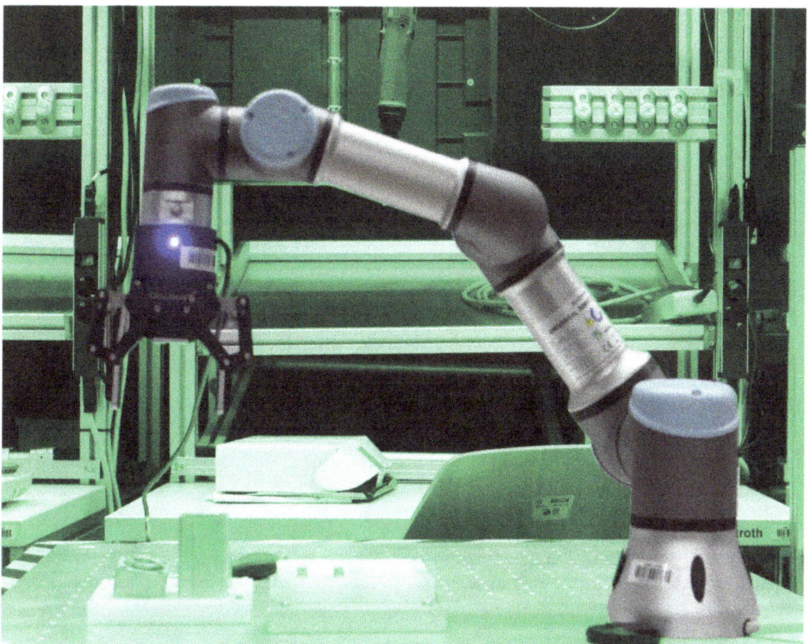

Fig. 12.9 Universal robot UR3 model

- Programming: Polyscope graphical user interface on touchscreen with mounting
- Collaboration operation: 15 advanced adjustable safety functions; TÜV NORD Approved Safety Function; tested in accordance with EN ISO 13849:2008 PL d.

12.4.3 Pneumatic Cylinder Collaborative Assembly

The workpiece which will be analyzed during the proposed case study is a medium-size pneumatic cylinder. The components (see Fig. 12.10) and related subassemblies (see Fig. 12.11) are summarized in Table 12.3.

The current manual assembly cycle is represented in Fig. 12.12.

The actual assembly cycle data are shown in Table 12.4.

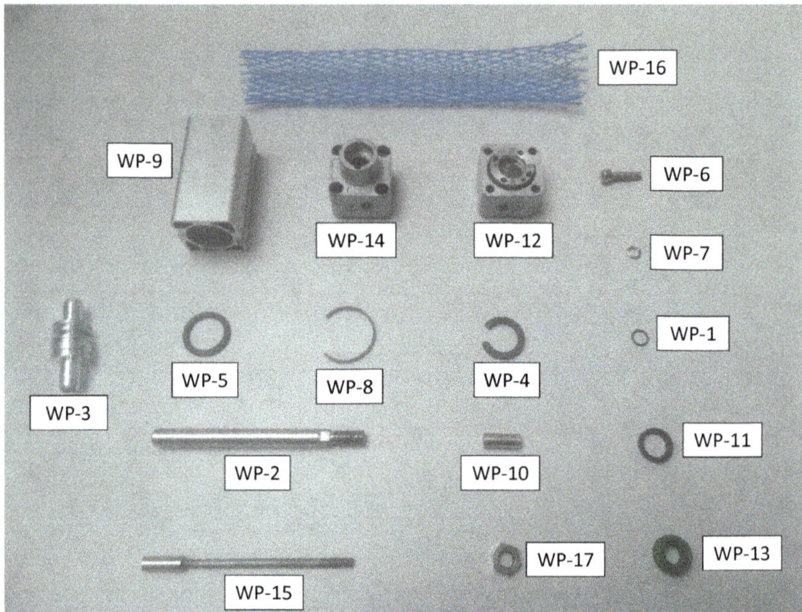

Fig. 12.10 Pneumatic cylinder components

12.4.4 Case Study Evaluation

a) **Technical evaluation**

According to the pneumatic cylinder description shown previously and by using the guidelines explained in Table 12.2, a preliminary technical evaluation was performed. All the mentioned feeding, handling and assembly critical issues were considered through a detailed visual and operational analysis. It is important to underline that it is of primary importance to really try to perform the tasks in order to understand all the assembly critical points in detail. Table 12.5 shows an example of technical critical issues examination through three feeding and handling tasks.

A further investigation could be performed in order to estimate the importance levels of the identified critical issues. In fact, there might be a difference between the impact of one critical issue with respect to

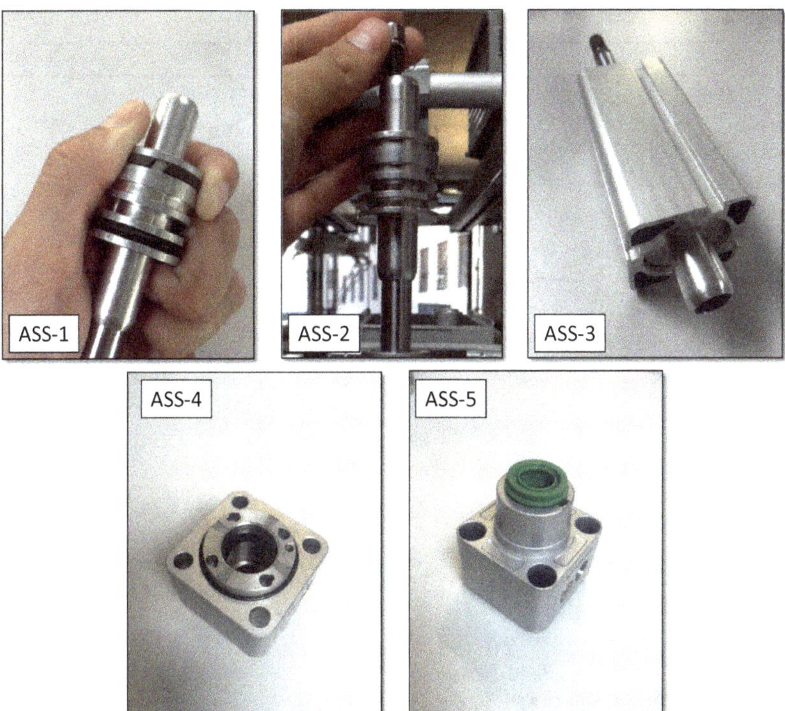

Fig. 12.11 Pneumatic cylinder subassemblies

another in terms of the possibility of using the robot for a certain task. Actually, the use of the same importance level for all the critical issues could be misleading in many cases. The definition of a scale of values could be a good idea for a more detailed technical evaluation. According to the main workflow explained in Fig. 12.5, Table 12.6 investigates the technical feasibility of the analyzed tasks.

As a result, task O5 is not adequate for robotic implementation since it is supposed that the related critical issues make the feeding and the handling of components too complex to be done using a proper amount of production resources in a suitable time. In fact, the ring is magnetic, potentially tangled and also flexible. Those conditions make the automatic gripping impossible since it will be necessary to adopt a

Table 12.3 Components and subassemblies list

	Nr	Code
Part name		
O-ring	1x	WP-1
Piston rod	1x	WP-2
Piston	1x	WP-3
Magnetic ring	1x	WP-4
Piston seal	1x	WP-5
Screw	1x	WP-6
Washer	1x	WP-7
Plastic ring	1x	WP-8
Cylinder	1x	WP-9
Nut 1 (Tie-rod)	4x	WP-10
Seal 1 (black)	2x	WP-11
Base cover	1x	WP-12
Seal 2 (green)	1x	WP-13
Head cover	1x	WP-14
Tie-rod	4x	WP-15
Mesh	1x	WP-16
Nut 2 (piston rod)	1x	WP-17
TOT PIECES		24
Subassembly name		
Piston + Magnetic ring + Piston seal	1x	ASS-1
ASS-1 + Piston rod + O-ring + Screw + Washer	1x	ASS-2
ASS-2 + Plastic ring + Cylinder	1x	ASS-3
Base cover + Seal 1 + Nut 1 (4x)	1x	ASS-4
Head cover + Seal 2 + Seal 1	1x	ASS-5
ASS-3 + ASS-4 + ASS-5 + Tie-rod (4x)	1x	ASS-6
TOT SUBASSEMBLIES		6

very complex solution to properly manage the component (for example a dedicated dispenser for single ring separation, placement, and supply). Task O8 could be a potential candidate. The main problem is that the seal is slightly flexible. In this case, that critical issue does not complicate the related feeding and handling since the utilized gripper can properly manage the component without the necessity of dedicated solutions. Finally, task O27 does not present any critical issues and as a result is a perfect candidate for robotic implementation. From the assembly point of view, Tables 12.7 and 12.8 investigate the technical feasibility of two assembly tasks.

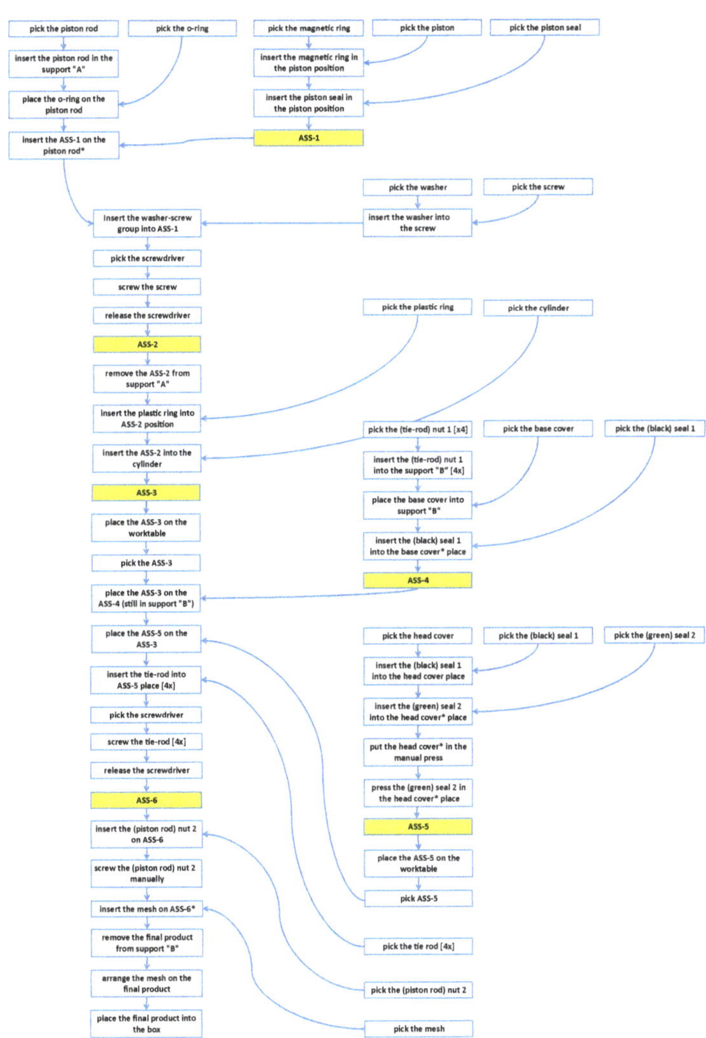

Fig. 12.12 Pneumatic cylinder—manual assembly cycle

O10 could be a potential candidate for robotic implementation if certain adjustments were integrated into the robot system. In fact, since the insertion tolerances are rather demanding, it is advisable to adapt the robotic system for the automatic recognition of the piston—rod

Table 12.4 Manual assembly cycle main data

Nr	Rip	Task	Average task time [s]	Stnd. Dev. [s]
O1	1x	pick the piston rod	2.54	0.35
O2	1x	insert the piston rod in the support "A"	0.80	0.19
O3	1x	pick the o-ring	2.14	1.02
O4	1x	place the o-ring on the piston rod	1.33	0.08
O5	1x	pick the magnetic ring	3.15	1.79
O6	1x	pick the piston	3.67	1.51
O7	1x	insert the magnetic ring in the piston position	4.89	0.87
O8	1x	pick the piston seal	1.61	0,12
O9	1x	insert the piston seal in the piston position	10.13	0.24
O10	1x	insert the ASS-1 on the piston rod*	2.00	0.00
O11	1x	pick the washer	3.31	1.47
O12	1x	pick the screw	1.37	0.64
O13	1x	insert the washer into the screw	1.57	0.26
O14	1x	insert the washer-screw group into ASS-1	1.27	0.18
O15	1x	pick the screwdriver	(/)	(/)
O16	1x	set the screwdriver head (if it is necessary)	(/)	(/)
O17	1x	screw the screw	3.28	1.12
O18	1x	release the screwdriver	(/)	(/)
O19	1x	remove the ASS-2 from support "A"	1.30	0.44
O20	1x	pick the plastic ring	2.59	0.30
O21	1x	insert the plastic ring into ASS-2 position	3.00	0.44
O22	1x	pick the cylinder	5.45	2.18
O23	1x	insert the ASS-2 into the cylinder	5.13	0.60
O24	1x	place the ASS-3 on the worktable	17.42	12.90
O25	4x	pick the (tie-rod) nut 1	8.21	0.61
O26	4x	insert the (tie-rod) nut 1 into the support "B"	9.18	12.1
O27	1x	pick the base cover	1.85	0.56
O28	1x	place the base cover into support "B"	1.62	0.79
O29	1x	pick the (black) seal 1	5.61	0.42
O30	1x	insert the (black) seal 1 into the base cover* place	6.15	2.13
O31	1x	pick the head cover	0.51	0.06
O32	1x	pick the (black) seal 1	4.44	0.86
O33	1x	insert the (black) seal 1 into the head cover place	4.03	1.11

(continued)

Table 12.4 (continued)

Nr	Rip	Task	Average task time [s]	Stnd. Dev. [s]
O34	1x	pick the (green) seal 2	3.33	2.06
O35	1x	insert the (green) seal 2 into the head cover* place	4.35	1.67
O36	1x	put the head cover* in the manual press	2.17	0.84
O37	1x	press the (green) seal 2 in the head cover* place	2.52	0.91
O38	1x	place the ASS-5 on the worktable	3.29	0.93
O39	1x	pick the ASS-3	1.56	0.27
O40	1x	place the ASS-3 on the ASS-4 (still in support "B")	2.86	1.48
O41	1x	pick the ASS-5	2.01	0.58
O42	1x	place the ASS-5 on the ASS-3	3.10	1.35
O43	4x	pick the tie-rod	5.99	1.23
O44	4x	insert the tie-rod into ASS-5 place	5.06	2.15
O45	1x	pick the screw driver	(/)	(/)
O46	1x	set the screwdriver head (if it is necessary)	(/)	(/)
O47	4x	screw the tie-rod	7.34	4.27
O48	1x	release the screwdriver	(/)	(/)
O49	1x	pick the (piston rod) nut 2	2.85	1.08
O50	1x	insert the (piston rod) nut 2 on ASS-6	2.47	1.53
O51	1x	screw the (piston rod) nut 2 manually	3.61	1.50
O52	1x	pick the mesh	1.96	0.39
O53	1x	insert the mesh on ASS-6*	1.95	1.04
O54	1x	remove the final product from support "B"	2.26	0.99
O55	1x	arrange the mesh on the final product	1.97	0.78
O56	1x	place the final product into the box	1.96	0.57

* means that the involved parts or subassemblies are partially assembled with other components

coupling in order to avoid insertion errors. There are two main possible solutions. The first one is the adoption of a vision system which allows the visual recognition of the components and the related insertion direction. That system could also be useful for other feeding and assembly applications. Unfortunately, the cost would be quite high and a detailed analysis is required in order to estimate the related return on investments. Another possibility could be the use of the robot power

Table 12.5 Examples of technical evaluation of feeding and handling tasks

Critical issue typology		Feeding critical issues		Handling critical issues					
Nr	Task	The component is magnetic or sticky	The component is a nest or tangle	The component has no symmetry axis	The component is fragile/delicate	The component is flexible	The component is very small or very big (referring to a human hand)	The component is light so that air resistance would create conveying problems	The component is slippery
05	Pick the magnetic ring	YES	YES	NO	NO	YES	NO	NO	NO
08	Pick the piston seal	NO	NO	NO	NO	YES	NO	NO	NO
027	Pick the base cover	NO	NO	NO	NO	NO	NO	NO	NO

Table 12.6 Technical evaluation of tasks O5, O8, and O27 according to main critical issues analysis and technical evaluation workflow

Criteria	05 Pick the magnetic ring	08 Pick the piston seal	027 Pick the base cover
Do the process or components present one or more feeding, handling and assembly critical issues?	YES	YES	NO
Referring to the robot system, does the identified critical issues make it absolutely impossible to feed, handle or assemble the components by using a proper amount of production resources in a suitable time?	YES	NO	(/)
Is it possible to solve the critical issues by using a simple and low-cost solution?	(/)	Not necessary	(/)
FINAL EVALUATION	Human only (not suitable for robot)	Suitable for robot with no modifications	Suitable for robot with no modifications

and force control system (which is an inherent peculiarity of collaborative robots) to delicately and systematically touch the rod with the gripped piston in order to find the suitable insertion direction. This solution does not require additional systems and as a consequence it will be totally free. On the other hand, it requires medium-high programming skills.

O9 is totally unsuitable for robotic implementation with the selected gripper. In fact, there are different critical issues which strongly limit an automatic assembly. For example, conditions like the need for high physical dexterity, the request for two hands for handling the components and the need to compress parts during the assembly operations

Table 12.7 Examples of technical evaluation of assembly tasks

Critical issues	Task	
	O9 Insert the piston seal in the piston position	O10 Insert the ASS-1 on the piston rod*
The component is magnetic or sticky	NO	NO
The component is a nest or tangle	NO	NO
The component has no symmetry axis	NO	NO
The component is fragile or delicate	NO	NO
The component is flexible	YES	NO
The component is very small or very big (in reference to a human hand)	NO	NO
The component is light so that air resistance would create conveying problems	NO	NO
The component is slippery	NO	NO
Components do not have a "datum surface" (reference surface) which simplifies precise positioning during the assembly	NO	NO
Components cannot be easily orientated	NO	NO
Components do not include features which allow self-aligning during the assembly	NO	NO
Components cannot be located before they are released	YES	NO
Components provide resistance to insertion	YES	NO
Components do not provide chamfers or tapers that help to guide and position the parts in the correct position	NO	NO
Components do not have a suitable base part on which to build the assembly	NO	NO
Components cannot be assembled in layer fashion from directly above (z-axis assembly)	YES	NO
The assembly is overconstrained	NO	NO
It is difficult to reach the assembly area / the components access for assembly operations is restricted or not easy to reach	NO	NO
The assembly requires high accuracy and/or demanding insertion tolerances	NO	YES
The component and/or the assembly sequence requires high physical dexterity	YES	NO
The assembly needs to reposition the partially completed sub-assembly, other components or fixtures	YES	YES
The assembly needs to reorient the partial assembly or to manipulate previously assembled parts	YES	YES
Components need to be compressed during the assembly	YES	NO
The component and/or the assembly sequence requires two hands for handling	YES	NO
The component and/or the assembly sequence require typical human skills (for example touch perception, hearing, ability to interpret situations…)	YES	NO

Table 12.8 Technical evaluation of task O9 and O10 according to main critical issues analysis and technical evaluation workflow

Criteria	O9 Insert the piston seal in the piston position	O10 Insert the ASS-1 on the piston rod*
Does the process or components present one or more feeding, handling and assembly critical issues?	YES	YES
Referring to the robot system, does the identified critical issues make it absolutely impossible to feed, handle or assemble the components by using a proper amount of production resources in a suitable time?	YES	NO
Is it possible to solve the critical issues by using a simple and low-cost solution?	YES	YES
FINAL EVALUATION	**Human only (not suitable for robot)**	**Suitable for robot with modifications**

* means that the involved parts or subassemblies are partially assembled with other components

make the components absolutely impossible to feed, handle, or assemble by using a proper amount of production resources in a suitable time. For these reasons, the analyzed task is reasonably performable only by an operator. The list of results for all the task technical evaluation is provided in Table 12.12 at the end of this section.

b) **Physical ergonomics evaluation**

For a preliminary analysis, the RULA method is selected. Considering the static muscle activity and the force caused on the upper limbs, this method is appropriate for the analysis of upper body activities and it involves body part diagrams integrated with the code for joint angles, body postures, load/force, coupling, and muscle activity. It investigates the exposure of individual workers to risk factors associated with work-related musculoskeletal disorders (Karwowski and Marras 2003). The outputs are risk level scores on a given scale to indicate the risk effects, as shown in Table 12.9.

In general, according to the results of the technical evaluation, if the RULA analysis of the selected tasks shows a value equal to or higher than three, it is necessary to deeply understand the problem's root-cause in order to provide a practical solution. In this case, the use of a collaborative robot should be a good option for improving related physical

Table 12.9 RULA action levels and relative task allocation

RULA value	Action level
1;2	**Action level 1:** the posture is acceptable if it is not maintained or repeated for long periods
3;4	**Action level 2:** further investigations are needed and changes may be required
5;6	**Action level 3:** investigations and changes are required soon
7 +	**Action level 4:** investigations and changes are required immediately

Table 12.10 List of tasks which present a RULA index value equal to or higher than three and which could potentially be performed by the robot from a technical point of view

Nr	Rip	Task	RULA INDEX
O1	1x	pick the piston rod	3
O3	1x	pick the o-ring	3
O6	**1x**	**pick the piston**	**4**
O8	1x	pick the piston seal	3
O10	1x	insert the ASS-1 on the piston rod*	3
O11	1x	pick the washer	3
O12	1x	pick the screw	3
O17	1x	screw the screw	3
O19	1x	remove the ASS-2 from support "A"	3
O20	1x	pick the plastic ring	3
O22	1x	pick the cylinder	3
O25	**4x**	**pick the (tie-rod) nut 1**	3
O27	1x	pick the base cover	3
O29	1x	pick the (black) seal 1	3
O31	1x	pick the head cover	3
O32	1x	pick the (black) seal 1	3
O34	1x	pick the (green) seal 2	3
O36	1x	put the head cover* in the manual press	3
O42	1x	place the ASS-5 on the ASS-3	3
O43	**4x**	**pick the tie-rod**	3
O44	**4x**	**insert the tie-rod into ASS-5 place**	3
O47	**4x**	**screw the tie-rod**	**4**
O49	1x	pick the (piston rod) nut 2	3
O52	1x	**pick the mesh**	**4**
O56	1x	place the final product into the box	3

ergonomics. The list of results for all the task physical ergonomics evaluation is provided in Table 12.12 at the end of this section. Starting from the tasks which could potentially be performed by the robot from a technical point of view, it is necessary to identify the ones with the highest priority from a physical ergonomic point of view. A first classification is provided in Table 12.10.

The tasks which are highlighted in red are the ones with the highest impact from a physical ergonomics point of view. In fact, the related

RULA index value is equal to four (O6 and O52) or equal to three but presenting a non-negligible number of repetitions per task (O25, O43, O44, and O47), a condition which can lead to long-term physical strain. Of course, a large part of the identified tasks could be solved using different kinds of organizational solutions (i.e., by changing the manual station layout—a probable valid solution for all the identified feeding tasks). On the other hand, in particular cases, the use of a robot could be very interesting. Task O47 presents a typical example of physical stress provided by screw operations. The number of repetitions combined with a medium RULA index makes that task an excellent candidate for automation. In addition, tasks O44 and O45 can be easily joined with task O47 in order to create an overall activity (pick, insert, and screw the tie-rods) which would be perfect for the use of the collaborative robot.

c) **Quality evaluation**

It is possible to preliminarily analyze process variability through the coefficient of variation (CV). *CV* is a parameter which can be used to measure and qualify production systems' variability starting from a set of data which are quite easy to obtain and commonly utilized in manufacturing process analysis and optimization. *CV* is defined as the ratio between the standard deviation (σ) and the mean value (*Xm*) (Nwanya et al. 2016):

$$CV = \frac{\sigma}{Xm}$$

According to the definition of *CV*, it is possible to have three different process variability categories: low process variability ($CV = 0 \div 0.75$), moderate process variability ($CV = 0.75 \div 1.33$) and high process variability ($CV > 1.33$). The list of results for all the task quality evaluation is provided in Table 12.12 at the end of this section. Starting from the tasks which could potentially be performed by the robot from a technical point of view, it is necessary to identify the tasks with high process variability. According to the collected data, there is only one high-variability process in the analyzed case study (O26). In fact, a large number of the activities present a value of *CV* lower than 0.50 (see Fig. 12.13), which means that the actual assembly is qualitatively under control according to this parameter. Further investigation could be undertaken by combining the *CV* values with the strategic importance of operations and/or tasks (i.e., by considering the components' economic value). Nevertheless, after a preliminary quality evaluation, O26 would be perfect for the use of the collaborative robot.

d) **Economic evaluation**

For a preliminary analysis, it is possible to categorize the cycle tasks as follows: grasping, handling, moving, positioning as NVA tasks; insertion, fastening, fixing, assembly as VA tasks. It is possible to recognize the tasks' typology just by a visual inspection. According to that classification and starting from the tasks, which could potentially be performed by the robot from a technical point of view, Table 12.11 summarizes the proposed economic division.

According to this classification, all the tasks which are classified as NVA are good candidates for the use of the collaborative robot. In fact, these activities will absorb production resources and/or time by generating unnecessary costs without providing perceived value and satisfaction to the final customer. That condition justifies the use of automation for the execution of these tasks.

e) **Final evaluation**

Finally, it is necessary to combine all the single evaluation results by using the hierarchical approach proposed in Fig. 12.5. This process allows the designer to have a preliminary and approximate estimation of the human–robot task division. After the validation of the task allocation, it will be possible to start the design of the collaborative workstation layout by using a set of structured data. Table 12.12 explains the overall evaluation results.

Actually, the final task allocation will be defined by the hierarchical contribution of every single evaluation. According to the proposed

Table 12.11 VA/NVA classification of tasks which could potentially be performed by the robot from a technical point of view

Nr	Task	Activity type	Classification
O1	pick the piston rod	grasping, handling, moving, positioning	NVA
O2	insert the piston rod in the support "A"	insertion, fastening, fixing, assembly	VA
O3	pick the o-ring	grasping, handling, moving, positioning	NVA
O6	pick the piston	grasping, handling, moving, positioning	NVA
O8	pick the piston seal	grasping, handling, moving, positioning	NVA
O10	insert the ASS-1 on the piston rod*	insertion, fastening, fixing, assembly	VA
O11	pick the washer	grasping, handling, moving, positioning	NVA
O12	pick the screw	grasping, handling, moving, positioning	NVA
O14	insert the washer-screw group into ASS-1	insertion, fastening, fixing, assembly	VA
O17	screw the screw	insertion, fastening, fixing, assembly	VA
O19	remove the ASS-2 from support "A"	grasping, handling, moving, positioning	NVA
O20	pick the plastic ring	grasping, handling, moving, positioning	NVA
O22	pick the cylinder	grasping, handling, moving, positioning	NVA
O24	place the ASS-3 on the worktable	grasping, handling, moving, positioning	NVA
O25	pick the (tie-rod) nut 1	grasping, handling, moving, positioning	NVA
O26	insert the (tie-rod) nut 1 into the support "B"	insertion, fastening, fixing, assembly	VA
O27	pick the base cover	grasping, handling, moving, positioning	NVA
O28	place the base cover into support "B"	insertion, fastening, fixing, assembly	VA
O29	pick the (black) seal 1	grasping, handling, moving, positioning	NVA
O31	pick the head cover	grasping, handling, moving, positioning	NVA
O32	pick the (black) seal 1	grasping, handling, moving, positioning	NVA
O34	pick the (green) seal 2	grasping, handling, moving, positioning	NVA
O36	put the head cover* in the manual press	grasping, handling, moving, positioning	NVA
O38	place the ASS-5 on the worktable	grasping, handling, moving, positioning	NVA
O39	pick the ASS-3	grasping, handling, moving, positioning	NVA
O40	place the ASS-3 on the ASS-4 (still in support "B")	insertion, fastening, fixing, assembly	VA
O41	pick the ASS-5	grasping, handling, moving, positioning	NVA
O42	place the ASS-5 on the ASS-3	insertion, fastening, fixing, assembly	VA
O43	pick the tie-rod	grasping, handling, moving, positioning	NVA
O44	insert the tie-rod into ASS-5 place	insertion, fastening, fixing, assembly	VA
O47	screw the tie-rod	insertion, fastening, fixing, assembly	VA
O49	pick the (piston rod) nut 2	grasping, handling, moving, positioning	NVA
O52	pick the mesh	grasping, handling, moving, positioning	NVA
O54	remove the final product from support "B"	grasping, handling, moving, positioning	NVA
O56	place the final product into the box	grasping, handling, moving, positioning	NVA

framework (see Fig. 12.5), the task allocation logic can be summarized in Table 12.13.

A further design stage would be to unify, in terms of use of resources (human or robot), different tasks which are related to the same activity. For example, task O1 (R) and O2 (H or R) are successive and related to the same component. In this case, it is reasonably advisable to perform these tasks by using the collaborative robot for both the operations. On the other hand, even if tasks O3 (H) and task O4 (R) are in the same condition as previous tasks, it is not useful to perform them separately for an efficiency reason. In fact, the exchange of the component

Table 12.12 Overall and final evaluation results

Nr	Task	Technical evaluation	Ergonomics evaluation		Quality evaluation		Economics evaluation		FINAL RESULTS (task allocation)
		Technical task allocation	RULA index value	Ergonomics task allocation	CV value	Quality task allocation	VA/NVA classification	Economic task allocation	
O1	pick the piston rod	H or R	3	H or R	0.14	H or R	N.V.A.	R	R
O2	insert the piston rod in the support "A"	H or R	2	H or R	0.24	H or R	V.A.	H or R	H or R
O3	pick the o-ring	H or R	3	H or R	0.48	H or R	N.V.A.	R	R
O4	place the o-ring on the piston rod	H	2	H or R	0.06	H or R	V.A.	H or R	H
O5	pick the magnetic ring	H	3	H or R	0.57	H or R	N.V.A.	R	H
O6	pick the piston	H or R	4	R	0.41	H or R	N.V.A.	R	R
O7	insert the magnetic ring in the piston position	H	2	H or R	0.18	H or R	V.A.	H or R	H
O8	pick the piston seal	H or R	3	H or R	0.07	H or R	N.V.A.	R	R
O9	insert the piston seal in the piston position	H	3	H or R	0.02	H or R	V.A.	H or R	H
O10	insert the ASS-1 on the piston rod*	H or R	3	H or R	0.00	H or R	V.A.	H or R	H or R
O11	pick the washer	H or R	3	H or R	0.44	H or R	N.V.A.	R	R
O12	pick the screw	H or R	3	H or R	0.47	H or R	N.V.A.	R	R
O13	insert the washer into the screw	H	2	H or R	0.16	H or R	V.A.	H or R	H
O14	insert the washer-screw group into ASS-1	H or R	2	H or R	0.14	H or R	V.A.	H or R	H or R
O15	pick the screwdriver	(/)	(/)	(/)	(/)	(/)	(/)	(/)	(/)
O16	set the screwdriver head (if it is necessary)	(/)	(/)	(/)	(/)	(/)	(/)	(/)	(/)
O17	screw the screw	H or R	3	H or R	0.34	H or R	V.A.	H or R	H or R
O18	release the screwdriver	(/)	(/)	(/)	(/)	(/)	(/)	(/)	(/)
O19	remove the ASS-2 from support "A"	H or R	3	H or R	0.34	H or R	N.V.A.	R	R
O20	pick the plastic ring	H or R	3	H or R	0.11	H or R	N.V.A.	R	R
O21	insert the plastic ring into ASS-2 position	H	2	H or R	0.15	H or R	V.A.	H or R	H
O22	pick the cylinder	H or R	3	H or R	0.40	H or R	N.V.A.	R	R
O23	insert the ASS-2 into the cylinder	H	2	H or R	0.12	H or R	V.A.	H or R	H

(continued)

Table 12.12 (continued)

O24	place the ASS-3 on the worktable	H or R	2	H or R	0.74	H or R	N.V.A.	R	R
O25(4x)	pick the (tie-rod) nut 1	H or R	3	H or R	0.07	H or R	N.V.A.	R	R
O26(4x)	insert the (tie-rod) nut 1 into the support "B"	H or R	2	H or R	1.31	R	V.A.	H or R	R
O27	pick the base cover	H or R	3	H or R	0.30	H or R	N.V.A.	R	R
O28	place the base cover into support "B"	H or R	2	H or R	0.49	H or R	V.A.	H or R	H or R
O29	pick the (black) seal 1	H or R	3	H or R	0.08	H or R	N.V.A.	R	R
O30	insert the (black) seal 1 into the base cover* place	H	2	H or R	0.35	H or R	V.A.	H or R	H
O31	pick the head cover	H or R	3	H or R	0.11	H or R	N.V.A.	R	R
O32	pick the (black) seal 1	H or R	3	H or R	0.19	H or R	N.V.A.	R	R
O33	insert the (black) seal 1 into the head cover place	H	2	H or R	0.28	H or R	V.A.	H or R	H
O34	pick the (green) seal 2	H or R	3	H or R	0.62	H or R	N.V.A.	R	R
O35	insert the (green) seal 2 into the head cover* place	H	2	H or R	0.38	H or R	V.A.	H or R	H
O36	put the head cover* in the manual press	H or R	3	H or R	0.39	H or R	N.V.A.	R	R
O37	press the (green) seal 2 in the head cover* place	H	3	H or R	0.36	H or R	V.A.	H or R	H
O38	place the ASS-5 on the worktable	H or R	2	H or R	0.28	H or R	N.V.A.	R	R
O39	pick the ASS-3	H or R	2	H or R	0.17	H or R	N.V.A.	R	R
O40	place the ASS-3 on the ASS-4 (still in support "B")	H or R	2	H or R	0.52	H or R	V.A.	H or R	H or R
O41	pick the ASS-5	H or R	2	H or R	0.29	H or R	N.V.A.	R	R
O42	place the ASS-5 on the ASS-3	H or R	3	H or R	0.44	H or R	V.A.	H or R	H or R
O43(4x)	pick the tie-rod	H or R	3	R	0.21	H or R	N.V.A.	R	R
O44(4x)	insert the tie-rod into ASS-5 place	H or R	3	R	0.43	H or R	V.A.	H or R	R
O45	pick the screw driver	(/)	(/)	(/)	(/)	(/)	(/)	(/)	(/)
O46	set the screwdriver head (if it is necessary)	(/)	(/)	(/)	(/)	(/)	(/)	(/)	(/)
O47(4x)	screw the tie-rod	H or R	3	R	0.58	H or R	V.A.	H or R	R
O48	release the screwdriver	(/)	(/)	(/)	(/)	(/)	(/)	(/)	(/)
O49	pick the (piston rod) nut 2	H or R	3	H or R	0.38	H or R	N.V.A.	R	R
O50	insert the (piston rod) nut 2 on ASS-6	H	3	H or R	0.62	H or R	V.A.	H or R	H
O51	screw the (piston rod) nut 2 manually	H	3	H or R	0.42	H or R	V.A.	H or R	H
O52	pick the mesh	H or R	4	R	0.20	H or R	N.V.A.	R	R
O53	insert the mesh on ASS-6*	H	3	H or R	0.53	H or R	V.A.	H or R	H
O54	remove the final product from support "B"	H or R	2	H or R	0.44	H or R	N.V.A.	R	R
O55	arrange the mesh on the final product	H	3	H or R	0.40	H or R	V.A.	H or R	H
O56	place the final product into the box	H or R	3	H or R	0.29	H or R	N.V.A.	R	R

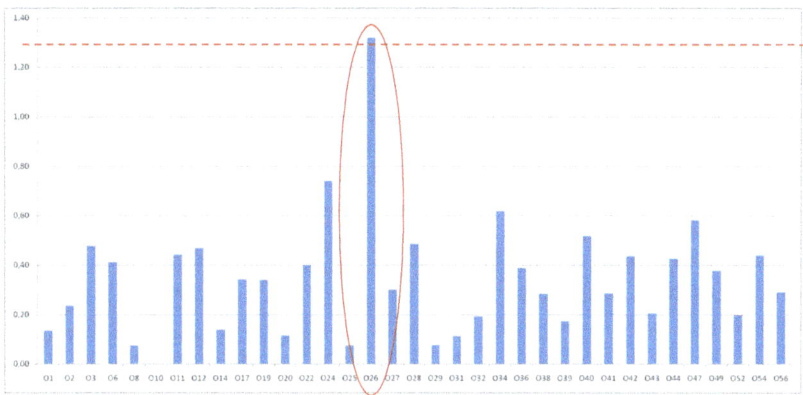

Fig. 12.13 CV value of tasks which could potentially be performed by the robot from a technical point of view

Table 12.13 Final task allocation logic according to the proposed framework and related examples

Technical evaluation result	Ergonomics evaluation result	Quality evaluation result	Economic evaluation result	FINAL RESULT	Task (example)
H	*irrelevant*	*irrelevant*	*irrelevant*	H	O4, O5, O7, O9 …
H or R	R	*irrelevant*	*irrelevant*	R	O6, O25, O43, O44 ….
H or R	H or R	R	*irrelevant*	R	O26
H or R	H or R	H or R	R	R	O11, O12, O19, O20 …
H or R	H or R	H or R	H or R	**H or R**	O2, O28, O40, O42 …

between the human and the robot will be useless and time-consuming. For this reason, it would be advisable for the operator to perform both tasks. This concept is summarized in Table 12.14.

12.5 Discussion and Hypothesis for Future Work

The following section aims to critically analyze the proposed approach in order to identify the main method critical issues for future developments and to investigate the main possibilities and innovations for collaborative assembly.

Table 12.14 Examples of task unification according to the use of resources

Task	INVOLVED COMPONENTS	FINAL RESULTS – task	FINAL RESULTS - activity
O1 - pick the piston rod	**piston rod**	R	R
O2 - insert the piston rod in the support "A"		H or R	
O3 - pick the o-ring	**o-ring**	H or R	H
O4 - place the o-ring on the piston rod		H	

12.5.1 Task Allocation Methodology: Future Developments

a. **Inclusion of dynamic task allocation methodologies for tasks which are classified as "H or R"**

The dynamic task allocation will be a core part of future collaborative workstations. In fact, a system where the operator can choose in real time and indiscriminately which task will be the next one according to his/her needs and wants could significantly improve cognitive ergonomics conditions, operators' psychological well-being, and production flexibility. Of course, this would be the perfect implementation of a human-centered design in the Industry 4.0 context. For this reason, it would be useful to add this possibility to future workstation development. Nevertheless, it is necessary to firstly identify which tasks of a sequence can be efficiently performed both by humans and robots by consider the real technical limitations of the robot system.

b. **Development of a multi-gripper technical evaluation**

The ability to properly pick up and manage a component strictly depends on the type of gripper which is intended for use for assembly activities. For this reason, it will be useful to further develop the proposed methodology by including multi-criteria evaluation of different kinds of gripper types in order to identify which one is the best solution for a certain assembly sequence. In this context, a recommended solution will be the development of a technical parameter which quantifies the percentage of success (in terms of robot usage) for a certain task according to the selected gripper type.

c. **Use of a more specific methodology for physical ergonomics evaluation (i.e., OCRA)**

The methodology for physical ergonomics evaluation proposed in this work is RULA. The selected method is simple and quick to use for different kinds of industrial applications. Nevertheless, this methodology presents some limitations especially because it does not consider in detail the tasks, workloads, and repetitions. For this reason, the RULA method is suggested for use for a preliminary postural evaluation; further investigation of the situations with dedicated approaches (i.e., NIOSH for lifting and carrying, Snook and Ciriello for the whole-body pushing and pulling, OCRA for the handling of low loads at high frequency..) is recommended for a proper physical ergonomics analysis.

d. **Integration of cognitive ergonomics considerations**

The sharing of workspaces and the physical interaction between humans and industrial robots could affect the cognitive ergonomics of the collaborative work. In this context, it would be mandatory to minimize mental stress and psychological discomfort, which could arise during hybrid operations. In fact, even if safety measures are well designed and implemented, the presence of the robot must not be perceived as a hazard or as a source of stress for humans. Designers should consider these kinds of cognitive ergonomics problems in order to develop anthropocentric and human-friendly collaborative workstations also from a psychological point of view.

e. **Inclusion of a method for the new assembly cycle definition according to the calculated task allocation**

Finally, the last consideration concerns the development of a quick and structured procedure for the new assembly cycle definition according to the planned task allocation. In this case, a new sequence should respect the defined human–robot task division and provide useful information for the design of the layout of the new collaborative workcell. The new cycle data should also support the designers in the comparison between the production performance of the actual and future workstation, in order to offer a clear overview of the costs and benefits that the introduction of the collaborative robot can provide to the overall production system.

12.5.2 Real-Time Allocation for Assembly

One of the most interesting and challenging features of collaborative assembly is the real-time allocation of tasks and responsibilities between the robot and the human operator. In these situations, the robot may behave with its own agency, i.e., goal-oriented initiative. Such agency endows the robot with the ability to negotiate the task owner with the operator and the order of the tasks. Researchers have shown that providing a machine or robotic agent with autonomous capabilities yields important benefits for human–robot team fluency (Gombolay et al. 2017). Furthermore, such agency is the basis for the emergence of smart interaction patterns with continuously distributed tasks among all contributors. In fact, while there are capabilities unique to both machines and humans, there is also a natural overlap. The objective of task allocation is to achieve an optimal sharing of these capabilities. For the dynamic task allocation, it is necessary to create a model of the assembly process and the sensing capacity to endow the collaborative robot with sufficient situational awareness. In fact, correct real-time task allocation is only possible with a sufficiently accurate virtual twin of the system. Such a virtual twin will be the object of analysis of the task-sharing system in conjunction with the feedback from situational awareness. Task allocation may be modified online by communication (verbal, nonverbal), by operator initiative or by another change in the system state. As can be seen, the dynamic task allocation problem is characterized by a degree of unpredictability. Such kinds of environments are called unstructured. Even in the simplest environment, the design of such interactions is not easy. All the challenges of motion planning in a dynamic environment are combined with the task allocation problem. This results in a combination of geometrically constrained motions in the space and ordered sequences of discrete tasks. For this reason, real-time task allocation problems are best modeled as hybrid systems. On the one hand, the task is part of a sequence of discrete states best modeled as a discrete event system (DES). Such a system evolves from the actual state in an undeterministic way due to the action of the operator. On the other hand, the motions executed by the robot are defined by the executed task and the traditional constrains of human–robot physical interaction. Such a combination of a DES with a motion planning system is called a hybrid system. Observe that this topic is characteristic for the fourth industrial revolution. Including the human operator as a CPS and making possible the dynamic integration of humans and machines, the coordination and orchestration of the smart factories become pervasive. This is only possible thanks to the correct integration of the required systems. In fact, the integration of heterogeneous digital systems is a must. In particular, advanced visual sensing systems must share high-structured information between CPS using the correct communication channels. Among these channels, physical contact may also be used to create an interaction interface and convey information to CPS. Traditionally, such problems have been attacked via high-level task planning where a sequencing of task sequences is computed to lower-level planning to compute the motions for the arm (Pellegrinelli et al. 2017). Initial research in task allocation perceived that, while there are capabilities unique to both machines and humans, there are also overlapping capabilities that provide the opportunity to variably assign tasks in accordance with resource availability (Ranz et al. 2017). Other systems based on Artificial Intelligence are based on a learning framework to construct an optimal task-sharing schedule. In works like Munzer et al. (2017) and Mitsunaga et al. (2008), the authors propose an online learning algorithm which adapts to the operator behavior during the task-sharing procedure. Given an initial task schedule, it is possible to adapt the robot's actions based on comfort and discomfort measurements gathered from the sensing system. Other approaches leverage the fact that people act not only as a response to external or internal stimuli, but also in order to achieve goals to design algorithms capable of predicting the intended actions of the operator in order to perform the task allocation (Demiris 2007). This is a feasible alternative to communication to understand the intentions in real time.

12.5.3 Cell Digitalization

We have already discussed how a collaborative robot can be introduced inside a manufacturing cell to simultaneously improve the cell's productivity and reduce the operator's strain. To this end, we have identified and allocated the optimal set of tasks that a collaborative robot can execute. Our approach, however, presents two fundamental limitations:

- Manual synchronization between the operator and the robot to conclude the assembly sequence
- The robot can only execute its allocated tasks in a predefined sequence.

Both limitations result from the fact that our approach only exploits static information or knowledge of the assembly process known a priori. Indeed, to allow higher levels of flexibility like dynamic task allocation, autonomous reaction, and adaption to unexpected situations, etc., it is necessary to enrich the "situational awareness" of the robot and to endow it with an autonomous or semi-autonomous decision-making mechanism. In other words, full cell digitalization in smart factories not only involves delegating known tasks to robots but also endows them with proper perceptive, cognitive, and control mechanisms for real-time monitoring and adaptation during the execution of the assembly process. The rest of the section is devoted to briefly introducing different technologies and approaches found in the state of the art to improve the situational awareness of the robot, including operator monitoring, and different inference mechanisms allowing autonomous adaptation during the assembly process.

12.5.4 Situational Awareness

The first step to increase the situational awareness of the robot is to provide it with some means of perception, not only to perceive the surrounding environment but also to measure its own internal states and to monitoring the operator's activities. Therefore, such means of perception cannot be defined only in terms of raw measurements of the physical world (e.g., provided by laser scanners, RGB cameras, RGB-D sensors, inertial measurement units, encoders, torque sensor, etc.) but also in terms of interactive HMI allowing bidirectional and natural information flows between the operator and the robot. This evolution from raw measurements to information flows defines the second step: understanding of the current situation, which includes the environment, robot, operator, and manufacturing process states. The third and last step consists of the prediction of future situations.

Key HMI interfaces in Industry 4.0 are the *automatic speech recognition*, the *gesture recognition*, the *enhanced reality* (either in terms of *augmented reality* or *virtual reality*), *physical HRI,* and the *prediction of operator's intentions* (Ruiz Garcia et al. 2019):

- *Automatic speech recognition* consists of the identification and recognition of patterns bearing the information content inside the speech waveform (O'Shaughnessy 2008). Although speech represents the most efficient method of human interaction, Lotterbach and Peissner (Lotterbach and Peissner 2005) state that voice user interfaces cannot represent a replacement for a classical graphic user interface (GUI) but can complement to them, so that under certain conditions and in certain contexts, they provide the most comfortable and efficient method of interaction.
- A gesture is defined by any expressive body motion capable of transmitting meaningful information to other entities in the workspace. Nowadays, thanks to the advent of RGB-D sensors, visual *gesture recognition* is one of the most widely used methods in the industry (Sansoni et al. 2009). A comprehensive review of applications and technologies can be found in Mitra and Acharya (2007).
- *Enhanced reality* consists of the enrichment of perceptive measurements of the physical world with digital information superimposed on top of it (Craig 2013).
- *Physical contact detection*, isolation, and reaction have been extensively explored in the robotics community, especially in the field of collaborative robotics (Haddadin et al. 2012; Ajoudani et al. 2018)

where the contact between the robot and the operators is expected to be frequent.
- *Prediction of operator's intentions* relies on monitoring techniques (Pirri et al. 2019; Mauro et al. 2018, 2019) and allows the enhancement of the effectiveness of collaboration between robots and humans, especially in industrial scenarios where safety greatly depends on the understanding between humans and robots.

12.6 Conclusions

HRC is a primary cyber-physical technology in Industry 4.0. There is no doubt that the global market of industrial collaborative robotics is extensively and continuously growing. It is reasonably possible to suppose that collaborative assembly will be a crucial application in the near future. This chapter aims to explain the main concepts about the introduction of industrial collaborative robots into manual assembly systems. The contents gave a general overview of the main features and requirements of human–robot collaborative assembly in the context of Industry 4.0, and discussed the opportunities and problems related to its design procedure. The main objectives of the adoption of collaborative systems into traditional manual assembly workstations is to improve operators' work conditions and production performances by combining inimitable human ability with smart machines' strengths. The main specific outcomes of this chapter based on Industry 4.0 applied to SMEs are:

- Identification of the main parameters for the possible adoption of collaborative systems into the assembly process
- Development of a structured framework for the evaluation of the technical possibility to use a collaborative robot for assembly activities
- Implementation of a multi-criteria method for human–robot task allocation in assembly activities by considering technical, ergonomic, organizational and economic principles
- Creation of the basis for the development of a digital tool for a guided self-evaluation.

In general, the proposed approach is based on hierarchical task allocation, which is able to define if a task can be performed efficiently by the operator (H), by the robot (R), or by both indiscriminately (H or R). The core part of that method is the technical evaluation, which analyzes if a generic feeding, handling or assembly task can be performed efficiently by a robot by using a proper amount of production resources in a suitable time according to product and process critical issues. The proposed methodology enables SMEs to carry out a preliminary feasibility analysis of the collaborative process. In the future, this methodology will be used as a basis for developing a digital tool for supporting SMEs technicians to self-evaluate the potential of collaborative systems in assembly processes. Such an application will help SMEs to a proper use of industrial collaborative robots and as a result, to improve assembly performances, operators' work conditions, and production quality. The chapter also introduced the SMF laboratory of the Free University of Bolzano-Bozen and explained how to apply intuitive HRI for assembly purposes through the description of a laboratory case study in a realistic industrial lab environment. For the proposed case study, a Universal Robot model UR3 equipped with a 2-finger Robotiq collaborative gripper is used for the collaborative assembly of a medium-size pneumatic cylinder. Finally, the current work in progress and future hypotheses for improvement are introduced and discussed by investigating the main possibilities and innovations for collaborative assembly.

References

Ajoudani, A., A.M. Zanchettin, S. Ivaldi, A. Albu-Schäffer, K. Kosuge, and O. Khatib. 2018. Progress and Prospects of the Human-Robot Collaboration. *Autonomous Robots*: 1–19. https://doi.org/10.1007/s10514-017-9677-2.

Boothroyd, G. 2005. *Assembly Automation and Product Design*. Boca Raton: CRC Press. https://doi.org/10.1201/9781420027358.

Boothroyd, G., P. Dewhurst, and W. Knight. 2010. *Product Design for Manufacture and Assembly*. CRC Press.

Chen, F., K. Sekiyama, J. Huang, B. Sun, H. Sasaki, and T. Fukuda. 2011. An Assembly Strategy Scheduling Method for Human and Robot Coordinated

Cell Manufacturing. *International Journal of Intelligent Computing and Cybernetics* 4 (4): 487–510. https://doi.org/10.1108/17563781111186761.
Chryssolouris, G. 2013. *Manufacturing Systems: Theory and Practice*. New York: Springer Science + Business Media. https://doi.org/10.1007/0-387-28431-1.
Craig, A.B. 2013. *Understanding Augmented Reality: Concepts and Applications*. Netherlands: Morgan Kaufmann, Elsevier. https://doi.org/10.1016/C2011-0-07249-6.
Crowson, R. (ed.). 2006. *Assembly Processes*. Boca Raton: CRC Press. https://doi.org/10.1201/9781420003666.
Darvish, K., F. Wanderlingh, B. Bruno, E. Simetti, F. Mastrogiovanni, and G. Casalino. 2018. Flexible Human-Robot Cooperation Models for Assisted Shop-Floor Tasks. *Mechatronics* 51: 97–114. https://doi.org/10.1016/j.mechatronics.2018.03.006.
Demiris, Y. 2007. Prediction of Intent in Robotics and Multi-Agent Systems. *Cognitive Processing* 8 (3): 151–158. https://doi.org/10.1007/s10339-007-0168-9.
Djuric, A.M., R.J. Urbanic, and J.L. Rickli. 2016. A Framework for Collaborative Robot (CoBot) Integration in Advanced Manufacturing Systems. *SAE International Journal of Materials and Manufacturing* 9 (2): 457–464. https://doi.org/10.4271/2016-01-0337.
Gombolay, M., A. Bair, C. Huang, and J. Shah. 2017. Computational Design of Mixed-Initiative Human-Robot Teaming That Considers Human Factors: Situational Awareness, Workload, and Workflow Preferences. *The International Journal of Robotics Research* 36 (5–7): 597–617. https://doi.org/10.1177/0278364916688255.
Gombolay, M.C., C. Huang, and J. Shah. 2015. Coordination of Human-Robot Teaming with Human Task Preferences. In *AAAI Fall Symposium Series on AI-HRI*.
Gualtieri, L., E. Rauch, R. Vidoni, and D.T. Matt. 2019. An Evaluation Methodology for the Conversion of Manual Assembly Systems into Human-Robot Collaborative Workcells. Paper presented at the International Conference in Flexible Automation and Intelligent Manufacturing (FAIM 2019), 24–28 June, Limerick, Ireland.
Gualtieri, L., R. Rojas, G. Carabin, I. Palomba, E. Rauch, R. Vidoni, and D.T. Matt. 2018. Advanced Automation for SMEs in the I4.0 Revolution: Engineering Education and Employees Training in the Smart Mini Factory Laboratory. In *IEEE International Conference on Industrial Engineering and Engineering Management (IEEM)*, 1111–1115.

Haddadin, S., S. Haddadin, A. Khoury, T. Rokahr, S. Parusel, R. Burgkart, … and A. Albu-Schäffer. 2012. On Making Robots Understand Safety: Embedding Injury Knowledge into Control. *The International Journal of Robotics Research* 31 (13): 1578–1602. https://doi.org/10.1177/0278364912462256.

International Organization for Standardization. 2003. Ergonomics—Manual Handling-Part 1: Lifting and Carrying (ISO Standard No. 11228-1). https://www.iso.org/standard/26520.html.

International Organization for Standardization. 2007a. Ergonomics—Manual Handling-Part 2: Pushing and Pulling (ISO Standard No. 11228-2). https://www.iso.org/standard/26521.html.

International Organization for Standardization. 2007b. Ergonomics—Manual Handling Handling of Low Loads at High Frequency (ISO Standard No. 11228-3). https://www.iso.org/standard/26522.html.

International Organization for Standardization. 2010. Ergonomics of Human-System Interaction-Part 210: Human-Centred Design for Interactive Systems (ISO Standard No. 9241-210). https://www.iso.org/standard/52075.html.

International Organization for Standardization. 2016. Robots and Robotic Devices—Collaborative Robots (ISO/TS Standard No. 15066). https://www.iso.org/standard/62996.html.

Kagermann, H., J. Helbig, A. Hellinger, and W. Wahlster. 2013. *Recommendations for Implementing the Strategic Initiative INDUSTRIE 4.0, Securing the Future of German Manufacturing Industry, Final Report of the Industrie 4.0 Working Group*. Frankfrut: Forschungsunion.

Karwowski, W., and W. Marras (eds.). 2003. *Occupational Ergonomics*. Boca Raton: CRC Press. https://doi.org/10.1201/9780203507933.

Liu, H., and L. Wang. 2017. Human Motion Prediction for Human-Robot Collaboration. *Journal of Manufacturing Systems* 44: 287–294. https://doi.org/10.1016/j.jmsy.2017.04.009.

Lotterbach, S., and M. Peissner. 2005. Voice User Interfaces in Industrial Environments. *GI Jahrestagung* 2: 592–596.

Lu, Y. 2017. Industry 4.0: A Survey on Technologies, Applications and Open Research Issues. *Journal of Industrial Information Integration* 6: 1–10. https://doi.org/10.1016/j.jii.2017.04.005.

Matthias, B., and T. Reisinger. 2016. Example Application of ISO/TS 15066 to a Collaborative Assembly Scenario. In *Proceedings of ISR 2016: 47st International Symposium on Robotics*, 1–5.

Mauro, L., E. Alati, M. Sanzari, V. Ntouskos, G. Massimiani, and F. Pirri. 2018. Deep Execution Monitor for Robot Assistive Tasks. In *Proceedings of the European Conference on Computer Vision (ECCV)*. https://doi.org/10.1007/978-3-030-11024-6_11.

Mauro, L., F. Puja, S. Grazioso, V. Ntouskos, M. Sanzari, E. Alati, and F. Pirri. 2019. Visual Search and Recognition for Robot Task Execution and Monitoring. *Frontiers in Artificial Intelligence and Applications* 310: 94–109. https://doi.org/10.3233/978-1-61499-929-4-94.

Mitra, S., and T. Acharya. 2007. Gesture Recognition: A Survey. *IEEE Transactions on Systems, Man, and Cybernetics, Part C (Applications and Reviews)* 37 (3): 311–324. https://doi.org/10.1109/tsmcc.2007.893280.

Mitsunaga, N., C. Smith, T. Kanda, H. Ishiguro, and N. Hagita. 2008. Adapting Robot Behavior for Human–Robot Interaction. *IEEE Transactions on Robotics* 24 (4): 911–916. https://doi.org/10.1109/TRO.2008.926867.

Monkman, G.J., S. Hesse, R. Steinmann, and H. Schunk. 2007. *Robot Grippers*. Wiley. https://doi.org/10.1002/9783527610280.

Munzer, T., M. Toussaint, and M. Lopes. 2017. Preference Learning on the Execution of Collaborative Human-Robot Tasks. In *2017 IEEE International Conference on Robotics and Automation (ICRA),* 879–885. https://doi.org/10.1109/icra.2017.7989108.

Nwanya, S.C., C.N. Achebe, O.O. Ajayi, and C.A. Mgbemene. 2016. Process Variability Analysis in Make-to-Order Production Systems. *Production & Manufacturing* 3 (1). https://doi.org/10.1080/23311916.2016.1269382.

O'Shaughnessy, D. 2008. Automatic Speech Recognition: History, Methods and Challenges. *Pattern Recognition* 41 (10): 2965–2979. https://doi.org/10.1016/j.patcog.2008.05.008.

Pedersen, M.R., L. Nalpantidis, R.S. Andersen, C. Schou, S. Bøgh, V. Krüger, and O. Madsen. 2016. Robot Skills for Manufacturing: From Concept to Industrial Deployment. *Robotics and Computer-Integrated Manufacturing* 37: 282–291. https://doi.org/10.1016/j.rcim.2015.04.002.

Pellegrinelli, S., A. Orlandini, N. Pedrocchi, A. Umbrico, and T. Tolio. 2017. Motion Planning and Scheduling for Human and Industrial-Robot Collaboration. *CIRP Annals* 66 (1): 1–4. https://doi.org/10.1016/j.cirp.2017.04.095.

Penas, O., R. Plateaux, S. Patalano, and M. Hammadi. 2017. Multi-Scale Approach from Mechatronic to Cyber-Physical Systems for the Design of Manufacturing Systems. *Computers in Industry* 86: 52–69. https://doi.org/10.1016/j.compind.2016.12.001.

Pirri, F., L. Mauro, E. Alati, V. Ntouskos, M. Izadpanahkakhk, and E. Omrani. 2019. Anticipation and Next Action Forecasting in Video: An End-to-End Model with Memory. *Computer Vision and Pattern Recognition*, 1–6.

Ranz, F., V. Hummel, and W. Sihn. 2017. Capability-Based Task Allocation in Human-Robot Collaboration. *Procedia Manufacturing* 9: 182–189. https://doi.org/10.1016/j.promfg.2017.04.011.

Ruiz Garcia, M.A., R. Rojas, L. Gualtieri, E. Rauch, and D. Matt. 2019. A Human-in-the-Loop Cyber-Physical System for Collaborative Assembly in Smart Manufacturing. Paper presented at the CIRP Conference on Manufacturing Systems (CIRP CMS 2019), June 12–14, Ljubljana, Slovenia.

Salvendy, G. (eds.). 2012. *Handbook of Human Factors and Ergonomics*. Wiley. https://doi.org/10.1002/9781118131350.ch8.

Sanchez-Salas, A., Y.M. Goh, and K. Case. 2017. Identifying Variability Key Characteristics for Automation Design—A Case Study of Finishing Process. *Proceedings of the 21st International Conference on Engineering Design (ICED17), 4,* 21–30.

Sansoni, G., M. Trebeschi, and F. Docchio. 2009. State-of-the-Art and Applications of 3D Imaging Sensors in Industry, Cultural Heritage, Medicine, and Criminal Investigation. *Sensors* 9 (1): 568–601. https://doi.org/10.3390/s90100568.

Shen, Y., G. Reinhart, and M.M. Tseng. 2015. A Design Approach for Incorporating Task Coordination for Human-Robot-Coexistence within Assembly Systems. In *2015 Annual IEEE Systems Conference (SysCon) Proceedings*, 426–431. https://doi.org/10.1109/SYSCON.2015.7116788.

Siciliano, B., and O. Khatib (eds.). 2016. *Springer Handbook of Robotics*. Berlin and Heidelberg: Springer. https://doi.org/10.1007/978-3-540-30301-5.

Swamidass, Paul M. 2000. Non-Value-Added Activities. In *The Encyclopedia of Production and Manufacturing Management*, ed. Pual M. Swamidass. Boston, MA: Kluwer Academic, Springer. https://doi.org/10.1007/1-4020-0612-8.

Universal Robot. 2019. UR3 Quick Facts. https://www.universal-robots.com/products/ur3-robot/. Accessed on Apr 2019.

Zhong, R.Y., X. Xu, E. Klotz, and S.T. Newman. 2017. Intelligent Manufacturing in the Context of Industry 4.0: A Review. *Engineering* 3 (5): 616–630. https://doi.org/10.1016/J.ENG.2017.05.015.

13

Axiomatic Design for Products, Processes, and Systems

Christopher A. Brown

13.1 The Axioms and Engineering Design as a Scientific Discipline

13.1.1 The Axioms

The axioms were developed in the late nineteen seventies by Prof. Nam P. Suh and his students at MIT. The project was supported by the US National Science Foundation. They studied design processes and the quality of design solutions.

They found that the best designs had two things in common. These became Suh's design axioms.

Axiom one: Maintain the **independence** of the functional elements.

Axiom two: Minimize the **information** content.

C. A. Brown (✉)
Department of Mechanical Engineering, Worcester Polytechnic Institute, Worcester, MA, USA
e-mail: brown@wpi.edu

D. T. Matt et al. (eds.), *Industry 4.0 for SMEs*,
https://doi.org/10.1007/978-3-030-25425-4_13

When these axioms are applied properly, the resulting design solution is the best it can be for a given set of functional requirements and candidate design parameters (Suh 1990). Beyond the proper application of the axioms, the value, quality, and virtue of the design solution, rests on the formulation of functional requirements and the generation of candidate design parameters, which are discussed below.

These design axioms can apply to all kinds of problem solving. This includes a daily agenda and an arrangement of items on a desk, an online electric vehicle (Suh and Cho 2017), a mobile harbor (Lee and Park 2010; Park and Suh 2011) or software, and organizations (Suh 1998).

Note that the word "design" can refer to a representation of an object that has been designed, which is also a design solution. "Design" can also refer to a process during which the design solution is developed.

Newton's laws also apply to solving a wide variety of problems, from planetary motion to apples falling on earth. Suh's axioms, like Newton's laws, and any other set of axioms or natural laws, cannot be proven. Their validity relies on the lack of observations to the contrary within their domains of applicability.

13.1.2 Three Parts and Six Elements

Axiomatic design has three essential parts: axioms, structure, and process. Each of these three parts can be decomposed into two elements, as shown in Table 13.1.

The axioms are applied during the design process. Design problems are formulated and design solutions developed, from abstract to detailed, down hierarchies of abstraction and across functional, physical and process domains, during a zigzagging decomposition process, so that the design solution is consistent with the axioms.

13.1.3 Axiom One: Maintain Independence

Adherence to the independence axiom assures that design solutions avoid unintended consequences, and that it will be adjustable and controllable. It does this by developing a design solution that avoids

Table 13.1 Parts and elements of design

Parts	Elements	Notes
1. Axioms	**1.1 Maintain the independence of the functional elements**	*Facilitates adjustment, controllability, and avoids iteration and unintended consequences*
	1.2 Minimize the information content	*Maximizes probability (p) of success.* $I = ln(1/p)$
2. Structures	**2.1 Design domains**	*Decomposes laterally by type of domains, customer, functional, physical, and process*
	2.2 Design hierarchies	*Decomposes vertically by level of detail in the domains*
3. Processes	**3.1 Zigzagging decomposition**	*Developing detail down through the hierarchical domains*
	3.2 Physical integration	*Integrating detailed physical components up through the physical domain*

unwanted coupling. When two functions are coupled, they cannot be adjusted or changed independently, i.e., a change in one influences the other, to bring about a change in it too. The result is that some non-productive iteration will generally be required to reach the desired result, if indeed a convergent adjustment strategy can be found. Coupling can occur when one physical component is used to satisfy two FRs. This frequently occurs as the result of well-meaning attempts to reduce the number of components, believing that this will save cost and simplify the design. The result can be just the opposite.

13.1.4 Axiom Two: Minimize the Information Content

Information content is the log of one divided by the probability of success $I = ln(1/p)$ Information content in design in this sense is not related to knowledge. The first axiom can be seen to be a special case of the first, because the independence promotes success. Axiom two is applied after the first. It is used to select between candidate solutions, favoring those with the greatest possibility of fulfilling the functions and satisfying the customer needs.

13.2 Structures

In order to apply the axioms, the developing design solution should be structured so unwanted coupling and information content are obvious and can be designed out. The need to apply the axioms to the developing design solution, during the design process, drives the design structure. The form of the design process is based on the functions required for the process of conceptualizing and developing design solutions. The axiomatic design process is therefore based on the axioms and is thereby self-consistent.

The completed design solution structure is a verbal description of the design solution. Its construction is essential for completing the design solution. The design structure is like a free body diagram for the design problem, a multidimensional, orthogonal description of the functions, and solutions.

The structure consists of design domains laterally, and hierarchies of detail within the domains, vertically. The design domains are different kinds of descriptions of the design. The functional requirements (FRs), items in the functional domain, are formulated to satisfy the Customer Needs (CNs), items in the customer domain. FRs are fulfilled by the physical solutions, i.e., the design parameters (DPs), which are items in the physical domain. The DPs are created with Process Variables (PVs), which are items in the process domain.

The design domains are listed in Table 13.2, along with the elements of which they consist, their abbreviations, and their descriptions. The items in the different design domains connect to each other.

Table 13.2 The design domains

Domains	Customer	Functional	Physical	Process
Items	Customer Needs	Functional Requirements	Design Parameters	Process Variables
Abbreviations	**CNs**	**FRs**	**DPs**	**PVs**
Grammatical form	Anything	Imperative: begins with a verb	Adjectives and nouns	Present participles
Notes	What is needed	What it does	How it looks	How to produce it
Other		Constraints **Cs**		

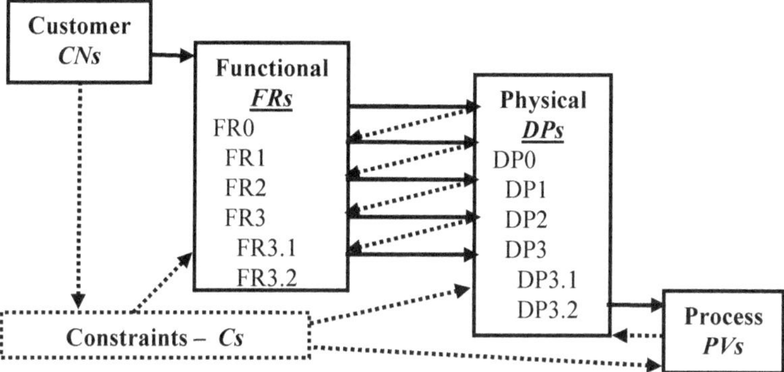

Fig. 13.1 Design domains, customer, functional, physical, and process, and the constraints, with their components. Zigzagging decomposition is indicated between the functional and physical domains. Upper level DPs constrain lower level FRs. The process domain and its connection to the physical domain is not elaborated here

These connections are evaluated by the axioms, particularly axiom one. The CNs describe what will be valued in the design. There is a value chain that runs from the Customer Domain to the Functional Domain, to the Physical Domain, and then on to the Process Domain (Fig. 13.1).

13.2.1 Design Domains and Constraints—Lateral Decompositions

The design domains (see Table 13.2) are spaces, or zones, on a field where a design game can be played. When the design decomposition is complete here is should be one-to-one correspondences between the items in the functional, physical, and process domains. Each FR needs its own DP, which should have a corresponding PV.

13.2.1.1 Customer Needs (CNs)

The Customer Domain describes what customers want, i.e., what they would value. Thompson (2013a) provides an excellent description of the process of identifying the customer and other stakeholders, e.g., manufacturing, transport, and salespeople. CNs might also contain constraints (Cs), things that must be avoided, and preferences that can become selection criteria (SCs) or optimization criteria (OCs) as described by Thompson (2013b).

In practice, CNs could be developed by a marketing group. Describing CNs can begin a design process. CNs have no special grammatical form. They should describe the fundamental needs, preferences, and constraints in a way that opens the solution space appropriately.

13.2.1.2 Functional Requirements (FRs)

The FRs should be an organized technical description of what the design must do, i.e., the functions it needs to provide. FRs are stated in the imperative. They begin with verbs. At the highest levels, FRs can be things like "transport people," or "deliver hot water."

Most often just the top-level FRs are developed from the CNs, because the CNs tend to be more abstract. The initial FRs should be the minimum list of functions that satisfy the CNs. The FRs should be collectively exhaustive with respect to the CNs. At any level, FRs should be mutually exclusive with respect to each other. Formulating good FRs is essential. A design solution cannot be better than its FRs. The top-level FRs establish the starting point, indicate the direction, and define when the destination is reached.

The FRs should state the design objective directly, for example, "transport people." A common mistake of novices is to make "design a bicycle" an FR, intending to design a new kind of bicycle. The operative verb is "design," so this suggests designing a system for designing a bicycle. If the intent is to design a new sort of bicycle then the FR might rather be "transport people." It could have constraints to be powered by the people being transported. Using a term like "bicycle" already suggests a physical solution, thereby limiting the solution space, and reducing the potential for a new, creative solution to the transport problem.

13.2.1.3 Design Parameters (DPs)

DPs describe what the design looks like. What eventually appears in a CAD file for mechanical parts, or lines of code for software. DPs begin with nouns. At the highest levels, DPs are things like "personal transportation device" or "hot water system." At the lower levels DPs can be things like "ball bearings" or "valves," or the design can start there, if you are designing rolling elements or fluid control devices.

The DPs are the physical items that fulfill the FRs. In order to comply with axiom one, the independence axiom, in the completed decomposition there should be one and only one DP corresponding to each FR. Ideally, that DP should only influence the FR that it is intended to fulfill.

The term "physical," referring to the physical domain and physical attributes, is not necessarily physical literally. DPs can also be "computer code," "education modules," or "slots of time in an agenda." These DPs would fulfill FRs like "calculate best fit line," "teach engineering design," or "meet with clients."

13.2.1.4 Process Variables (PVs)

PVs describe how the DPs are produced. PVs can be manufacturing processes, like machining, injection molding, and assembly. The PVs can be developed so that the PVs have a one-to-one relation with the DPs. This can be necessary when functional and physical tolerances are difficult to achieve (Brown 2018). Often it is not necessary to develop this kind of detail in the Process Domain. When it is time to do process design, then it can be advisable to begin a new design problem with FRs and DPs relating to the process (Brown 2014).

13.2.1.5 Constraints (Cs)

Cs interface with the domains, although they are not in the value chain. Cs are limits or restrictions on the items in the other domains. Cs sometimes cannot be decoupled from the other items in the domains.

They are things that must be avoided rather than fulfilled. Cs differ from FRs in that Cs do not have DPs. Another important difference is that while each of the FRs should be independent of all other FRs, Cs can influence all the FRs and DPs, e.g., weight and cost limitations.

13.2.2 Design Hierarchies—Vertical Decomposition

When more than one level of detail is necessary to describe a design solution to the point where the solution is obvious, then the design should be decomposed vertically. This decomposition can continue increasing in detail, from parent to child, through several generations, or levels, until the solution is obvious. The upper level FRs describe main branches, subsequent decomposition of the children define branches that are more detailed. A decimal notation is used for the branches. The top-level FR is FR0. The next level are FR1, FR2, FR3, etc. The children of FR1 are FR1.1, FR1.2, FR1.3, etc.

13.3 The Design Process

The design process seeks first to generate items comprising a design solution by systematically decomposing abstract design concepts into progressively more detailed items until the decomposed solution is obvious. Then the design decomposition is physically assembled to produce a complete, integrated solution. The decomposition and integration are both done so as to assure that the resulting design solution complies with the axioms. A complete solution should include functional and physical metrics with dimensioning and tolerancing. Steps in the design process are shown in Table 13.3.

13.3.1 Zigzagging Decomposition

The decomposition is generated by zigzagging between domains at one level, then proceeding to the next, more detailed level (Fig. 13.2). Generally, this is done from the functional to the physical domain,

Table 13.3 Steps for zigzagging decompositions to develop FRs and select the best DPs

1. Identify the needs of the customers and stakeholders (CNs) (Thompson 2013a)
2. Develop FR0 based on CNs
   - 2.1 Select DP0 candidates
   - 2.2 Select best DP0 based on minimizing information content (Axiom 2)
3. Develop next level or branch of FRs, based on CNs where appropriate
   - 3.1 Check for non-FRs, OCs, SCs (Thompson 2013b) and violation of constraints
   - 3.2 Check FRs for CEME decomposition (Axiom 1), and correct if necessary
4. Select corresponding DP candidates for each FR, and check for violation of constraints
5. Select the best DP for each FR from the candidates
   - 5.1 Select those that maintain independence (Axiom 1—use design matrix)
   - 5.2 Select best for minimum information content (Axiom 2—use $I = log(1/p)$)
6. If the solution is obvious, then leave the decomposition and go to physical integration, if not, go to step 3

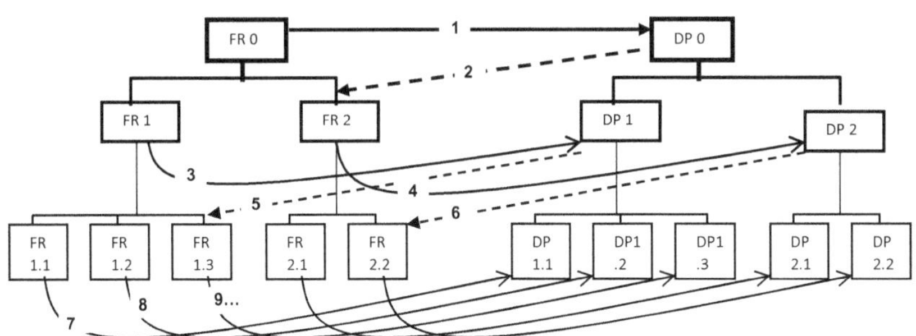

Fig. 13.2 Zigzagging decomposition symbolic example. The arrows show the direction of the flow and the order is indicated by the numbers on the arrows. There must be at least two children at each level of the decomposition in each branch

although it can be extended to the process domain as well. Each parent must have at least two children in its domain.

The children contain more detail than the parent does. The children should combine to equal the parent, that is, they must be collectively exhaustive (CE) with respect to the parent. If not, the decomposition is incomplete. The children at each level in each branch must be mutually exclusive (ME) with respect to each other, in order to comply with axiom one, independence. An appropriate decomposition is said to be CEME-min, indicating in addition the minimum number of FRs.

In these hierarchal decompositions upper level DPs constrain lower level FRs. If, for example, a pneumatic solution has already been selected at a higher level, then all subsequent lower-level solutions should fit with the higher-level choice, for example, pneumatic valves and actuators.

13.3.1.1 Checking for Independence (Axiom One)

At each level and on each branch of the decomposition the proposed solutions should be checked for compliance with the axioms. This is done by examining the design matrix, which, minimally, shows which DPs influence which FRs (Tables 13.4, 13.4, 13.5 and 13.6).

Table 13.4 Uncoupled basic design matrix, diagonal, and therefore fully independent

	DP1	DP2	DP3
FR1	X	0	0
FR2	0	X	0
FR3	0	0	X

Table 13.5 Decoupled basic design matrix, triangular, and therefore quasi-independent

	DP1	DP2	DP3
FR1	X	0	0
FR2	X	X	0
FR3	X	X	X

Table 13.6 Coupled basic design matrix, full, and therefore not independent

	DP1	DP2	DP3
FR1	X	X	X
FR2	X	X	X
FR3	X	X	X

The Xs in the matrices indicate that the DP in the upper row influences the FR in the left column. The nature of the relation could be described in any way. The exact nature of the relation need not be known, only that an imaginable change in the DP could be enough to take the FR out of tolerance.

An uncoupled design matrix is diagonal. Each DP influences only one FR. The DPs can be adjusted in any order to satisfy all FRs without any non-productive iterations. This is the best.

A decoupled design matrix has one or more off-diagonal interactions on the same side of the diagonal (Table 13.5). There are certain orders for adjusting DPs that can fulfill the FRs without non-productive iterations. In the case of a fully triangular matrix there is only one order of adjustment of the DPs to fulfill the FRs. In Table 13.5, which is lower triangular, the adjustment order is 1, 2, 3. The order of adjustment for an upper triangular matrix would be 3, 2, 1. Some matrices, which are not triangular, can be rearranged to be triangular using linear algebra. Triangular is acceptable. Favor those solutions with the most off-diagonal zeros in their design matrices.

A coupled matrix (Table 13.6) has at least one interaction (X) on both sides of the diagonal and no way to rearrange to the matrix to triangular. There is no order of adjustment without iteration. Even with iteration, there is no guarantee of convergence on an acceptable solution. This matrix indicates a failure to be adjustable, controllable, and to avoid unintended consequences. Another solution should be sought with more independence.

13.3.1.2 Decomposition Themes

Themes can be used to facilitate CEME decompositions. Different themes can be used at different levels. Energy, work, and partition, by location or by sequences in time, are useful themes in manufacturing. In testing for being CE, each form of energy and work, location or instance of time can be accounted for to insure the decomposition is exhaustive. For example, in response to a CN to improve sustainability in manufacturing an FR might be to recover waste in machining. One branch might be to recover thermal energy. This can be partitioned into

Table 13.7 Example of an FR-DP, top-level decomposition in response to a CN for sustainability in metal cutting

FR0 recover waste heat in machining	DP0 heat recovery system
FR1 Recover heat from the tool	DP1 Thermoelectric generator
FR2 Recover heat from the chips	DP2 Gaseous heat exchanger in chip handler
FR3 Recover from the workpiece	DP3 Liquid heat exchanger with cutting fluid

locations that contain excess thermal energy as the result of the work during machining, the chips, the tool, and the workpiece (Table 13.7).

Location specific heat recovery systems could be selected as corresponding DPs. Selected themes can influence the process, as well as the ability to apply the axioms. Different themes, or themes selected in a different sequence at different levels, can result in the same elemental items in the decomposition at the lowest levels. An alternative decomposition could be made by the temperature of the heat, which would have had to have been addressed eventually in a decomposition beginning with a location theme, because of the fundamental thermodynamics of heat recovery.

Another theme for decomposition can be directional, e.g., Cartesian. For example, limit or control movement could be decomposed into controlling or limiting movement in three orthogonal directions. Another kind of theme can be provided by terms used in the FRs. The FRs are at least a verb-noun pair. Often it is possible to decompose the verb and the noun into two or more components each.

13.3.1.3 Metrics and Equations

Metrics and equations can assist with the decomposition process, although frequently the decomposition is done qualitatively, at least initially. The FRs benefit from functional metrics that indicate the extent to which the function is satisfying a CN. The DPs need physical metrics, which can be measures of physical properties. The PVs need process metrics, which are adjustments in the process domain.

Two kinds of equations can be useful in the decomposition. Equations relating two different domains. The FRs and DPs are related by the design equations (13.1). The DPs and the PVs are related by the process equations (13.2).

$$\text{FRi} = \text{f(DPj)} \tag{13.1}$$

$$\text{DPj} = \text{g(PVk)} \tag{13.2}$$

The functions f and g can be described in detailed design and process matrices. The FRs, DPs, and PVs can be described as column vectors. When written in differential form, which is useful because the interest is in how changes in the DPs change the FRs, the design and process matrices are the Jacobians, describing these changes.

Equations can be used to relate the parents and the children. These could be used to verify that the decomposition is CEME.

$$\text{FRi} = \text{FRi.1} + \text{FRi.2} + \text{FRi.3} \ldots \tag{13.3}$$

They do not have to be related by a summation, although this is the simplest relation. Any kind of combination of the children to achieve the parent would suffice.

13.3.2 Physical Integration

When the decomposition is complete, all the physical items for the design have been selected. Then for a complete solution, the physical items need to be physically integrated. This could be a drawing or some other representation of the juxtaposition of the physical parts of the design solution. Note that physical integration can be in space or time.

The branches and themes in the physical integration can differ from those in the decomposition. Some items need to be proximate to provide the intended function. In a car, for example, the braking system could be one branch in the decomposition and be dispersed in the physical structure of the car after the physical integration.

It is good practice to include the FR-DP numbers in the representation, so each physical item (DP) can be traced back to an FRs in the decomposition and the intent can be identified. In this way, the consequences of modifications can be quickly traced and unintended consequences avoided.

To facilitate the physical integration, a DP-DP matrix or a PV-PV matrix can be created. It can show the interactions, or lack of them, with Xs and 0s, like the design or process matrices. The diagonal is superfluous, because it indicates that an item interacts with itself. The off-diagonal Xs indicate physical interactions, in space or time that are introduced by the integration. These need to be checked to be sure the independence is not compromised (Axiom 1) and that the information is not increased unnecessarily beyond some equally functional physical configurations (Axiom 2).

13.4 Additional Applications and Industry 4.0

This section and the following subsections discuss some applications and aspects of AD that could be of particular interest to people doing work related to Industry 4.0 (I4.0).

As we look to integrate Industry 4.0 and AD, we note that the current descriptions of Industry 4.0 seem to fit best as items in the physical domain, as DPs. Industry 4.0 seems to comprise things like cyber-physical production systems (CPPS), connectivity for data exchange and automated decision-making for routine situations. Industry emphasizes solutions in the physical domain, i.e., DPs. The task then is understanding the appropriate functions, FRs, and the CNs that would use these DPs. In other words, engineers should decide how these interesting tools should be used ethically for the benefit of humanity.

13.4.1 Analyzing Existing Designs

The AD method as described above is intended to be used on new design problems. It can be difficult or impossible to fix an existing design to make it adjustable, controllable, robust, and avoiding unintended consequences. Nonetheless, there are techniques that can be used to study and improve existing designs.

Design thinking starts with knowing or formulating the intended functions, i.e., the design intent. Often problems result from not adequately considering and articulating the functions. The functions must satisfy needs defined by the customers and stakeholders. Improving an existing design solution or portion of it, by using AD, generally requires doing a decomposition, for at least of some part of the design solution.

Problems can also arrive in achieving a solution when a design effort has been partitioned too early in the design process. This can result from an assumption that independence of branches can be maintained during subsequent hierarchal decomposition and integration of these branches. The solutions in some branches can be more challenging than in others, and these branches might need a higher priority in allocation of constraints, such as cost, space, and weight. This assumes that it will be easier to adapt the less challenging branches to comply with system-wide constraints.

Many poor designs can be linked to specific problems (Table 13.8). These problems can be related to failing some aspect of the axioms, which can be linked to shortcomings in the process. References in Table 13.8 suggest some solutions.

In some cases, new CNs are introduced, as with items 1 and 2 in Table 13.8. This requires the formulation of new FRs, and maybe nFRs (Cs, OCs, and SCs, Thompson 2013a), and returning to procedures in Table 13.3.

Items 3, 4, and 5 in Table 13.8 require the selection of new DPs and returning to procedures in Table 13.3. Sometimes, if there are

Table 13.8 Some typical problems in a poor design

1. Poor understanding of the CNs and stakeholder needs (Thompson 2013a)
2. Not developing appropriate FRs and nFRs (Thompson 2013b)
3. Having fewer DPs than the FRs (violates axiom one)
4. Selecting DPs that influence too many FRs (violates Axiom one)
5. Difficulties achieving solutions consistently (low yield, violates Axiom two)
6. Not knowing the correct order of adjustment (imaginary complexity according to Suh 2005)
7. Violation of constraints (Axiom two)
8. Poor physical integration introduces coupling (Axiom one), or reduces probabilities of success (Axiom two)

difficulties finding DPs, then the FRs might need modifications to enlarge the solution space as described in Sect. 13.4.2 below.

Imaginary complexity (item 6 in Table 13.8) can be addressed by constructing a design matrix and decoupling using linear algebra to show the correct order of adjustment (Suh 1990).

There are times when the existing design solutions should be modified to reduce cost, weight, or volume, a constraint violation (item 7 Table 13.8). An approach is to develop the decomposition. Starting with the CNs, including all the stakeholders, clearly establish what provides the required value. Look at each item in the solution FRs, DPs, and PVs, moving through the value chain and identify those things that add cost, weight, or volume and do not add value, and then eliminate them. Then examine DPs and PVs that add to the constrained quantities, i.e., Cs, and look for alternative solutions that add less and still comply appropriately with the axioms.

Item 8 in Table 13.8 suggests a return to the physical integration procedure, and an examination of the physical integration matrices (DP-DP, or PV-PV). If the necessary independence or information content cannot be achieved, then this indicates a return to the FRs. Perhaps some FRs need to be modified to increase the solution space (Sect. 13.4.2), while still satisfying the CNs, and new DPs or PVs found. Maybe the CNs need to be better understood (see Sect. 13.4.2).

13.4.2 Recognizing Opportunities for Creativity and Innovation

Being creative is having good, useful ideas. Innovation is making them viable. The decomposition and integration processes are a kind of functional modeling. These can help to make a good idea viable.

Working in the best solution space is important for creativity. FRs need to be solution neutral. This starts with understanding the CNs to formulate good FRs.

Henry Ford supposedly said that if he had asked people what they wanted, they would have said a faster horse. The issue is then to understand that the desire was for faster transportation. There is a tendency to go with known solutions, like a horse for transportation at the beginning of the twentieth century.

The FRs should not limit the DPs unduly. One test for a good FR is to see how many candidates DPs can fit with the FR. Be skeptical of FRs that do not provide enough solution space for several candidate DPs. In these cases check to see if the FR suggests a physical solution. Sometimes, when the FR is too close to the DPs, the FR can be moved closer to the CN and provide more solution space for the DPs.

Another approach is to ask why that FR is needed. Often in the answer to this question is another FR that will also satisfy the same CNs, although in a more fundamental way that provides a larger solution space for selecting candidate DPs.

13.4.3 Application for Industry 4.0

For enterprises, a good CN and FR0 can be a return on investment (ROI). This suggests a decomposition for based on return and on investment (Suh 2001). It can be tempting to make the next level FRs maximize return and minimize investment. Maximizing and minimizing only make good FRs when there is a system or program for achieving the maximum or minimum, which could be a DP.

The question for I4.0 might be, do the DP heavy descriptions fit into solutions for ROI. The temptation to use a I4.0 type solution without looking to see if there is another solution that works and requires less investment. The decision to use a I4.0 type solution should be justified in terms of ROI, and in terms of satisfying the axioms.

New technologies can also be used to fulfill FRs that could not previously be considered. It is in this vein that the best use might be made of the I4.0 type solutions. In the pursuit of Industry 4.0 there are CNs and resulting FRs that should be considered now, which were not previously, because there appeared to be no physical solution. Industry 4.0 can enable new design solutions for production systems that were previously unattainable. These new design solutions might help to reverse climate change, and provide appropriate ROIs, through improved productivity.

13.5 Concluding Remarks

Considering the CNs, for all the stakeholders in Industry 4.0, engineers must hold paramount the safety, health, and welfare of the public (first fundamental canon of ethics for engineers). Engineers must act according to the canons of ethics, otherwise what they are doing is not engineering. Industry 4.0 must consider sustainability because it is essential to the safety, health, and welfare of the public.

In 1953, a few years after the invention of the transistor as the first digital large computers appeared Kurt Vonnegut published Player Piano (Vonnegut 1953). The title appears to a reference to John Parson's first numerical controls for machine tools with punch tapes and pneumatic actuators. This novel describes a highly automated, yet highly disturbing, society, which is in the grips of another industrial revolution with some similarities to Industry 4.0. It mentions that this later industrial revolution devalues human thought, just as the first industrial revolution devalued human labor.

Axiomatic Design is powerful. It can empower tools that are available and being developed for Industry 4.0. AD and Industry 4.0 must be used for the benefit of humanity. Sustainability and the preservation and enhancement of freedom and dignity, must be included in the CNs. If Industry 4.0 diminishes these things, and serves mainly to enrich further those who are already wealthy, then it has failed.

References

Brown, C.A. 2014. Axiomatic Design of Manufacturing Processes Considering Coupling. In *Proceedings of the 8th International Conference on Axiomatic Design*, 149–153, Sept 24–26, Campus de Caparica, Portugal.

Brown, C.A. 2018. Specification of Surface Roughness Using Axiomatic Design and Multiscale Surface Metrology. *Procedia CIRP* 70: 7–12. https://doi.org/10.1016/j.procir.2018.03.094.

Lee, T., and G.J. Park. 2010. Managing System Design Process Using Axiomatic Design: A Case on KAIST Mobile Harbor Project. *SAE International Journal of Materials and Manufacturing* 3 (1): 125–132. https://doi.org/10.4271/2010-01-0278.

Park, N.K., and S.C. Suh. 2011. Modal Shifting from Road to Coastal Shipping Using a Mobile Harbor. *The Asian Journal of Shipping and Logistics* 27 (3): 447–462. https://doi.org/10.1016/S2092-5212(11)80021-0.

Suh, N.P. 1990. *The Principles of Design*. New York: Oxford University Press.

Suh, N.P. 1998. Axiomatic Design Theory for Systems. *Research in Engineering Design* 10 (4): 189–209. https://doi.org/10.1007/s001639870001.

Suh, N.P. 2001. *Axiomatic Design: Advances and Applications*. MIT-Pappalardo Series in Mechanical Engineering. New York: Oxford University Press.

Suh, N.P. 2005. *Complexity: Theory and Applications*. New York: Oxford University Press.

Suh, N.P., and D.H. Cho. 2017. *The On-Line Electric Vehicle: Wireless Electric Ground Transportation Systems*. Cham: Springer.

Thompson, M.K. 2013a. Improving the Requirements Process in Axiomatic Design Theory. *CIRP Annals* 62 (1): 115–118. https://doi.org/10.1016/j.cirp.2013.03.114.

Thompson, M.K. 2013b. A Classification of Procedural Errors in the Definition of Functional Requirements in Axiomatic Design Theory. In *Proceedings of the 7th International Conference on Axiomatic Design*, June 27–28, Worcester, MA, USA.

Vonnegut, K. 1953. *Player Piano: A Novel*. London: Macmillan.

Index

D. T. Matt et al. (eds.), *Industry 4.0 for SMEs*,
https://doi.org/10.1007/978-3-030-25425-4

GPSR Compliance
The European Union's (EU) General Product Safety Regulation (GPSR) is a set of rules that requires consumer products to be safe and our obligations to ensure this.

If you have any concerns about our products, you can contact us on

ProductSafety@springernature.com

In case Publisher is established outside the EU, the EU authorized representative is:

Springer Nature Customer Service Center GmbH
Europaplatz 3
69115 Heidelberg, Germany

www.ingramcontent.com/pod-product-compliance
Lightning Source LLC
LaVergne TN
LVHW052124070326
832902LV00038B/1742